Structure and Properties of Cell Membranes

Volume II

Molecular Basis of Selected Transport Systems

Editor

Gheorghe Benga, M.D., Ph.D.
Head, Department of Cell Biology
Faculty of Medicine
Medical and Pharmaceutical Institute
Cluj-Napoca, Romania

CRC Press, Inc.
Boca Raton, Florida

Library of Congress Cataloging in Publication Data
Main entry under title:

Molecular basis of selected transport systems.

(Structure and properties of cell membranes; v. 2)
Includes bibliographies and index.
1. Biological transport. 2. Molecular biology.
I. Benga, Gheorghe. II. Series. [DNLM: 1. Cell
Membrane-physiology. 2. Cell Membrane—ultrastructure.
3. Biological Transport. QH 601 S9285]
QH509.M65   1985   574.87'5   84-19942
ISBN 0-8493-5765-9

Direct all inquiries to CRC Press, Inc., 2000 Corporate Blvd., N.W., Boca Raton, Florida, 33431.

© 1985 by CRC Press, Inc.
International Standard Book Number 0-8493-5765-9

Library of Congress Card Number 84-19942
Printed in the United States

PREFACE

In recent years it has become apparent that many essential functions of living cells are performed by membrane-associated events. Membranes are highly selective permeability barriers that impart their individuality on cells and organelles (Golgi apparatus, mitochondria, lysosomes, etc.) by forming boundaries around them and compartmentalizing specialized environments. By receptor movement and responses to external stimuli, membranes play a central role in biological communication. Owing to various enzymes attached to or embedded into membranes they are involved in many metabolic processes. The two important energy conversion processes, photosynthesis in chloroplasts and oxidative phosphorylation in mitochondria, are carried out in membranes. To understand all these processes, which are essential for living organisms, it is necessary to understand the molecular nature of membrane structure and function.

The main purpose of this book is to provide in-depth presentations of well-defined topics in membrane biology, focusing on the idea of *structure-function relationships at the molecular level*. The book consists of three volumes.

Volume 1 covers general aspects of structure-function relationships in biological membranes. Attention has been paid both to protein and lipid components of cell membranes regarding the interactions between these components, mobility of proteins and lipids, as well as to the physiological significance of membrane fluidity and lipid-dependence of membrane enzymes. Since some molecular components of the plasma membrane appear to function in concert with some component macromolecules of basement membranes a review of this expanding topic has been included.

Volume 2 is devoted to models and techniques which allow molecular insights into cell membranes. After the first chapter describes quantum chemical studies of proton translocation, several chapters present the most extensively used model systems (monomolecular films, planar lipid bilayers, and liposomes) in relation to biomembranes as well as the reconstitution of membrane transport systems. The description of some biophysical techniques (X-ray, spin labeling ESR, NMR) is focused on their use in studying stucture-function relationships in cell membranes. The remaining chapters in this volume are devoted to the physiological significance of surface potential of membranes and of the dietary manipulation of lipid composition.

Volume 3 covers transport at the molecular level in selected systems. The first chapters present basic kinetics and pH effects on membrane transport, while subsequent chapters focus on the effect of membrane lipids on permeability in prokaryotes or on Ca^{2+} permeability. Three chapters describe structure-function relationships in mitochondrial H^+-ATPase, cytochrome oxidase, and adenine nucleotide carrier. The last chapter is devoted to exocytosis, endocytosis, and recycling of membranes, which are distinct, albeit overlapping, cellular processes.

From this survey it is obvious that by application of biochemical and biophysical techniques it is possible to explain membrane phenomena at the molecular level in a meaningful way. Moreover, it is now clear that the study of cell membranes at the molecular level is important for understanding the alterations leading to abnormal cells or the understanding of drug and pesticide action. The multidisciplinary approach of research in this area and the permanent need for information regarding the recent advances require new books on cell membranes. The present collection of reviews is by no means a comprehensive treatise on all aspects of ''membranology'', rather a sampling of the status of selected topics. The volumes, providing contributions for reference purposes at the professional level, are broadly aimed at biochemists, biologists, biophysicists, physicians, etc., active investigators working on cell membranes and hopefully will also be of great help to teachers and students at both the undergraduate or postgraduate levels.

THE EDITOR

Dr. Gheorghe Benga, M.D., Ph.D., is the Head of the Department of Cell Biology at the Medical and Pharmaceutical Institute, Cluj-Napoca, Romania. He is also heading the Laboratory of Human Genetics of the Cluj County Hospital.

In 1967, Dr. Benga received an M.D. with academic honors from the Medical and Pharmaceutical Institute. After 3 years of internship (1966 to 1969) in basic medical sciences (biochemistry, microbiology), he studied for a Ph.D. in medical biochemistry from 1969 to 1972 under Prof. Ion Manta, Department of Biochemistry, at the same Institute. In 1972, Dr. Benga received a B.Sc. in Chemistry and in 1973, an M.Sc. in Physical Chemistry of Surfaces from the University of Cluj.

From 1972 to 1978, Dr. Benga was Lecturer and Senior Lecturer in the Department of Biochemistry at the Medical and Pharmaceutical Institute. In 1974 he was awarded a Wellcome Trust European Travelling Fellowship and spent 1 year in England as a postdoctoral research worker under Prof. Dennis Chapman, Department of Chemistry and Biochemistry, University of Sheffield and Chelsea College University of London. In 1978, Dr. Benga was appointed to head the newly formed Department of Cell Biology at the Medical and Pharmaceutical Institute. He is currently teaching cell biology to medical students.

In addition to his other duties, Dr. Benga has spent several 1- to 3-month periods as a Visiting Scientist at many British and American universities and in 1983 was a Visiting Professor at the University of Illinois at Urbana-Champaign.

Dr. Benga has attended several international courses on biomembranes and has presented numerous papers at international and national meetings, as well as guest lectures at various universities and institutes in Romania, England, the U.S., the Netherlands, and Switzerland. He has taken an active part in the organization of three international workshops on biological membranes (1980, Cluj-Napoca — Romanian-British; 1981, Cluj-Napoca — Romanian-American; 1982, New York City — American-Romanian) and has published over 80 papers to date.

Dr. Benga is the author of several text books of cell biology for medical students and of the book, *Biologia moleculară a membranelor cu aplicații medicale,* published by Editura Dacia, Cluj-Napoca, 1979. He is the co-author of *Metode biochimice în laboratorul clinic,* Editura Dacia, 1976; co-editor of *Biomembranes and Cell Function,* New York Academy of Sciences, 1983; and co-editor of *Membrane Processes: Molecular Biology and Medical Applications,* Springer-Verlag, New York, 1984.

His major interests in the field of biological membranes include the characterization of molecular composition and functional properties of human liver subcellular membranes, the molecular interactions (lipid-protein, lipid-sterol, and drug effects) in model and natural biomembranes, and the investigation of water diffusion through red blood cell membranes.

Dr. Benga is President of the Cluj-Napoca Section of the Romanian National Society of Cell Biology and Vice-President of this Society. He is on the board of the Subcommission of Biochemistry of the Romanian Academy and is on the editorial board of *Clujul Medical.*

CONTRIBUTORS

Angelo Azzi
Medizinisch-Chemisches Institut
Universität Bern
Bern, Switzerland

Kurt Bill
Medizinisch-Chemisches Institut
Universität Bern
Bern, Switzerland

Marc R. Block
Laboratoire de Biochimie
Departement de Recherche Fondamentale
Centre d'Etudes Nucléaires
Grenoble, France

Reinhard Bolli
Medizinisch-Chemisches Institut
Universität Bern
Bern, Switzerland

François Boulay
Laboratoire de Biochimie
Departement de Recherche Fondamentale
Centre d'Etudes Nucléaires
Grenoble, France

Gérard Brandolin
Laboratoire de Biochimie
Departement de Recherche Fondamentale
Centre d'Etudes Nucléaires
Grenoble, France

Robert P. Casey
Medizinisch-Chemisches Institut
Universität Bern
Bern, Switzerland

Zdeněk Drahota
Director
Institue of Physiology
Czechoslovak Academy of Sciences
Prague, Czechoslovakia

Ross P. Holmes
Senior Food Scientist
Burnsides Research Laboratory
University of Illinois
Urbana, Illinois

Josef Houštěk
Institute of Physiology
Czechoslovak Academy of Sciences
Prague, Czechoslovakia

Jan Kopecký
Institute of Physiology
Czechoslovak Academy of Sciences
Prague, Czechoslovakia

Arnošt Kotyk
Institute of Physiology
Czechoslovak Academy of Sciences
Prague, Czechoslovakia

Guy J.-M. Lauquin
Laboratoire de Biochimie
Departement de Recherche Fondamentale
Centre d'Etudes Nucléaires
Grenoble, France

Ronald N. McElhaney
Professor
Deparment of Biochemistry
University of Alberta
Edmonton, Alberta
Canada

D. James Morré
Professor
Departments of Medicinal Chemistry and
  Biological Sciences, and Purdue Cancer
  Center
Purdue University
West Lafayette, Indiana

Katarzyma A. Nałęcz
Medizinisch-Chemisches Institut
Universität Bern
Bern, Switzerland

Maciej J. Nałęcz
Department of Cellular Biochemistry
Nencki Institute of Experimental Biology
Warsaw, Poland

Paul O'Shea
Medizinisch-Chemisches Institut
Universität Bern
Bern, Switzerland

Pierre V. Vignais
Laboratoire de Biochimie
Departement de Recherche Fondamentale
Centre d'Etudes Nucléaires
Grenoble, France

Lech Wojtczak
Department of Cellular Biochemistry
Nencki Institute of Experimental Biology
Warsaw, Poland

STRUCTURE AND PROPERTIES OF CELL MEMBRANES

Gheorghe Benga

Volume I

The Evolution of Membrane Models
Protein-Protein Interactions in Cell Membranes
Lateral Mobility of Proteins in Membranes
Lateral Diffusion of Lipids
Topological Asymmetry and Flip-Flop of Phospholipids in Biological Membranes
Membrane Fluidity: Molecular Basis and Physiological Significance
Lipid Dependence of Membrane Enzymes
Protein-Lipid Interactions in Biological Membranes
Basement Membrane Structure, Function, and Alteration in Disease

Volume II

Basic Kinetics of Membrane Transport
pH Effects on Membrane Transport
The Effect of Membrane Lipids on Permeability and Transport in Prokaryotes
The Influence of Membrane Lipids on the Permeability of Membranes to Ca^{2+}
Molecular Aspects of Structure-Function Relationship in Mitochondrial H^+-ATPase
Molecular Aspects of the Structure-Function Relationship in Cytochrome c Oxidase
Molecular Aspects of the Structure-Function Relationships in Mitochondrial Adenine
Nucleotide Carrier
Exocytosis, Endocytosis, and Recycling of Membranes
The Surface Potential of Membranes: Its Effect on Membrane-Bound Enzymes and
Transport Processes

Volume III

Quantum Chemical Approach to Study the Mechanisms of Proton Translocation Across
Membranes Through Protein Molecules
Monomolecular Films as Biomembrane Models
Planar Lipid Bilayers in Relation to Biomembranes
Relation of Liposomes to Cell Membranes
Reconstitution of Membrane Transport Systems
Structure-Function Relationships in Cell Membranes as Revealed by X-Ray Techniques
Structure-Function Relationships in Cell Membranes as Revealed by Spin Labeling EPR
Structure and Dynamics of Cell Membranes as revealed by NMR Techniques
The Effect of Dietary Lipids on the Composition and Properties of Biological Membranes

TABLE OF CONTENTS

Volume II

Chapter 1

BASIC KINETICS OF MEMBRANE TRANSPORT

Arnošt Kotyk

TABLE OF CONTENTS

I. INTRODUCTION

In this chapter, we shall deal with the rates of movement of substances across membranes during their transport or translocation. At the outset, let us accept some restrictions and simplifications for the purposes of this treatment, as they are common throughout transport literature.

1.   The membrane is "infinitely" thin so that no intramembrane gradients are to be considered.
2.   The solutions separated by the membrane are sufficiently mixed so that there is a unique concentration of solute at any time in each of them.
3.   The transport system resides in the membrane and cannot leave it so that its "concentration" in the membrane does not change during the experiment.

Two fundamentally different mechanisms of membrane transport exist: nonspecific permeation (often called simple diffusion) and specific, saturable transport (often termed carrier or porter transport).

II. NONSPECIFIC PERMEATION

In the first type of transmembrane movement, no specific reactions take place between the permeant solute and the permeable membrane, although nonspecific interactions, such as stripping of hydration water, attraction by opposite electrical charges, and the like, are possible and, in fact, likely.

The rate of such transport (in kmol s^{-1} m^{-2}) is defined from Fick's First Law of diffusion for movement along one of the coordinates

$$J_s = -D(S_{II} - S_I)/\delta = P(S_I - S_{II}) \tag{1}$$

where S_I and S_{II} are solute concentrations (kmol m^{-3}) at the starting and at the target sides of the membrane, respectively, D is the diffusion coefficient across the membrane in $m^2 s^{-1}$, δ the membrane thickness in m, and P the permeability constant in m s^{-1}.

D itself may be defined as RTU where R is the gas constant (8.314 J mol^{-1} K^{-1}), T the absolute temperature (in K), and U the mobility (in m^2 s^{-1} J^{-1} mol). This suggests that diffusion processes have a poorly expressed temperature dependence, something like 3% increase in D per 10°C rise in the physiological temperature range. This is, however, not true for transmembrane crossing. We must consider here a different definition of P, which must be taken as phenomenological, only as it holds both for passage through membrane lipid domains as well as for movement through hydrophilic channels. It states that

$$P = K D_m/\delta \tag{2}$$

where D_m is the diffusion coefficient of the solute within the membrane and K the partition coefficient between the membrane and the exterior aqueous solution. It is in this quantity that processes such as stripping of hydration water molecules are included, and these may possess much steeper temperature dependences, corresponding to apparent activation energies of 50 to 80 kJ mol^{-1}. This, by the way, is the range of activation energies commonly found with the specific types of transport discussed below.

Equation (1) predicts that: (1) the net movement ceases ($J_s = 0$) when $S_I = S_{II}$, so that, for an uncharged solute not adsorbed within the cell (such as urea in human erythrocytes), the equilibrium concentrations are the same in and out of the cell; (2) initial transport rate (when $S_{II} = 0$) as well as unidirectional flux, followed with the aid of labeled substance, is linearly related to concentration over the entire reasonable concentration range.

It should be noted, however, that if an ion is translocated by the same type of simple diffusion, it will distribute itself according to the existing membrane potential so that, in equilibrium,

$$S_{II} = S_I e^{-nF\Delta\psi m/RT} \tag{3}$$

where $\Delta\psi_m$ is the membrane potential in V, F is the Faraday constant (96.5 kC mol^{-1}), and n is the number of positive charges of the ion. Thus, for a membrane potential of -100 mV, passively distributed K$^+$ will reach an S_{II}/S_I of 46, while Mg^{2+} might attain a S_{II}/S_I ratio of 2120 at 30°C.

III. SPECIFIC TRANSPORT

Characteristics of this type of transport are its specificity and its saturability, both indicating that a finite number of membrane proteins are involved in the process.

Although the molecular mechanisms of this type of membrane translocation are many (cf. other chapters of this book), they all involve binding of the transported solute to a receptor (binding) site in the membrane, its movement across the membrane, and dissociation toward the other membrane face.

There is an important distinction between a situation where the binding site is simultaneously accessible from both sides of the membrane (we may call this a channel or pore) and one where it is accessible alternately from the one and from the other side (we may call this a carrier). The limited accessibility in this last case may be due to isomerization of the carrier protein during the transport process or to the shifting of a "gate" which is part of the surrounding membrane, presumably another associated protein.

A. Specific Channels or Pores

This case may be formally depicted as follows:

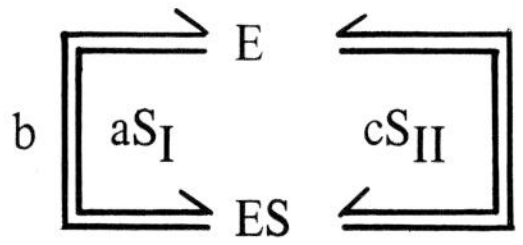

where a, b, c, d are rate constants for the association and dissociation of solute with channel receptor site E. The site is accessible both to S_I and to S_{II}, although the rates of association a and c may be different (say, for steric reasons).

The equation for rate of transport of S is easily derived to be

$$J_s = E_t \frac{adS_I - bcS_{II}}{b + d + aS_I + cS_{II}} \tag{4}$$

which, for unidirectional flux (as followed by a tracer) becomes

$$J_{s(I \rightarrow II)} = E_t \frac{adS_I}{b + d + aS_I + cS_{II}} \tag{5}$$

For initial flux

$$J_{so} = E_t \frac{ad\ S_I}{b + d + aS_I} = \frac{d\ E_t S_I}{(b + d)/a + S_I} \tag{6}$$

from which the J_{max} is equal to dE_t and the half-saturation constant K_T is equal to $(b + d)/a$. E_t represents the total receptor site concentration.

In analogy with enzymological practice, Equation (4) may be converted to the form

$$J_s = \frac{J_{max}^{\rightarrow} (S_I - S_{II}/K_{eq})}{K_{T_1} (1 + S_{II}/K_{T_2}) + S_I} \tag{7}$$

where K_{eq}, the equilibrium constant (for a simple channel equal to 1), is defined by ad/bc; $J_{max}^{\rightarrow}$, the maximum rate going from side I to side II, is equal to dE_t; K_{T_1}, the half-saturation constant going from I to II, is $(b + d)/a$; and K_{T_2}, the half-saturation constant going from II to I, is $(b + d)/c$. Finally, $J_{max}^{\leftarrow} = K_{T_2} J_{max}^{\rightarrow}|K_{T_1} K_{eq}$.

The channel described here does not operate in an active manner and it does not show the phenomenon of countertransport (see below).

B. Specific Carriers

The specific carrier may be formally represented as follows:

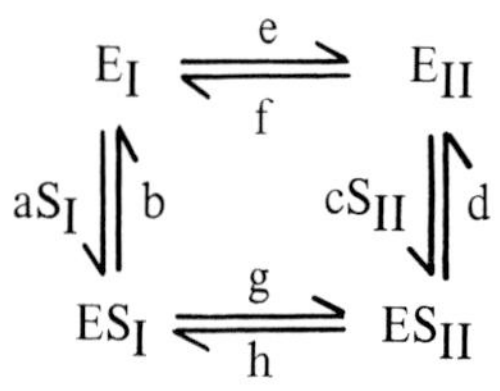

where the two membrane sides are distinct with respect to the location of the carrier.[1] (Phenomenologically equivalent expressions are obtained from a model where ES_I and ES_{II} are identical but E_I and E_{II} distinct.)[2]

Using a steady-state approach, such that the individual carrier "concentrations" are constant during the measurable reaction time but none of the steps of scheme II is rate-limiting, one obtains the following equation for the rate of transport of S:

$$J_s = E_t \frac{adfgS_I - bcehS_{II}}{(e + f)(bh + bd + dg) + a[d(f + g) + f(g + h)]S_I + c[h(b + e) + e(b + g)]S_{II} + ac(g + h)S_I S_{II}} \tag{8}$$

the unidirectional flux from side I to side II is then

$$J_{s(I \rightarrow II)} = E_t adgS_I[f + bchS_{II}/(bh + bd + dg)]/D \tag{9}$$

D being the same denominator as in Equation (8). The opposite unidirectional flux may be obtained by replacing each of the rate constants with its opposite counterpart from the schematic model above.

The initial rate of reaction from side I to side II is then

$$J_{s0} = E_t \frac{adfgS_I}{(e + f)(bh + bd + dg) + a[d(f + g) + f(g + h)]S_I}$$

so that $J_{max} = E_t dfg/[d(f + g) + f(g + h)]$

and $K_T = (e + f)(bh + bd + dg)/a[d(f + g) + f(g + h)]$ \quad (10)

Like in the case of channel transport, the equations can be readily converted to forms containing only measurable kinetic constants, thus, Equation (8) becomes

$$J_s = \frac{N_1 S_1 - N_2 S_{II}}{C + D_1 S_1 + D_2 S_{II} + D_{12} S_1 S_{II}} \tag{11a}$$

and

$$J_s = \frac{\overrightarrow{J_{max}} (S_1 - S_{II}/K_{eq})}{K_{T_1} (1 + S_{II}/K_{T_2}) + S_1 (1 + S_{II}/K_i)}$$

with $K_{eq} = N_1/N_2$, $K_{T1} = C/D_1$, $K_{T2} = C/D_2$, $\overrightarrow{J_{max}} = N_1/D_1$ $\qquad$ (11b)

and, a new expression not present in Equation (7), $K_i = D_1/D_{12}$. Inversion of this equation leads to

$$1/J_s = [K_{T_1} (1 + S_{II}/K_{T_2})/\overrightarrow{J_{max}} + S_1 (1 + S_{II}/K_i)/\overrightarrow{J_{max}} K_{eq}]$$

$$\times \; 1/(S_1 - S_{II}/K_{eq}) + (1 + S_{II}/K_i)/\overrightarrow{J_{max}} \tag{12}$$

A plot of $1/J_s$ against $1/(S_1 - S_{II}/K_{eq})$ for different S_{II} leads to a set of $1/J_{max_{app}}$ which are related to true $\overrightarrow{J_{max}}$ by

$$1/J_{max_{app}} = S_{II}/\overrightarrow{J_{max}} K_i + 1/\overrightarrow{J_{max}} \tag{13}$$

A plot of $1/J_{max_{app}}$ against S_{II} will yield $1/\overrightarrow{J_{max}}$ as the intercept with the ordinate and $-K_i$ as the intercept with the abscissa.

The plot of $1/J_s$ against $1/(S_1 - S_{II}/K_{eq})$ above also yields a family of $K_{T_{app}}$ which is related to the true K_T's as follows:

$$K_{T_{app}} = K_{T_1} (1 + S_{II}/K_{T_2})/(1 + S_{II}/K_i) + S_{II}/K_{eq} \tag{14}$$

Now K_{T_1} is known from measurement of initial rate of uptake at different S_1; K_i is known from the $1/J_{max_{app}}$: S_{II} plot; K_{eq} is equal to the equilibrium accumulation ratio (unity for nonenergized transports); then K_{T_2} can be calculated from any $K_{T_{app}}$ at an appropriate S_{II}. $\overleftarrow{J_{max}}$ is then derived from the condition that $K_{eq} = K_{T_2} \overrightarrow{J_{max}}/K_{T_1} \overleftarrow{J_{max}}$.[3]

There are a number of ways to distinguish between the "channel" and the "carrier" mechanism, based on measuring unidirectional fluxes under different conditions.[4] In the (1) zero-trans procedure (zt), we measure initial rates ($S_{II} = 0$); in the (2) equilibrium exchange (ee), we measure flows when $S_1 = S_{II}$; in the (3) infinite-trans (it) procedure, we set up a very high S_{II} and follow rates of flow of different S_1; in the (4) infinite-cis procedure (ic), we follow the rate of flow from extremely high S_1 toward various small S_{II}'s. Procedures (1) and (2) do not qualitatively distinguish between the concepts, but in (3) and (4) finite values for K_T are obtained only for the carrier mechanism so that the channel can be rejected. If $1/J_{max_{ee}} = 1/J_{max_{zt (I \to II)}} + 1/J_{max_{zt (II \to I)}}$, the simple channel concept is valid.

Another important diagnostic feature is the occurrence of the so-called countertransport which is caused by the bilateral "competition" of S_1 and S_{II} for the same carrier. Although this is predictable for all the complex carrier mechanisms we shall demonstrate it on a simplified version of Equation (8) for two competing solutes (most conveniently S and R as the unlabeled and labeled forms of the same compound, respectively), in a mediated diffusion system. Equation (8) can be drastically simplified if pseudoequilibria of the association-dissociation reactions can be assumed so that $b/c = d/c = K$ and $e = f$ and $g = h$ (and the same for S and R). Let $S/K = S'$ and $R/K = R'$. The appropriate equation for the rate of flow of R in the presence of S is

$$J_{R(S)} = 2gE_t \frac{R_I'(gS_{II}' + gR_{II}' + e) - R_{II}'(gS_I' + gR_I' + e)}{(gS_I' + gR_I' + e)(1 + S_{II}' + R_{II}') + (gS_{II}' + gR_{II}' + e)(1 + S_I' + R_I')}$$

(15)

In an experimental arrangement where we preincubate cells with R until $R_I = R_{II} = R$ and then add S_I, at that moment (when $S_{II} = 0$)

$$J_{R(S)} = J_{max} \frac{-gR'S_I'}{(gS_I' + gR' + e)(1 + R') + (gR' + e)(1 + S_I' + R')}$$

(16)

Since, by convention, movement into cells is taken as positive solute R will move out of cells against its concentration gradient at the expense of a flow of S into cells down its concentration gradient.[5]

In a channel mechanism the countertransport phenomenon does not occur. If Equation (4) is expanded to include movement of S so that

$$J_{R(S)} = E_t(adR_I/S_I - bcR_{II}/S_{II})/(b + d + aR_I/S_I + cR_{II}/S_{II})$$

under the above condition that $S_{II} = 0$ and $R_I = R_{II} = R$, $J_{R(S)} = -b$ has no meaning.

Coming back to Equation (8) we shall see that net flow will cease when

$$S_{II}/S_I = adfg/bceh$$

(17)

If no energy flows into the system through a coupled reaction S_{II} will reach the value of S_I in equilibrium and hence $adfg = bceh$. This applies to the so-called mediated or facilitated diffusion. If, however, energy is used to drive the transport of S uphill, then in equilibrium (for nonmetabolized molecules or ions!) $S_{II} > S_I$.

Then one or more of the constants in the numerator will be increased by this energy coupling, say, with the ATP-splitting reaction (in one type of primary transport) or with the dissipation of the gradient of H^+ or Na^+ (in the so-called secondary active transport).

The primary active transport systems so far known concern the transport of ions and, again, mainly the transport of cations, and most of them are rather complicated molecules or molecular complexes. It has not been possible to ascribe a molecular-level meaning to the rate constants in such transports so as to satisfy Equation (17), although one may envisage a "phosphorylation potential", an "oxidative pressure", "photon pressure", and the like, to affect the rate constants driving the process uphill.

The situation is somewhat clearer with the so-called secondary active transport which is "energized" by mediating a downhill (both concentration- and potential-wise) transport of a driving ion, either H^+ or Na^+.

However, to derive the rate equations one must consider the various possibilities of association of solute S and ion A with the carrier (E + S + A, E + A + S) and of its dissociation (ESA − S − A, ESA − A − S), as well as the fact that some of the complexes may not move across the membrane.

The schematic model, in rapid equilibrium

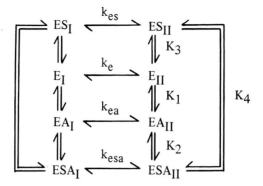

or in steady state

$$
\begin{array}{ccc}
ESA_I \xrightleftharpoons[f]{e} ESA_{II} & ESA_I \xrightleftharpoons[f]{e} ESA_{II} & ESA_I \xrightleftharpoons[f]{e} ESA_{II} \\
c \updownarrow d \quad h \updownarrow g & c \updownarrow d \quad h \updownarrow g & c \updownarrow d \quad h \updownarrow g \\
EA_I \quad EA_{II} & ES_I \quad ES_{II} & EA_I \quad ES_{II} \\
a \updownarrow b \quad j \updownarrow i & a \updownarrow b \quad j \updownarrow i & a \updownarrow b \quad j \updownarrow i \\
E_I \xrightleftharpoons[k]{l} E_{II} & E_I \xrightleftharpoons[k]{l} E_{II} & E_I \xrightleftharpoons[k]{l} E_{II}
\end{array}
$$

yields rate equations of considerable complexity. The initial rate equations are somewhat simpler and the kinetic parameters K_T and J_{max} can be derived from them (Table 1).

Table 1 of the section on pH effects lists the influences that increasing A_I, A_{II}, and $\Delta\psi$ will have on the parameters of these and similar models. Other predictions are made if a negatively charged carrier is assumed to exist.[6]

A somewhat more complicated situation arises if the steady-state approach is applied to the so-called random-sequence of binding, i.e., $E + A + S$ as well as $E + S + A$. The general expression for the initial rate here is of the form

$$ J_{so} = (\alpha S_I^2 + \beta S_I)/(\gamma S_I^2 + \delta S_I + \epsilon) \tag{18} $$

and it is known about such rate expressions that, in a Lineweaver-Burk arrangement, a biphasic convex plot is obtained if $\alpha\delta > \beta\delta$ and $\alpha\epsilon < \beta\delta$, while various types of concave plots are obtained for other combinations of these restrictive conditions.[7] It seems to hold for this situation that the convex plot is obtained for relatively low concentrations of the activator ion, while a concave one is obtained as A is increased.

The maximum accumulation ratio predicted by the above models generally relates S_{II}/S_I to A_I/A_{II} and the membrane potential as exponent of e. In the simplest cases, where only one loop of the carrier scheme is present, such as where k_{es} and k_{ea} are equal to zero, the expression is

$$ S_{II}/S_I = (ion_I/ion_{II})^n \, e^{-nF\Delta\psi/RT} = acegik/bdfhjl \tag{19} $$

where n is the number of cations H^+ or Na^+ bound to the carrier for efficient solute transport. For all other models the S_{II}/S_I ratio is less than shown by Equation (19).[8]

It should be noted that even if the system is tightly coupled (kinetically, this means that k_{ea} and k_{es} are zero) the ratio predicted by Equation (19) is not reached at higher concentrations

Table 1
KINETIC PARAMETERS FOR VARIOUS SECONDARY TRANSPORT MODELS

Model	J_{max}/E_t	K_T
Rapid equilibrium E + S + A	$\dfrac{\zeta^{1/2}k_{csa}(k_cK_1 + \zeta^{-1/2}k_aA_{II})}{k_{csa}\zeta^{1/2}(K_1 + A_{II}) + K_1k_c + \zeta^{-1/2}k_aA_{II}}$	$\dfrac{K_2[(K_1 + A_{II})(k_cK_1 + k_a\zeta^{1/2}A_I) + (K_1 + A_I)(k_cK_1 + k_a\zeta^{-1/2}A_{II})]}{A_I[k_{csa}\zeta^{1/2}(K_1 + A_{II}) + K_1k_c + \zeta^{-1/2}k_aA_{II}]}$
E + A + S	$\dfrac{k_c(k_{cs}K_4 + k_{csa}\zeta^{1/2}A_I)}{k_{cs}K_4 + k_{csa}\zeta^{1/2}A_I + k_c(K_4 + A_I)}$	$\dfrac{2K_3K_4k_c}{k_{cs}K_4 + k_{csa}\zeta^{1/2}A_I + k_c(K_4 + A_I)}$
E + S + A and E + A + S	$\dfrac{(k_cK_1 + k_a\zeta^{-1/2}A_{II})(k_{cs}\zeta^{1/2}A_IK_3 + k_{cs}K_1K_2)}{(k_{cs}K_1K_2 + k_{csa}\zeta^{1/2}A_IK_3)(K_1 + A_{II}) + (K_1K_2 + A_IK_3)(k_cK_1 + k_{csa}\zeta^{-1/2}A_{II})}$	$\dfrac{(K_1 + A_{II})(k_cK_1 + k_{cs}\zeta^{1/2}A_I) + (k_cK_1 + k_{csa}\zeta^{-1/2}A_{II})(K_1 + A_I)}{(k_{cs}K_1K_2 + k_{csa}\zeta^{1/2}A_IK_3)(K_1 + A_{II}) + (K_1K_2 + A_IK_3)(k_cK_1 + k_{csa}\zeta^{-1/2}A_{II})}$
Steady state E + A + S	$\dfrac{aegik\zeta^{1/2}A_I}{aA_I[fik\zeta^{-1/2} + gik + e\zeta^{1/2}(ih + gi + gk + gi)] + egi(k + 1)}$	$\dfrac{b(df + dg + eg\zeta^{1/2})(ik + il + jlA_{II}) + aikA_I(df + dg + eg\zeta^{1/2})}{aA_I[fik\zeta^{-1/2} + gik + e\zeta^{1/2}(ih + gi + gk + gi)] + egi(k + 1)}$
E + S + A	$\dfrac{cegik\zeta^{1/2}A_I}{cA_I[f\zeta^{-1/2}(hkA_{II} + ik) + e\zeta^{1/2}(hkA_{II} + ihA_{II} + gi + gk) + gik] + dfk\zeta^{-1/2}(hA_{II} + i) + gik(d + c\zeta^{1/2})}$	$\dfrac{b(k + 1)[df\zeta^{-1/2}(hA_{II} + i) + gi(d + e\zeta^{1/2}) + cegi\zeta^{1/2}A_I(a + k + 1)}{a cA_I[f\zeta^{-1/2}(hkA_{II} + ik) + e\zeta^{1/2}(hkA_{II} + ihA_{II} + gi + gk) + gik] + dfk\zeta^{-1/2}(hA_{II} + i) + gik(d + c\zeta^{1/2})}$
E + A + S, in E + S + A, out	$\dfrac{aegik\zeta^{1/2}A_I}{egi\zeta^{1/2}(k + 1) + aA_I[f\zeta^{-1/2}(hj + hk + ik) + e\zeta^{1/2}(hj + hk + gi) + gik]}$	$\dfrac{b(k + 1)(df + dg + eg\zeta^{1/2}) + aA_Id[f\zeta^{-1/2}(hk + hj + ik + gik + egik\zeta^{1/2}] + bdfh\zeta^{-1/2}A_{II}(k + 1)}{cegi\zeta^{1/2}(k + 1) + acA_I[f\zeta^{-1/2}(hj + hk + ik) + e\zeta^{1/2}(dj + hk + gi) + gik]}$

Note: In all cases $\zeta^{1/2} = e^{-nF\Delta\psi/2RT}$ and $\zeta^{-1/2} = e^{nF\Delta\psi/2RT}$

of S. In fact, S_{II}/S_I may be less than zero at very high concentrations. This situation is predictable on the basis of the following assumption. If the source of energy for the active transport, Q — e.g., the pH difference — is seen by the transport system only as it is limited within the membrane, its local "concentration" will be $Q = Q_t - ES$, Q_t being the total concentration and ES the concentration of the carrier complex to which Q can be bound. If the carrier concentration is high enough S may be bound to form a large concentration of ES so that Q will decrease substantially. If Q is part of one of the numerator constants of Equation (19) it can under these conditions decrease so that $S_{II}/S_I < 1$.[9]

In all the previous derivations it was assumed that one solute molecule is bound per carrier. If two molecules or ions are bound per carrier the binding of the second one or movement across the membrane of the doubly bound carrier will be either supported by the first molecule on the carrier (positive cooperativity) or it will be impaired (negative cooperativity). The corresponding reciprocal plots will then be concave or convex, respectively, corresponding to the general rapid-equilibrium equation for such a case which will have the form of Equation (18) and $\alpha\delta > \beta\gamma$, $\alpha\epsilon > \beta\delta$ for positive cooperativity and $\alpha\delta > \beta\gamma$, $\alpha\epsilon < \beta\delta$ for negative cooperativity.

From positive cooperativity in transport one can predict an interesting kinetic phenomenon that has been termed cotransport.[10] In a situation like that used for the study of counter-transport, if equilibrium is reached with a low concentration of solute, addition of another low concentration of the same or related solute will evoke a transient increase of the pre-equilibrated intracellular concentration, due to the parabolic dependence of rate on concentration in the low concentration range. If we assume a rapid-equilibrium system in which K_{es} is the dissociation constant of the first molecule, K_{ess} that of the second molecule, then preequilibrated labeled R will move inward (cotransport) if $1 > S(S + R)/K_{es}K_{ess}$.

REFERENCES

1. **Kotyk, A. and Janáček, K.,** *Membrane Transport — An Interdisciplinary Approach,* Plenum Press, New York, 1977, 207.
2. **Stein, W. D. and Lieb, W. R.,** A necessary simplification of the kinetics of carrier transport, *Isr. J. Chem.,* 11, 325, 1973.
3. **Cuppoletti, J. and Segel, I. H.,** Kinetic analysis of active membrane transport systems: equations for net velocity and isotope exchange, *J. Theor. Biol.,* 53, 125, 1975.
4. **Stein, W. D.,** Concepts of mediated transport, in *Membrane Transport,* Bonting, S. L. and de Pont, J., Eds., Elsevier/North-Holland, Amsterdam, 1981, chap. 5.
5. **Rosenberg, T. and Wilbrandt, W.,** Uphill transport induced by counterflow, *J. Gen. Physiol.,* 41, 289, 1957.
6. **Kotyk, A.,** Coupling of secondary active transport with $\bar{\mu}_{H^+}$, *J. Bioenerg. Biomembr.,* 15, 307, 1983.
7. **Segel, I. H.,** *Enzyme Kinetics,* Wiley-Interscience, New York, 1975.
8. **Kotyk, A.,** Critique of coupled vs. noncoupled transport of nonelectrolytes, in *5th Winter School on Biophysics of Membrane Transport,* Vol. 2, Agricultural University Press, Wroclaw, 1979, 49.
9. **Kotyk, A and Stružinský, R.,** Effect of high substrate concentrations on active transport parameters, *Biochim. Biophys. Acta,* 470, 484, 1977.
10. **Kotyk, A. and Janáček, K.,** *Cell Membrane Transport — Principles and Techniques,* 2nd ed., Plenum Press, New York, 1975, 89.

Chapter 2

pH EFFECTS ON MEMBRANE TRANSPORT

Arnošt Kotyk

TABLE OF CONTENTS

I. KINETIC EFFECTS

The majority of transport mechanisms and practically all those that have to do with the physiologically important solutes (both nonelectrolytes and ions) include a specific interaction of the transported solute with a membrane protein. Such membrane proteins resemble enzymes known from the aqueous ''soluble'' phase both in their molecular structure[1] and in their kinetic behavior,[2] the ''product'' of the reaction being identical with the substrate but translocated to the other membrane side.

One of the characteristic properties of enzymes, as well as membrane carrier proteins, is the dependence of their catalyzed reaction rate on pH which, in the vast majority of cases, is bell-shaped, there being an optimum at intermediate pH values. This has been analyzed for soluble enzymes[3,4] and shown convincingly to be due to protonation of the enzyme-active site which is functional only if it is half protonated, i.e., it does not function when no protons are bound, as well as when two protons are bound, as shown in Scheme I.

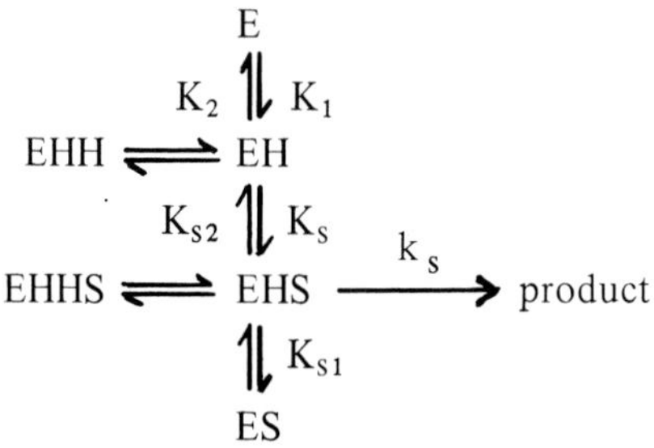

$$
\begin{array}{c}
\text{E} \\[2pt]
\text{K}_2 \Big\updownarrow \text{K}_1 \\[2pt]
\text{EHH} \rightleftharpoons \text{EH} \\[2pt]
\text{K}_{s2} \Big\updownarrow \text{K}_s \qquad k_s \\[2pt]
\text{EHHS} \rightleftharpoons \text{EHS} \longrightarrow \text{product} \\[2pt]
\Big\updownarrow \text{K}_{s1} \\[2pt]
\text{ES}
\end{array}
\qquad\qquad (\text{I})
$$

(E is the enzyme, S the substrate, and H hydrogen ion.)

The expression for initial rate based on the rapid-equilibrium assumption (i.e., all the forms above may be considered to be in a virtual equilibrium with one another) states that

$$
v_0 = \frac{E_t \cdot k_s}{1 + [H^+]/K_{s2} + K_{s1}/[H^+]} \cdot \frac{S}{K_s(1 + [H^+]/K_2 + K_1/[H^+])/(1 + [H^+]/K_{s2} + K_{s1}/[H^+]) + S} \qquad (1)
$$

From this, the apparent maximum rate V is easily derived as the first fraction of the expression, the apparent half-saturation constant K_m as the fraction appearing in the denominator of the right-hand-side compound expression.

Plotting the logarithm of V, K_m and their ratio against pH yields the curves shown in Figure 1.

In a transport reaction, the pH dependence is similar but somewhat more complicated, as one must take into account the fact that the membrane protein is exposed alternately to the extracellular and to the intracellular pH.

Assuming again that it is only the properly protonated carrier form which is productive, i.e., which moves across the membrane, and introducing the simplification that the system is intrinsically symmetrical, i.e., all the various dissociation constants are identical with their counterparts on the other membrane side, we may write the kinetic scheme as follows:

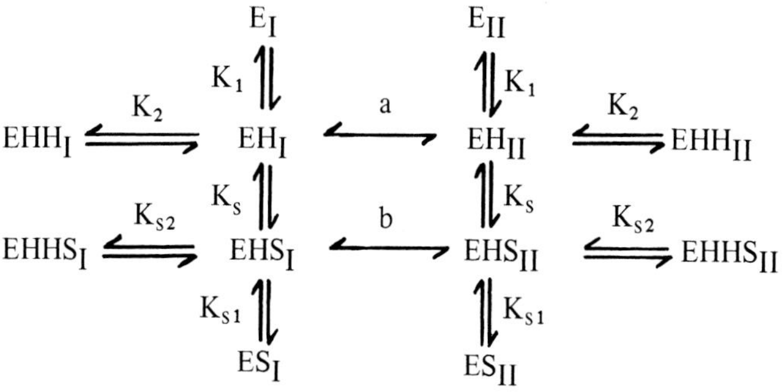

$$
\begin{array}{ccccc}
E_I & & & E_{II} & \\
\Big\updownarrow K_1 & & & \Big\updownarrow K_1 & \\
EHH_I \xrightleftharpoons{K_2} EH_I & \xrightarrow{a} & EH_{II} & \xrightleftharpoons{K_2} EHH_{II} & \\
\Big\updownarrow K_s & & & \Big\updownarrow K_s & \\
EHHS_I \xrightleftharpoons{K_{s2}} EHS_I & \xrightarrow{b} & EHS_{II} & \xrightleftharpoons{K_{s2}} EHHS_{II} & \\
\Big\updownarrow K_{s1} & & & \Big\updownarrow K_{s1} & \\
ES_I & & & ES_{II} &
\end{array}
\tag{II}
$$

The parameters of the initial rate equation (for $V = J_{max}$; $K_m = K_T$) are as follows:

$$
J_{max} = \frac{2E_t b}{1 + [H^+]_I/K_{s2} + K_{s1}/[H^+]_I + (b/a)(1 + [H^+]_{II}/K_{s2} + K_1/[H^+]_{II})}
\tag{2a}
$$

$$
K_T = \frac{K_s(2 + K_1/[H^+]_I + [H^+]_I/K_2 + K_1/[H^+]_{II} + [H^+]_{II}/K_2)}{1 + [H^+]_I/K_{s2} + K_{s1}/[H^+]_I + (b/a)(1 + [H^+]_{II}/K_{s2} + K_1/[H^+]_{II})}
\tag{2b}
$$

A plot analogous to that in Figure 1A, but taking into account intracellular pH, is shown in Figure 2. Spreading of the intersections of the straight lines with respect to Figure 1 indicates that in assigning the apparent pH values to the amino acid residues within the carrier-active site, one must be aware of the intracellular pH exerting its effect.[5]

II. THERMODYNAMIC EFFECTS

In the so-called secondary active transport, a cation (H^+ or Na^+, but only very exceptionally interchangeably) is cotransported with another solute, usually an organic molecule, such as sugar or amino acid. In fact, the concentration difference of this ion across the membrane, plus the membrane potential, constitutes the driving force for transporting such solutes uphill. If the cotransported ion is H^+ we can describe effects of pH at each of the membrane sides on kinetic parameters, as well as of ΔpH across the membrane on the equilibrium distribution of the transported solute. The kinetic equations involved derive from a transport scheme, such as

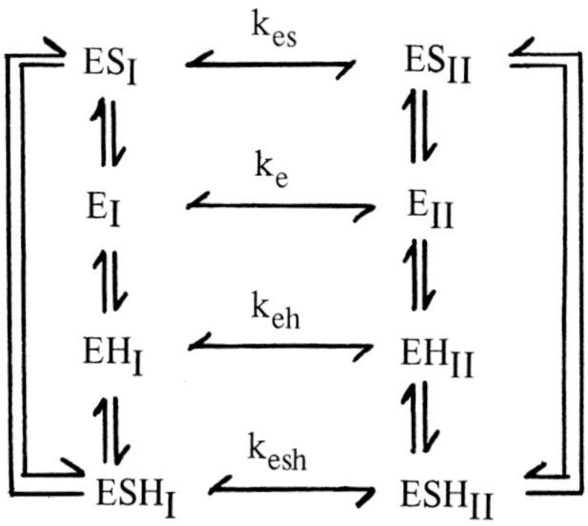

$$
\left[
\begin{array}{ccc}
ES_I & \xrightarrow{k_{es}} & ES_{II} \\
\Big\updownarrow & & \Big\updownarrow \\
E_I & \xrightarrow{k_e} & E_{II} \\
\Big\updownarrow & & \Big\updownarrow \\
EH_I & \xrightarrow{k_{eh}} & EH_{II} \\
\Big\updownarrow & & \Big\updownarrow \\
ESH_I & \xrightarrow{k_{esh}} & ESH_{II}
\end{array}
\right]
\tag{III}
$$

which includes all sequences of binding of solute S and activating ion H to the carrier enzyme E. Depending on whether the system is in rapid equilibrium (like above in the effect of pH

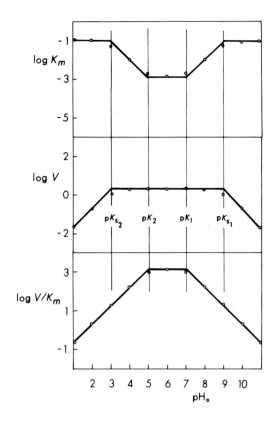

FIGURE 1.   Plot to determine intrinsic dissociation constants of an enzyme-catalyzed reaction. Calculated on the
basis of $K_1 = 10^{-7} M$, $K_2 = 10^{-5} M$, $K_s = 10^{-3} M$, K_{s1}
$= 10^{-9} M$, $K_{s2} = 10^{-3} M$, $k_s = 1$.

on rates), whether the sequence of binding is E + S + H or E + H + S, and on whether
all the complexes involved are mobile in the membrane, one may derive more or less complex
equations such as shown in the section on transport kinetics. It should be noted that all the
complexes that move across the membrane, including the free carrier, are in fact half-
protonated as discussed in the earlier section. However, the protons shown in the present
scheme are ligands that are transported across and released at the other side of the membrane,
whereas the protons involved in the catalytic effects on rates remain bound to the carrier
unless ambient pH changes considerably during the transport process.

Since the binding of a proton to the carrier increases its positive charge, the movement
of the complex will be favored in the direction into cells by the existing transmembrane
potential, which is under practically all conditions oriented with its negative side inward. It
will be seen in Table 1 how changes of pH_{out}, pH_{in}, and of the membrane potential affect
the kinetic parameters of proton-driven transports, for a variety of models, both in rapid
equilibrium and in steady state.

It should be observed that these effects may obscure or enhance the catalytic pH influences
discussed above.

Now as the ΔpH is part of the driving force it will be reflected in the accumulation ratio
of a H^+-symported solute as may be derived either from the appropriate rate equation or
from the thermodynamic expression for the electrochemical potential gradient of protons
across the membrane[6]

Table 1
EFFECTS OF THE DRIVING ION CONCENTRATION AT THE STARTING SIDE H_{out}^+ AND AT THE TARGET SIDE H_{in}^+ AND OF THE MEMBRANE POTENTIAL ON THE MAXIMUM RATE OF TRANSPORT J_{max} AND THE HALF-SATURATION CONSTANT K_T

Kinetic scheme, rapid equilibrium	Increase of					
	H_{out}^+		H_{in}^+		$\Delta\psi$	
	J_{max}	K_T	J_{max}	K_T	J_{max}	K_T
E + H + S, full	0	−	−	−	+	−
E + H + S, $k_{ch} = 0$	0	−	−	−	+	−
E + S + H, full	+	−	0	0	+	−
E + S + H, $k_{cs} = 0$	+	−	0	0	+	−
E + H + S and E + S + H	+	−	−	−	+	−
Steady state						
E + H + S, $k_{ch} = 0$	+	−	−	−	+	+
E + S + H, $k_{cs} = 0$	+	−	−	−	+	−
E + H + S in E + S + H out	+	−	0	+	+	−

Note: +, increase; −, decrease; 0, no effect.

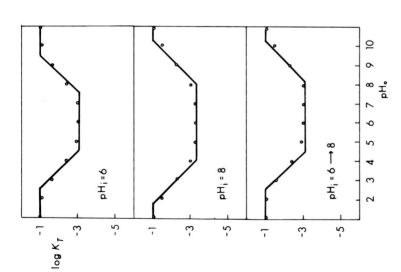

FIGURE 2. Plots generated on the basis of assuming the same constants as in Figure 1 but taking into account the effect of intracellular pH. This was equal to 6 above, to 8 in the center, and it changed from 6 to 8 (at pH$_o$ equal to 7) below. The dissociation constants are the same as in Figure 1, but a = 1 and b = 2.

$$\Delta \tilde{\mu}_{H^+} = \Delta \tilde{\mu}^{\circ}_{H^+_{in}} - \tilde{\mu}^{\circ}_{H^+_{out}} - RT \ln([H^+]_{in}/[H^+]_{out}) - F\Delta\psi \qquad (3)$$

where the standard zeroed potentials are concentration- and potential-independent and taken to be equal; R is the gas constant ($8.314 \ J \ mol^{-1}K^{-1}$), T is the absolute temperature (in K), F is the Faraday constant ($96.5 \ kC \ mol^{-1}$), and $\Delta\psi$ the membrane potential (in V). It is often assumed that in Scheme III only k_{esh} and k_e are nonzero, the argument being that if k_{eh} were appreciable it would short-circuit the H^+ concentration difference at the two membrane sides uselessly, and if k_{es} were mobile this would represent an intrinsic leak such that the system would operate less efficiently. Be it as it may, a system thus defined (termed "tightly coupled") is much easier to describe kinetically and it is the only variant that can be handled by a thermodynamic approach, proceeding from Equation (3). The maximum accumulation ratio of a H^+-driven solute is then derived by assuming that the Gibbs free energy available from the electrochemical potential gradient of H^+ is

$$\Delta G = n\Delta \tilde{\mu}_{H^+} \qquad and \qquad -\Delta G = RT \ln ([solute]_{in}/[solute]_{out}) \qquad (4)$$

where n is the number of protons accompanying the solute during transport. Hence, in equilibrium

$$[solute]_{in}/[solute]_{out} = ([H^+]_{out}/[H^+]_{in})^n \ e^{-nF\Delta\psi/RT} \qquad (5)$$

It is often difficult to determine the true ΔpH just across the membrane, not to mention uncertainties of estimating the actual membrane potential,[6] so that more often than not the accumulation of solute calculated from macroscopic assays of ΔpH and $\Delta\psi$ does not tally with experimentally determined $\Delta\mu_{solute}$. Moreover, the maximum attainable value of $\Delta\mu_{solute}$ is observed only at low concentrations of solute, as if the saturation of the carrier with solute at its high concentration were not adequately met by the supply of H^+.[7]

III. OTHER pH-DEPENDENT TRANSLOCATIONS

A. Dissociable Molecules

It should be observed that some molecules capable of dissociating or associating a proton but existing also in the undissociated form, such as weak organic acids or weak bases, will distribute themselves unevenly across biological membranes, according to the following simple principle. If the neutral form is the only one that can permeate freely — and this is most often the case in the above-mentioned examples — it will reach identical concentrations in the two membrane-bathing solutions, but then will dissociate or associate the proton according to the ambient pH. From measuring the analytical concentrations of the substances (neutral plus charged) at both membrane sides, one may in fact deduce the intracellular pH, using the following formulas:

$$pH_{in} = pH_{out} + \log \left[\frac{[solute]_{in}}{[solute]_{out}} (1 + 10^{pKa - pHout}) + 10^{pKa - pHout} \right] \qquad (6)$$

(for weak acids, used if $pH_{in} > pH_{out}$), and

$$pH_{in} = pH_{out} - \log \left[\frac{[solute]_{in}}{[solute]_{out}} (1 + 10^{pHout + pKb - 14}) + 10^{pHout + pKb - 14} \right] \qquad (7)$$

(for weak bases, used if $pH_{in} < pH_{out}$). Here pK_a and pK_b are the negative logarithms of the dissociation constants of the acid and the base, respectively.

B. Indirect Effects

There are various indirect effects of pH on membrane transport parameters that have to do with the local interactions of membrane constituents, especially charged phospholipids. Among the more important ones one should mention the surface potential ψ_s, generally believed to be caused by the negative charges predominating on membrane surfaces, due especially to phosphate anionic residues. It causes the concentration of a charged solute at the membrane (S_m) to be different from the "analytical" bulk concentration (S_b) as follows:

$$S_m = S_b\, e^{-nq\psi_s/kT} \tag{8}$$

where n is the number of positive charges on the solute, q is the charge of the electron (0.1602 fC), and k is the Boltzmann constant $(1.38 \cdot 10^{-23}$ J K$^{-1})$. From this expression it is useful to extract the quantity $e^{-q\psi_s/kT}$ and set it equal to, say, y. Then $S_m = S_b y^n$.

This potential exerts a variety of effects on the transport of ions and ion-accompanied transports of other solutes.[8] Thus, in a symport of a cation with an anion the K_T of anion uptake (in the simplest case of random binding) will be $K_a \cdot y$, the K_T of cation uptake will be K_c/y. The J_{max} of the anion will be given by $J_{max,0} \cdot S_c y/(K_c + S_c y)$, that of the cation by $J_{max,0} \cdot S_a/(K_a y + S_a)$. (Here subscripts a refer to anion, subscripts c to cation concentrations or constants.)

Because cations decrease the surface potential low pH will do the same and thus affect the K_T and J_{max} of a variety of transports.

IV. CONCLUSION

It will have become apparent that ambient pH can affect various transports in different ways.

1. Every saturable transport will be less effective (lower J_{max} and higher K_T) at the ends of the pH scale because of improper protonation of the carrier-active site.
2. Most transports of cations will have both K_T and J_{max} increased and most transports of anions will have the two parameters decreased at low pH because of decreased surface potential.
3. Transports of solutes in symport with H^+ will be affected in a variety of ways by low pH, invariably by a decrease of K_T and, in most cases, by an increase of J_{max}.

REFERENCES

1. **Carafoli, E. and Scarpa, A.,** Transport ATPases, *Ann. N.Y. Acad. Sci.,* 402, 207, 1982.
2. **Kotyk, A. and Janáček, K.,** *Cell Membrane Transport — Principles and Techniques,* 2nd ed., Plenum Press, New York, 1975.
3. **Dixon, M. and Webb, E. C.,** *Enzymes,* Longmans, Green, London, 1958, 116.
4. **Michaelis, L.,** *Die Wasserstoffionenkonzentration,* Springer-Verlag, Berlin, 1922, 48.
5. **Kotyk, A. and Horák, J.,** Effects of pH and temperature on saturable transport processes, in *Water and Ions in Biological Systems,* Vasilescu, V., Pullman, B., Packer, L., and Leahu, L., Eds., Plenum Press, New York, 1985.
6. **Kotyk, A.,** Coupling of secondary active transport with μ_{H^+}, *J. Bioenerg. Biomembr.,* 15, 307, 1983.
7. **Kotyk, A. and Stružinský, R.,** Effect of high substrate concentrations on active transport parameters, *Biochim. Biophys. Acta,* 470, 484, 1977.
8. **Roomans, G. M. and Borst-Pauwels, G. W. F. H.,** Co-transport of anions and neutral solutes with cations across charged biological membranes. Effects of surface potential on uptake kinetics, *J. Theor. Biol.,* 73, 453, 1978.

Chapter 3

THE EFFECT OF MEMBRANE LIPIDS ON PERMEABILITY AND TRANSPORT IN PROKARYOTES

Ronald N. McElhaney

TABLE OF CONTENTS

I. INTRODUCTION

The cells of prokaryotic microorganisms, because of their relative genetic, morphological, and biochemical simplicity, possess a number of inherent advantages over the more complex eukaryotic cells for studies of the structure and function of biological membranes generally, and for the investigation of membrane permeability and transport processes in particular. In addition to their relatively rapid generation times and ease of cultivation, most prokaryotes contain only a single membrane, the limiting or plasma membrane, as well as an extracellular wall structure. In contrast, most eukaryotic cells contain a variety of cytoplasmic and organelle membrane systems in addition to their plasma and nuclear membranes. Thus, the isolation of substantial quantities of pure plasma membrane for biochemical or biophysical study is a fairly straightforward process for most prokaryotic microorganisms,[1,2] particularly for the mycoplasmas, which lack a cell wall.[3] On the other hand, the isolation of a particular membrane from eukaryotic cells in relatively pure form and in usable amounts can be a formidable task, especially in the case of plasma membranes.[4] Finally, the lipid polar head group and fatty acid compositions of prokaryotic membranes are generally considerably simpler than those of eukaryotic membranes. In prokaryotes, as few as three major phospho- and/or glycolipid types may be present, and each of these typically possesses only a relatively small number of saturated, monounsaturated, branched-chain, or alicyclic fatty acids. In addition, sterols are generally absent in prokaryotes, although pentacyclic triterpenes of the hopanoid family, which are structurally and functionally analogous to sterols, may be present.[5,6] In eukaryotes, a larger number of phospholipids are present, as well as sphingolipids and sterols. In addition, a number of polyunsaturated as well as saturated and monounsaturated fatty acids are found. Thus, a prokaryotic plasma membrane may contain as few as ten major lipid molecular species (chemically unique combinations of polar head groups and fatty acyl chains), while a typical eukaryotic plasma membrane possesses at least ten times as many.[7]* It is thus not surprising that our knowledge of the structure and function of prokaryotic plasma membranes generally greatly exceeds that of eukaryotic plasma membranes.[1,3,9] The human erythrocyte membrane may be a partial exception to this generalization, since red blood cells, which are highly differentiated and simplified eukaryotic cells, possess many of the advantages offered by prokaryotes.

In addition to possessing relatively simple membrane lipid polar head group and hydrocarbon chain compositions, in many cases considerable alterations in membrane lipid structure can be induced in prokaryotic plasma membranes through genetic and/or environmental manipulation. In a number of *Escherichia coli* fatty acid auxotrophs,[1,2] for example, or in several mycoplasma species,[3] marked changes in membrane lipid fatty acid composition can be produced by culturing cells in the presence of appropriate exogenous fatty acids, which are readily incorporated to high levels. In several mycoplasmas, notably *Mycoplasma mycoides*[10] and *Acholeplasma laidlawii B*,[11] "fatty acid-homogeneous" membranes (i.e., membranes whose lipids contain essentially only a single species of fatty acyl group) can even be obtained. Moreover, substantial amounts of cholesterol and other sterols can be incorporated into the plasma membranes of prokaryotes such as *E. coli*[12] and *A. laidlawii*,[9] even though these organisms normally completely lack these constituents. Conversely, certain sterol-requiring mycoplasmas, which normally possess high levels of cholesterol in their plasma membranes, can be adapted to grow with very low levels of cholesterol.[13,14] The

---

*   The protein composition of prokaryotic plasma membranes, however, may be quite complex. Thus, the bacterium, *Escherichia coli,* has been estimated to contain at least 300 different polypeptides in its plasma membrane, while the simpler mycoplasma, *Acholeplasma laidlawii* B, has at least 140.[8] This situation doubtlessly arises from the fact that in most prokaryotes *all* membrane-associated processes must be localized on a single membrane system, whereas a number of specialized membranes exist in most eukaryotic cells.

fluidity and phase state of the membrane lipids of these and other microorganisms can thus be varied in a marked yet controlled manner through variations in the chemical composition of the hydrophobic core of their lipid bilayers. Of course, changes in the fatty acid composition and sterol content of the plasma membranes of some eukaryotic cells are also possible, but the degree of compositional alteration obtainable is quite limited in comparison to prokaryotic systems. Moreover, the presence of sterols and of biochemical compensatory mechanisms in most eukaryotic cells further restricts one's ability to significantly alter the physical properties of the membrane lipids.

The membrane lipid polar head group composition of prokaryotic organisms can also be markedly altered by environmental manipulations or by the utilization of biochemically defined mutants conditionally defective in various portions of the phospho- or glycolipid *de novo* biosynthetic pathways. In *E. coli* mutants, for example, not only can the proportions of the "normal" membrane lipids (phosphatidylethanolamine [PE], phosphatidylglycerol [PG], and cardiolipin) be considerably varied, but membranes containing substantial quantities of metabolic intermediates (phosphatidic acid [PA], phosphatidylserine [PS], or diglyceride), which are normally present in the membrane only in trace quantities, can be produced.[7] Moreover, the membrane lipid/protein ratio can be altered using conditional mutants defective at early steps in the lipid biosynthetic pathway. These mutants should prove very useful in studies of the relationship between lipid bilayer surface properties (chemical composition, degree of hydration, charge, etc.) and various membrane-associated processes, including membrane permeability and transport. Moreover, since the fluidity and phase state of the membrane lipids depend on the structure of their polar head group as well as on the structure and length of their hydrocarbon chains,[15] the physical properties of the *E. coli* membrane lipids can be altered without changing fatty acid or sterol composition, a possibly advantageous feature in certain types of studies. Again, although a small number of biochemically defined choline, inositol, and sterol auxotrophs have been isolated from eukaryotic microorganisms and from a few mammalian cell lines, these have proven less useful for membrane studies, since the increased biochemical complexity and the presence of compensatory metabolic pathways limit the polar head group variability which can be obtained with fully viable cells.[7]

Prokaryotic microorganisms also possess several important advantages over eukaryotic cells, specifically for investigations of membrane permeability and transport processes. Because prokaryotic cells normally lack cytoplasmic and organelle membranes, they usually behave essentially as one-compartment systems bounded by a single permeability barrier, the plasma membrane. In contrast, eukaryotic cells typically contain a number of intracellular compartments, each of which is bounded by its own permeability barrier. Moreover, in prokaryotic cells all protein-mediated transport systems are, of course, associated exclusively with the plasma membrane, whereas in eukaryotic cells a given molecule or ion may be a substrate for several different transport systems associated with different cytoplasmic or organelle membranes in addition to the plasma membrane. Finally, the ability to construct and isolate transport mutants makes it possible to selectively alter, delete, or enrich prokaryotic plasma membranes with the various protein components of active transport systems and, thereby, to gain additional insight into the function of these components, whereas this is more difficult with eukaryotic cells. For these and related reasons, the actual molecular mechanisms of facilitated diffusional and active transport processes and the effect of membrane lipid physical properties on these processes are better understood in bacteria than in other cell types.[16,17]

In this chapter I will attempt to summarize and critically evaluate recent studies from this and other laboratories on the effect of membrane lipids on the movement of molecules across the membranes of prokaryotic microorganisms. More specifically, this contribution will focus primarily on the relationship between membrane lipid fluidity and phase state and passive permeability and various protein-mediated transport processes. Before embarking on this analysis, however, brief discussions of the definition and measurement of membrane lipid fluidity and phase state and of the interpretation of diffusional and transport experiments are in order.

II. DEFINITION OF MEMBRANE LIPID FLUIDITY

The terms "viscosity" and "fluidity" were originally developed by physical chemists to describe the flow properties of macroscopic fluid systems. Viscosity is a measure of the frictional resistance that a fluid offers to an applied shearing force, that resistance resulting from a transfer of momentum from one layer of the moving liquid to the next. The fluidity of a fluid is simply the reciprocal of its viscosity and is thus a measure of the tendency of that fluid to flow in response to an applied force. The viscosity (or fluidity) of a system can be determined by measuring the rate of settling of a sphere in a fluid, as the frictional force retarding the movement of the sphere is directly proportional to the viscosity of that fluid. For most fluids, the viscosity decreases with increasing temperature and increases with increasing pressure. However, under a given set of environmental conditions, a single coefficient of viscosity is sufficient to fully describe a particular fluid system, at least macroscopically.

The terms fluidity and viscosity are often used in membrane research, usually in a rather qualitative way, to indicate something about the tightness of packing and/or about the relative mobilities of the lipid and protein components in the membrane. Although these terms are useful, they are inherently ambiguous when applied in a general sense to a highly anisotropic, asymmetric, two-dimensional structure such as a biological membrane, which, in addition, is only a few molecules thick.[18,19] Clearly, the organization and mobilities of lipids and proteins will be quite different when measured from within the plane of the membrane than when measured across the membrane plane. Not only does the effective or "average" fluidity of a biological membrane differ within and across the membrane plane, but also within the membrane structure itself. This is a consequence of the existence of gradients of motion and orientational order within the lipid[22-26] and protein[27,28] molecules themselves. It is clear from these and other considerations that no single "coefficient of viscosity" can adequately describe the fluid properties of a biological membrane.

The local or "microviscosity" of the lipid and protein components of model and biological membranes are usually determined by one or more of a variety of spectroscopic techniques. In considering the results of such studies, one should remember that these techniques often measure primarily either *average orientational order* or *average rates of motion* (again only within certain time scales); often single techniques do not provide reliable measures of both of these components of fluidity simultaneously. Although orientational order (usually expressed as an order parameter, S) and rates of motion (usually expressed as a relaxation time, T, or correlation time, τ) seem usually to be inversely related, as in the case of the phospholipid bilayer gel to liquid-crystalline phase transition, a simple inverse relationship between order and motion may not always obtain. For example, the presence of integral membrane proteins in membranes has been reported by several techniques to decrease both the motional rates and the orientational order of phospholipid hydrocarbon chains[9,29,30] (but see References 23, 24, and 31). Still, it seems most likely that changes in temperature, fatty acid composition, and cholesterol content do generally affect the order and motional rates of various portions of the phospholipid molecule in opposite ways, although this has by no means been rigorously established, even for simple model systems. In this chapter, the term "fluidity" will be used to describe both the relative degree of orientational order and relative rates of motion, although in most studies both of these parameters may not have been rigorously determined. Moreover, the term "fluidity" will be used only to describe the biologically relevant liquid-crystalline state of the lipid bilayer. I will thus differentiate between changes in membrane lipid *fluidity* (within the liquid-crystalline state) and changes in membrane lipid *phase state*, although, of course, orientational order and rates of motion are profoundly affected by lipid phase transitions.

III. ARRHENIUS PLOTS OF PERMEABILITY AND TRANSPORT PROCESSES

Studies of the temperature dependence of chemical, enzymatic, or transport processes can often provide information valuable for an understanding of the molecular mechanisms of these processes. The effects of temperature on the rate constants characterizing a chemical or a biological process are often analyzed in terms of an empirical activation energy (E_a) according to the Arrhenius equation, which may be written as

$$\log k = - \frac{E_a}{2.3R} \cdot \frac{1}{T}$$

where k is the rate constant of interest, R the gas constant, and T the absolute temperature. In practice the numerical value of E is determined from the slope of a plot of log k vs. $1/T$ (the Arrhenius plot). For most chemical reactions, and for chemical reactions catalyzed by most soluble enzymes, Arrhenius plots are linear over the accessible temperature range, although in both cases exceptions are known.[32,33] In these cases, the enthalpies of activation of these processes, which differ from their apparent activation energies only by a quantity equal to RT (about 0.6 kcal/mol), are essentially temperature invariant. For membrane-associated transport systems and enzymes, however, nonlinear Arrhenius plots are often obtained.[17,34] In most publications Arrhenius plots consisting of a relatively sharp break between two (or sometimes more) straight line segments are reported, while in a few cases Arrhenius plots of membrane-associated functions are depicted as smooth curves. Normally, even abrupt changes in the apparent activation energies of membrane transport or enzymic processes are not accompanied by a significant change in the reaction rate, although in several cases actual jump discontinuities in Arrhenius plots of membranous enzymes have been reported.[35] It has been proposed that true jump discontinuities can only arise as a thermodynamic consequence of a phase change.[36] However, a break or change in slope in an Arrhenius plot, which is not accompanied by a marked change in reaction rate, can arise from a number of causes.[32,33]

One should be aware that the use of the simplified form of the empirical Arrhenius equation given above involves a number of assumptions about the process of interest. Specifically, it assumes that the reaction geometry and activation entropy, as well as the activation enthalpy, are independent of temperature! These assumptions are not always met, even for relatively simple chemical reactions.[33] Of course, if a simple linear relationship between the logarithm of k and $1/T$ is indeed obtained for a particular process, then these assumptions are probably valid, at least to a first approximation. However, if a validly constructed Arrhenius plot departs substantially from a simple linear relationship, then at least one (and perhaps more) of these factors must be temperature dependent. If this is so, then it is no longer valid to employ a simplified form of the Arrhenius equation and the slope of the Arrhenius plot is no longer a reliable indicator of the reaction enthalpy. Since it is generally very difficult experimentally to evaluate activation enthalpies, activation entropies, and reaction geometries and their temperature dependencies, particularly for complex biological processes, in practice a valid mechanistic interpretation of an Arrhenius plot of a biological process that deviates from a simple linear relationship is usually not possible. These points have recently been stressed by Bagnall and Wolfe[37] in their useful critique of the use of Arrhenius plots in plant research.

It should also be pointed out that an Arrhenius plot of an enzymic reaction or transport process will normally be linear only if a single species of catalyst is responsible for the chemical process under study, and only if one particular step in the overall reaction is rate-limiting over the entire temperature range examined. In biological membranes, several enzymes or transport systems may simultaneously participate in a given process. Moreover, membrane enzymic reactions and transport processes are generally complex, multistep processes, and each partial reaction in the overall process may have a different temperature dependence (and a different lipid dependence).[34] Dixon and Webb[38] and Han[32] have discussed the effects of temperature on the rate of enzymic reactions generally and have formulated experimental approaches to recognize and correct for some of the more trivial effects of temperature. These effects can include such things as unrecognized temperature-induced changes in the pH of aqueous buffers, in solution viscosity, and in substrate-binding affinity (K_m). In addition, Han[32] has analyzed other factors that may produce nonlinearity in Arrhenius plots of enzymic reactions, and classified them into two categories: (1) thermodynamic factors, including all secondary equilibrium reactions that modify the elementary process being catalyzed, and (2) kinetic factors, attributed to changes in the rate-limiting step occurring within the experimental temperature range. Although the above treatments were developed for soluble enzymes, they apply to membrane enzymes and transport systems as well. In much of the transport work reviewed in subsequent sections, it appears that the factors noted above have usually not been explicitly considered, with the result that at least some of the conclusions reached must be regarded as tentative.

Several attempts have been made to develop a quantitative and systematic analysis of the mechanistic basis for nonlinear Arrhenius plots in membranous systems, taking into account the unique properties of these systems due to their existence in a lipid environment, the fluidity and phase state of which can also vary with temperature. Wynn-Williams[39] has proposed that the sudden change in the apparent activation energy of membrane enzymes could be due to the simultaneous presence of pure lipid (in the gel state) and of enzyme-lipid phases (in a fluid state) in the membrane. If enzyme activity depends on the composition of the enzyme-lipid phase, the temperature dependence of lipid solubility in the enzyme-lipid phase can lead to a sudden change in the apparent enzyme activation energy within the lipid phase transition temperature range, without the activity of the functional enzyme molecules undergoing an actual discontinuity. This is because the actual enthalpy of the "activated state" of the enzymic reaction is no longer equal to the slope of the Arrhenius plot of enzyme activity within the phase transition range. If this treatment proves generally valid for membrane enzymes and transport systems, it removes a theoretical difficulty, since it is no longer necessary to assume that a marked change in the activation enthalpy is exactly compensated for by a change in the activation entropy at the break temperature. However, the existence of an apparent inherent energy-entropy compensation in the passive permeation of simple lipid bilayers at their phase transition temperatures has recently been presented.[40] Thilo et al.[41] have provided evidence for membrane lipid phase transition-induced changes in the activities of several transport systems that are compatible with the proposal of Wynn-Williams, in that a differential partitioning of the transport system into gel and liquid-crystalline domains is postulated (see Section VI.A). However, the recent work of Silvius and McElhaney[42] on the $Mg^{2+} + Na^+$-ATPase of the *A. laidlawii* B plasma membrane is not supportive of this proposal, since the *physical state* of the boundary lipid surrounding this enzyme seems to determine its activity, rather than the composition of the annular lipid. Finally, Silvius and McElhaney[43] have systematically derived the rate-temperature relationships for a variety of physical models of membrane rate processes in order to predict the Arrhenius plot shape appropriate to each. Interestingly, only a few models predict Arrhenius plots with the "biphasic linear" form most commonly reported in studies of membrane enzymes and transport systems. Instead, most models predict Arrhenius plots consisting of

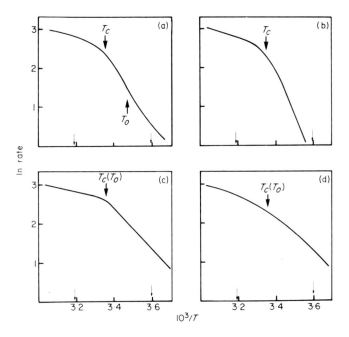

FIGURE 1.   Representative theoretical Arrhenius plots corresponding to some of the equations derived for several classes of physically realistic models of lipid-protein interactions (for details, see Silvius and Mc-Elhaney[43]. In all cases the lipid phase transition midpoint (denoted T_c or $T_c[T_o]$) has been arbitrarily set at 25°C (298.2 K). The small arrows along the x-axis indicate the usual limits of temperature variation (5 to 40°C or 278.2 to 313.2 K) used in most biological studies.

smooth curves (see Figure 1). However, many of the models yield plots which can be fit to two intersecting straight lines with a quite modest experimental error, particularly if the slope change around the "break" temperature corresponds to a change in apparent activation energy of less than 15 to 20 kcal/mol, and the temperature range examined is not large (Figure 2). These findings indicate a need for rigorous analysis of Arrhenius plot data in terms of graph shapes other than sets of intersecting straight lines and for a cautious interpretation of the physical basis of Arrhenius plot "breaks". The need for accurate determinations of the true maximum rates of the membrane-associated process of interest, and at a large number of experimental temperatures, for the valid interpretation of Arrhenius plots has been stressed recently by several groups,[44,45] and Silvius et al.[46] have demonstrated how unrecognized temperature-dependent variations in K_m can produce Arrhenius plot artifacts. Also, Keleti[47] has recently called attention to numerical errors common in the improper use of Arrhenius and van't Hoff plots. Moreover, the fitting of Arrhenius plot data points by eye, as is usually done, may lead to controversies over whether the author's subjective representation is the most correct one. In particular, a tendency to draw two straight lines through data points which actually fall on a single, continuously curving line is often evident. Recently, Sprague et al.[45] have developed statistical methods of assessing the goodness of fit of Arrhenius plot data points by various types of curves, as well as by straight-line segments. This is an important development, since of course two straight lines will always fit a set of curvilinear points better than a single straight line. To utilize these approaches effectively, however, a number of determinations of the rate of the process of interest at each experimental temperature must be available, and the variance between replicate measurements must be determined. Unfortunately, in little of the present literature

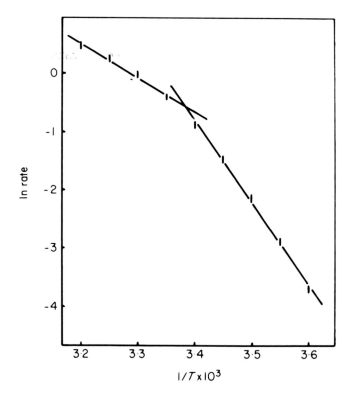

FIGURE 2. The fit of a curved Arrhenius plot with two intersecting straight lines. The data points plotted were computer generated from an equation giving activation enthalpies of 30 kcal/mol at 10°C and 10 kcal/mol at 35°C and producing a smooth, curved plot over the entire temperature range. The vertical line segments represent 5% error bars centered on the computed curve. This curved Arrhenius plot can actually be quite well fit, within experimental error, by two straight-line segments having apparent activation enthalpies of 11.6 kcal/mol above and 28.1 kcal/mol below the "break" temperature. However, no discrete break in this curve actually exists. (With permission from Silvius, J. R. and McElhaney, R. N., *J. Theor. Biol.*, 88, 135, 1980. Copyright: Academic Press Inc. (London) Ltd.)

is this information provided. Thus, in many published studies, evidence for the existence of a discrete Arrhenius plot "break" is equivocal, and this has important consequences for the interpretation of the experimental results.

IV. MEMBRANE LIPIDS AND PASSIVE DIFFUSION

A. Nonelectrolytes

The relationship between membrane passive permeability to nonelectrolytes and membrane lipid composition has been most systematically investigated with the simple prokaryote *A. laidlawii* B. In a series of studies by McElhaney and co-workers,[48-53] the fatty acid composition and cholesterol content of the *A. laidlawii* B plasma membrane were varied and the rates at which a number of nonelectrolytes passively diffuse into or out of intact cells, and into and out of liposomes prepared from the total membrane lipid, were studied as a function of temperature. Since alterations in membrane fatty acid composition or cholesterol content do not alter the qualitative or quantitative distribution of membrane

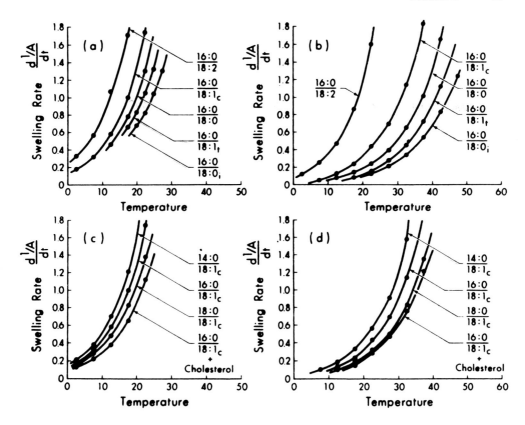

FIGURE 3. (a and c) Initial swelling rates in isotonic glycerol of intact cells of *A. laidlawii* B grown in the presence of different combinations of fatty acids, with or without cholesterol, as a function of temperature; (b and d) initial swelling rates in isotonic glycerol of liposomes, prepared from the total membrane lipids of *A. laidlawii* B, as a function of temperature. Under the experimental conditions employed here, the relative initial swelling rates, measured optically, are proportional to the relative rate of passive glycerol entry into cells and liposomes. (From McElhaney, R. N., De Gier, J., and van der Neut-Kok, E. C. M., *Biochim. Biophys. Acta*, 298, 500, 1973. With permission.)

proteins, and have only modest effects on lipid polar head group distribution,[9] it was possible to selectively study the effects of variations on the nature of the hydrophobic core of the plasma membrane of this organism on its passive permeability properties. The nonelectrolyte permeabilities of intact cells and derived liposomes were found to be markedly dependent on the chemical structure and chain length of the membrane lipid fatty acids. The biosynthetic incorporation of branched-chain or unsaturated fatty acids, or fatty acids of reduced chain length, increased nonelectrolyte permeability to a similar extent in both cells and liposomes (see Figures 3 and 4, respectively). The nonelectrolyte permeability of both plasma and liposomal membranes above their phase transition temperatures was also reduced by the incorporation of cholesterol. The mean activation energy values calculated for the permeation of several nonelectrolytes into intact cells and into liposomes were the same, within experimental error, and did not depend on the fatty acid composition or cholesterol content of the membrane (see Table 1). Mean activation energy values did, however, depend on permanent structure in such a way as to suggest that most nonelectrolytes permeate both natural and artificial membranes as single, fully dehydrated molecules. The passive permeabilities of a series of membranes of different fatty acid compositions were inversely related to their gel to liquid-crystalline membrane lipid phase transition temperatures, suggesting that permeation rates increase with increases in membrane lipid fluidity. Generally, similar

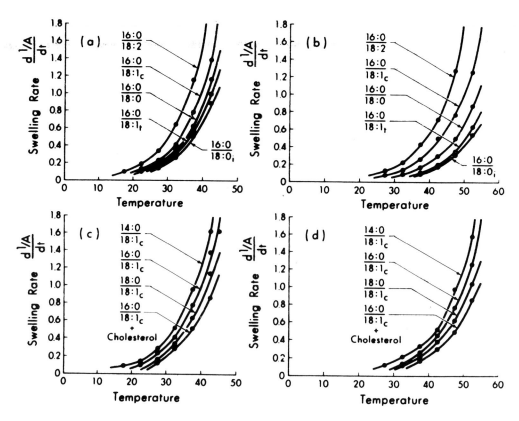

FIGURE 4.   (a and c) Initial swelling rates in isotonic erythritol of intact cells of *A. laidlawii* B, grown in the presence of different combinations of fatty acids, with and without cholesterol, as a function of temperature; (b and d) initial swelling rates in isotonic erythritol of liposomes, prepared from the total membrane lipid of *A. laidlawii* B, as a function of temperature. Initial swelling rates are measures of the rate of passive erythritol entry into cell and liposomes. (From McElhaney, R. N., De Gier, J., and van der Neut-Kok, E. C. M., *Biochim. Biophys. Acta, 298*, 500, 1973. With permission.)

results have been reported for the lipid dependence of glycerol passive permeation in an *E. coli* unsaturated fatty acid auxotroph by Eze and McElhaney.[54]

It has also been established that small changes in the chemical structure of the sterol molecule present in the *A. laidlawii* B membrane can significantly alter the passive permeability of intact cells by altering the physical state of the membrane lipid. De Kruijff et al.[51] have demonstrated that the presence of cholesterol itself decreased the fluidity of the hydrocarbon chains of the membrane lipids and increased the tightness of packing in the lipid bilayer, thus reducing the nonelectrolyte permeability of both intact cells and liposomes. The incorporation of epicholesterol, the 3α-hydroxy epimer of cholesterol, had no detectable effect on membrane lipid fluidity or packing and also did not alter the passive permeability of model or biological membranes. On the other hand, the presence of cholest-3-one, a keto analog of cholesterol which actually increased the fluidity of the membrane lipids and resulted in a more loosely packed bilayer structure, increased the nonelectrolyte permeability of both cells and liposomes. These studies confirmed that the physical properties of the membrane lipids, as determined primarily by their fatty acid and sterol compositions, determine the passive nonelectrolyte permeability properties of the *A. laidlawii* plasma membrane.

The intrinsic passive nonelectrolyte permeability of *A. laidlawii* B and *E. coli* plasma membranes appears to be very low when their membrane lipids exist entirely in the gel state.[51,55,56] However, at temperatures near or just below the membrane lipid gel to liquid-

Table 1
ACTIVATION ENERGIES CALCULATED FOR THE
PERMEATION OF GLYCEROL AND ERYTHRITOL
INTO *A. LAIDLAWII* B CELLS AND LIPOSOMES

	Activation energy (kcal/mol)			
	Glycerol		Erythritol	
Fatty acids added	Cells	Liposomes	Cells	Liposomes
16:0 + 18:2	18.2	20.2	20.5	20.7
16:0 + 18:1$_{cis}$	18.5	18.9	21.8	21.0
16:0 + 18:1$_{trans}$	19.4	17.3	21.5	23.8
16:0 + 18:0$_{iso}$	19.0	16.7	20.3	21.9
16:0 + 18:0	16.8	17.7	21.1	21.6
18:1$_{cis}$ + 14:0	18.6	18.6	21.7	21.2
18:1$_{cis}$ + 16:0	18.5	18.9	21.8	21.0
18:1$_{cis}$ + 18:0	17.3	17.6	22.6	20.3
18:1$_{cis}$ + 16:0 + cholesterol	17.8	17.6	21.2	21.3
Mean ± S.D.	18.2 ± 0.9	18.1 ± 1.1	21.3 ± 0.7	21.5 ± 1.1

crystalline phase transition midpoint temperature, these membranes can become quite leaky to nonelectrolytes.[51,53,55,56] Similar local increases in passive permeability near the phase transition temperature have also been reported for a number of model membranes composed of single or binary mixtures of phospholipids, and it thus appears to be an intrinsic property of lipid bilayers. It has been proposed that structural defects or mismatches in molecular packing at the fluid and solid lipid phase domain boundaries, an increased lateral compressibility, or an increased magnitude of structural fluctuations near the lipid phase transition temperature are responsible for this behavior. Whatever its exact molecular basis, this phenomenon must be considered carefully in the interpretation of protein-mediated transport experiments as well (see Section VI). Moreover, *A. laidlawii* and *E. coli* cells containing predominantly gel state lipid become very susceptible to cell lysis if subjected to mechanical or osmotic stress.[51-53,56]

B. Ions

The relationship between the ionic permeability of prokaryotic plasma membranes and their lipid compositions and physical states has received relatively little systematic study. In fatty acid-homogeneous *A. laidlawii* B cells, passive sodium ion permeability was found by Silvius et al.[53] to be influenced by fatty acid structure and chain length in much the same way as passive nonelectrolyte permeability. In *M. mycoides* cells depleted of cholesterol, passive sodium and potassium permeabilities were reported by Le Grimellec and Leblanc[57] to increase relative to cholesterol-rich cells of the same fatty acid composition. Moreover, in both *A. laidlawii* B[53] and in an unsaturated fatty acid auxotroph of *E. coli*,[55] cells become leaky to sodium and potassium ions, respectively, near their phase transition temperatures. It thus appears that the ionic permeability of prokaryotic plasma membranes can be markedly influenced by the fluidity and phase state of the membrane lipids in a manner similar to that observed for nonelectrolyte permeability.

No studies of the lipid dependence of the proton and hydroxide ion passive permeabilities of prokaryotic plasma membranes appear to have been carried out. This is unfortunate, as these ions play crucial roles in energizing various active transport processes (see Section VI).

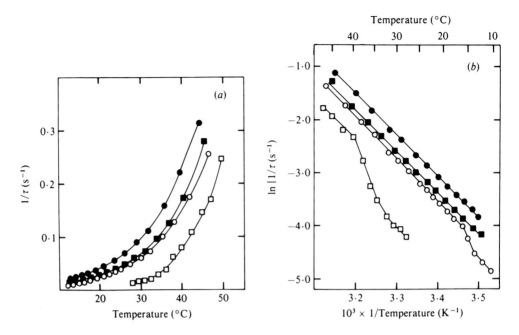

FIGURE 5.   (a) Temperature dependence of the rate of passive glycerol permeation into *E. coli* K1060 grown with xylose plus various unsaturated fatty acids. The rates of passive glycerol entry were measured as the reciprocal relaxation times ($1/\tau$ of cell swelling in hypertonic glycerol). Unsaturated fatty acids: ●, linoleic acid ($18:2_{c,c}$); ■, palmitoleic acid ($16:1_c$); ○, oleic acid ($18:1_c$); □, elaidic acid ($18:1_t$). (b) Arrhenius plots of the data in (a). (From Eze, M. O. and McElhaney, R. N., *J. Gen. Microbiol.*, 124, 299, 1981. With permission.)

V. MEMBRANE LIPIDS AND FACILITATED DIFFUSION

Eze and McElhaney[54] studied the influence of membrane lipid fatty acid composition and temperature on the passive permeation and facilitated diffusion of glycerol in an *E. coli* unsaturated fatty acid auxotroph. Cells grown on glycerol as the sole carbon and energy source contained the components of the *glp* regulon, including a membrane-associated glycerol permease, a protein capable of catalyzing the nonconcentrative passage of glycerol across the plasma membrane. In contrast, the *glp* regulon was repressed in cells grown on glucose or xylose, so that the glycerol permease was absent and glycerol entry occurred exclusively by passive diffusion. It was found that the relative rates of protein-mediated glycerol entry varied fairly markedly with membrane lipid fatty acid composition, just as observed for the passive entry of glycerol; the relative rates of glycerol entry decreased in the order linoleic > palmitoleic > oleic > elaidic acid-enriched cells (see Figures 5a and 6a). An inverse relationship was observed between the rate of glycerol-facilitated diffusion and the gel to liquid-crystalline phase transition temperature of the membrane lipids, suggesting that the rate of protein-mediated glycerol permeation increases with increasing membrane lipid fluidity, again as observed for the passive entry of glycerol into *E. coli* cells of similar fatty acid compositions. Finally, Arrhenius plots of the rates of glycerol-facilitated diffusion were roughly linear and of similar slopes, irrespective of membrane lipid fatty acid composition (see Figure 6b). The apparent activation energies for glycerol facilitated entry varied from 9.5 to 11.3 kcal/mol. In particular, no abrupt breaks were observed near the lipid phase transition temperatures of oleate- or elaidate-enriched cells. In contrast, Arrhenius plots of passive glycerol permeation exhibited breaks near the membrane lipid phase transition temperature in cells enriched in oleic and elaidic acids, and the apparent

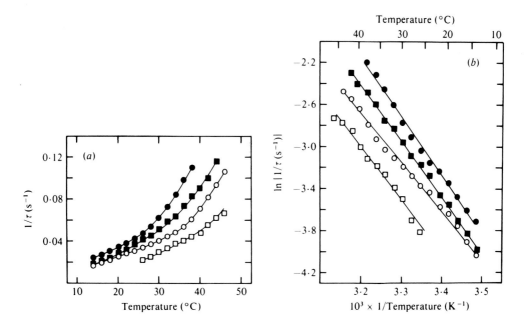

FIGURE 6.   (a) Temperature dependence of the rate of entry of glycerol by facilitated diffusion into *E. coli* K1060 grown in the presence of glycerol plus various unsaturated fatty acids. The rates of glycerol facilitated diffusion were measured as the difference in reciprocal relaxation times ($1/\tau$) of cell swelling between glycerol- and xylose-grown cells. Unsaturated fatty acids: ●, linoleic acid (18:2$_{c,c}$); ■, palmitoleic acid (16:1$_c$); ○, oleic acid (18:1$_c$); ■, elaidic acid (18:1$_t$). (b) Arrhenius plot of the data in (a). (From Eze, M. O. and McElhaney, R. N., *J. Gen. Microbiol.*, 124, 299, 1981. With permission.)

activation energy for glycerol passive diffusion across fluid membranes was higher than for glycerol-facilitated diffusion, being 15 to 16 kcal/mol (see Figure 5b). The apparent lack of an effect of the lipid phase transition on the function of the glycerol permease may indicate that this protein functions as a membrane channel. However, the apparent influence of membrane fluidity on the rates of mediated glycerol permeation would be difficult to explain if this were the case, unless the membrane lipid fatty acid composition influences the number of function carriers in the *E. coli* membrane. Alternatively, the boundary lipid immediately adjacent to the glycerol carrier protein may remain in a fluid or semifluid state at temperatures where the bulk membrane lipid is solid. The ability of intrinsic membrane proteins to disorder gel state lipid,[58] and of membrane enzymes and transport systems to retain activity below the bulk lipid phase transition temperature[42,58] has recently been demonstrated in several systems.

This study is the only one the author is aware of in which the relationship between membrane lipid fluidity and phase state and the function of a prokaryotic facilitated diffusion system has been investigated. Although the results presented above are probably valid, it would be desirable if these experiments could be repeated using a more direct assay of glycerol entry. The stopped-flow spectroscopic swelling rate assay utilized by Eze and McElhaney[59] is necessarily indirect and may be subject to some technical uncertainties.

VI. MEMBRANE LIPIDS AND ACTIVE TRANSPORT

A. Sugar Transport

The relationship between phospholipid biosynthesis and the induction of a functional lactose transport system was originally investigated by Fox[60] and Hsu and Fox;[61] in the first study, phospholipid synthesis was inhibited by the starvation of an *E. coli* unsaturated fatty acid auxotroph for unsaturated fatty acids, and in the second study by starvation of a glycerol auxotroph for glycerol. In both cases inhibition of phospholipid synthesis resulted in a marked reduction of the ratio of β-galactoside transport induction to β-galactosidase induction, suggesting that, although the products of lactose operon were being produced in the absence of phospholipid biosynthesis, the assembly of a functional transport system required normal phospholipid synthesis. Nunn and Cronan[62] and Weisberg et al.,[63] however, using a generally similar approach, reported that the induction of lactose transport is not preferentially inhibited in the absence of unsaturated fatty acid or phospholipid synthesis, respectively. Moreover, Overath et al.,[64] using an inhibitor of unsaturated fatty acid biosynthesis in a wild-type strain, also found normal rates of β-galactoside transport induction initially, although after half a generation the cells became leaky, indicating membrane damage. Similarly, Robbins and Rotman[65] reported that inhibition of unsaturated fatty acid biosynthesis in *E. coli* resulted in less than a 50% inhibition of lactose transport induction, if assayed before general cellular damage due to unsaturated fatty acid deprivation was evident, although the induction of methylgalactoside transport was markedly inhibited by this treatment. Finally, Mindich[66] and Willecke and Mindich[67] have demonstrated that the lactose permease in *Staphylococcus aureus* and citrate transport in *Bacillus subtilis,* respectively, can be induced in the absence of phospholipid synthesis, although the lactose permease system functions with a reduced efficiency (30 to 50% of normal) under such conditions. It thus appears that, in general, the induction and assembly of functional lactose and at least some other transport systems can occur in the absence of unsaturated fatty acid or phospholipid synthesis. This is not unexpected, since growth studies,[68] as well as estimates of the amount of "boundary lipid" present in prokaryotic plasma membranes,[1,2,9] indicate that most membranes contain significant amounts of fluid bilayer-phase lipid that at any given instant is not interacting directly with membrane proteins. Thus, sufficient "nonboundary" lipid would, presumably, be available for the solvation of newly synthesized integral membrane proteins, even in the temporary absence of continuing glycerolipid synthesis. In fact, McIntyre and Bell[69] and McIntyre et al.[70] have demonstrated that a 40% decrease in the phospholipid/protein ratio of the cytoplasmic and outer membranes can be induced by the inhibition of phospholipid synthesis in glycerol auxotrophs of *E. coli* before cell growth ceases.

Tsukagoshi and Fox,[71] utilizing an *E. coli* unsaturated fatty acid auxotroph supplemented with various fatty acids, have also reported that the induction of lactose transport is abortive in cells maintained at temperatures below the membrane lipid gel to liquid-crystalline phase transition temperature. This conclusion was again based on the decreased ratio of β-galactoside transport induction to β-galactosidase induction observed below characteristic temperatures, which depended on the fatty acid composition of the cell. Although this conclusion may be valid, one should remember that *E. coli* cells are unable to grow at temperatures below their lipid phase transition midpoints and, at least in the case of elaidic acid-enriched cells, rapidly lose viability.[68] Since Tsukagoshi and Fox induced their cultures under these conditions, it is possible that the reduced ratio of β-galactoside transport to β-galactosidase induction observed may simply reflect a differential decline in membrane function, generally (in relation, for example, to soluble protein biosynthesis), which accompanies this loss of viability. Also, since *E. coli* cells are known to become "leaky" below their lipid phase transition temperatures (see earlier discussion, Section IV), one must also consider the possibility that the transmembrane electrochemical proton gradient could become at least partially collapsed under these conditions, either due to an inhibition of substrate oxidation, a defective coupling of substrate oxidation to proton pumping, by an excessive proton passive leakage, or by a combination of two or more of these factors. Since the electrochemical proton gradient is known to drive *E. coli* transport systems, either directly or indirectly,[16] a dissipation of this gradient could result in an inhibition of the transport function even in

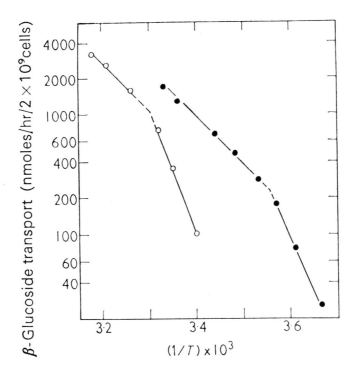

FIGURE 7.   Temperature dependence of PNP-glucose transport. Cultures of strain 30E of *E. coli* K12, an elaidate-positive variant of strain 30⁻, were grown from a small inoculum in standard medium supplemented with 0.1 mM TPglu (an inducer of the β-glucoside system) and 0.02% linoleic (— ● — ● —) or elaidic (— ○ — ○ —) acids. At a culture density of 10^9 cells per milliliter, choloramphenicol was added at 50 μg/mℓ, and the cells were processed for the assay of β-glucoside transport. (With permission from Wilson, G. and Fox, C. F., *J. Mol. Biol.*, 55, 49, 1971.

the presence of completely integrated and potentially fully functional membrane transport systems. The tacit assumption made in most studies of the role of lipids in microbial transport, that transport rates reflect only the behavior of the transport system itself, is at present largely unsubstantiated, since very little work has been done on the effect of alterations in membrane lipid fluidity and phase state on membrane energization.

The association of the lactose transport system with the lipids of the cytoplasmic membrane was also studied by Wilson and Fox by taking advantage of the fact that the rate-temperature profile of lactose transport is dependent on the fatty acid supplement provided to unsaturated fatty acid auxotrophs of *E. coli*.[72] As first demonstrated by Schairer and Overath[73] and later by Wilson et al.,[74] Arrhenius plots for transport could be fit (at least approximately) by two straight lines, the lower temperature line having a considerably greater slope than the higher temperature line, as illustrated in Figure 7. The "break" or inflection temperature observed in each Arrhenius plot depended on the fatty acid composition in such a way as to suggest that the break was related to the phase transition temperature of the membrane lipids. Thus, the activity-temperature profile of the lactose transport system reflected the physical properties of its lipid environment in the membrane. By growing cells in the presence of one exogenous fatty acid before the induction of the lactose transport system and then shifting to a second fatty acid upon induction, Wilson and Fox[72] reported that the rate-temperature profile of transport reflected the fatty acid present during the induction period, even though in some cases this was not the major fatty acid present in the total membrane lipid. They thus concluded that newly synthesized transport proteins are preferentially associated, in a relatively long-lived manner, with newly synthesized lipid. In contrast, Overath et al.,[64] using a generally similar experimental approach, found that the rate-temperature profiles of the lactose transport system reflected the average membrane lipid fatty acid composition and not that of the phospholipids synthesized during induction of the transport system, suggesting a relatively rapid randomization of the membrane phospholipids. In a later paper, Tsukagoshi and Fox[75] presented data tending to confirm the findings of Overath and co-workers, and ascribed their initial observations to an inadequate characterization of the lactose transport rate-temperature profiles. However, Tsukogoshi and Fox[75] then reported that if transport induction is carried out at 25°C rather than 37°C, triphasic Arrhenius plots are observed, with one break corresponding to the unsaturated fatty acid present during growth at 37°C and the second break corresponding to the fatty acid present during induction at 25°C. These authors concluded that newly synthesized transport proteins are, indeed, preferentially associated with newly synthesized lipid, but that the relatively rapid randomization of the membrane lipid phase at 37°C simply precludes its observation, although at lower temperatures (25°C) randomization is not complete and some preferential association is still observable in the rate-temperature profiles. This conclusion appears to be a tenuous one, however, for a number of reasons. First, there is uncertainty about the validity of transport measurements performed at temperatures below the phase transition temperature, due to the fact that *E. coli* cells lose viability and become leaky under these conditions. Second, it is far from clear that the experimental data presented is uniquely fit by the three straight-line-segment Arrhenius plots drawn; in fact, the experimental points actually fall on smooth, continuously curving lines. Third, later studies from the same laboratory report the presence of two or even three breaks in Arrhenius plots of lactose transport in cells cultured on only a single fatty acid.[76-79] Finally, recent ^{2}H nuclear magnetic resonance and saturation transfer electron spin resonance studies of *E. coli* membranes[79-83] and of *A. laidlawii* B membranes,[9] and of several reconstituted membrane protein-lipid model membrane systems,[84-86] have demonstrated that rapid exchange ($> 10^4$/sec) must be occurring between boundary and bulk lipid populations, even at temperatures near 0°C. Thus, unless the lactose transport system is quite atypical, no preferential association of newly synthesized membrane lipids with newly induced transport proteins should be detectable in rate-temperature profile experiments of the type described, irrespective of the temperature at which induction was performed.

Although it seems that the biogenesis of the lactose transport system is not dependent on concomitant phospholipid synthesis, in general, nor upon the synthesis of specific molecular species of phospholipid, in particular, these studies did establish that the function of the lactose transport system is dependent in some manner on the gel to liquid-crystalline membrane lipid transition, despite the discrepancies reported in the number and position of the Arrhenius plot breaks and lipid phase transition temperatures by Fox and co-workers[72,74,76-79] and by Overath and co-workers.[64,73,87-89] In particular, these investigations demonstrated that the function of the lactose transport system was partially inhibited by the presence of gel-state lipid in the *E. coli* membrane. Thilo et al.[41] have recently provided a reasonably explicit and physically plausible explanation for the observed temperature dependence of β-galactoside (and β-glucoside) transport in unsaturated fatty acid auxotrophs of *E. coli*, based on a recent reinvestigation of the transport rate-temperature profiles, paying particular attention to the determination of transport rates at low temperatures. These workers now observe triphasic Arrhenius plots consisting of two linear regions of similar and relatively shallow slopes occurring at both high and low temperatures, separated by a linear region of much steeper slope at intermediate temperatures (see Figure 8). The first fairly gradual, downward change in slope generally occurred between the lipid phase transition midpoint and upper boundary, as determined by a fluorescent probe, while the second, upward change in slope correlated well with the lower boundary of the phase transition. Thilo et al. interpreted

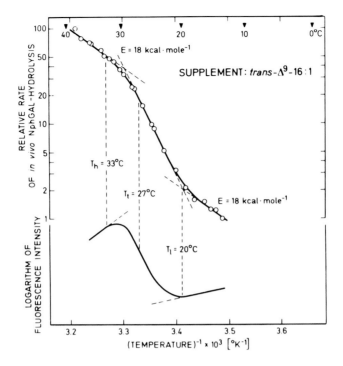

FIGURE 8. The temperature dependence of the rate of β-galactoside transport, measured as the in vivo rate of *O*-nitrophenyl β-Dαlactopyranoside (NphGAL) hydrolysis. For palmitelaidic acid-supplemented *E. coli* cells (upper curve) in comparison with the membrane lipid phase transition as measured by *N*-phenyl-1-naphthylamine fluorescence in these same cells (lower curve). (Reprinted with permission from Reference 41. Copyright 1977 American Chemical Society.)

these results in terms of a lateral partitioning of these transport proteins between the fluid and ordered lipid domains of the *E. coli* membrane, with these sugar transport proteins exhibiting a 10- to 25-fold higher activity in the liquid-crystalline than in the gel domains, but functioning by similar mechanisms in both lipid phases (hence, the similar activation enthalpy values exhibited at temperatures both above and below the lipid phase transition). The apparent lateral partition coefficient varied with the membrane lipid composition (which would explain why the exact relationship of the upper-temperature inflection to the position of the phase transition often appears to be variable), but in all cases the partitioning of the transport proteins into the fluid parts of the membrane appeared to be favored. The conclusion that the lactose transport system retains appreciable activity in a gel-state lipid environment should be accepted with caution, however, despite the authors' great care in determining transport rates at low temperatures, since these cells become highly permeable to the lactose analog utilized to measure transport at temperatures near the lower boundary of the lipid phase transition. It is thus unclear whether or not the rates of hydrolysis of the lactose analog, which are only 15 to 30% above the high background levels observed at these low temperatures, accurately reflect the activity of the lactose transport system functioning via its "normal" mechanism.

The concept of a selective partitioning of the lactose transport proteins into the fluid lipid domains of the membrane within the lipid phase transition temperature range has been supported by the studies of Therisod et al.[90] and of Letellier et al.[91] These workers utilized the fluorescence changes purportedly accompanying the energy-dependent *binding* (but not the transport) of dansyl galactoside to study the number of functional *lac* carriers as a function of temperature and of the membrane lipid phase transition. They concluded that the changes in slope observed for Arrhenius plots of lactose transport are due primarily to a change in the number of functional carriers with temperature, rather than to a change in the rate at which a constant number of *lac* carriers translocate substrate. Shechter and co-workers[95] confirmed that the *lac* carrier proteins do, indeed, segregate preferentially into the liquid-crystalline as opposed to the gel lipid domains, but found that about half of the carrier proteins in the fluid domains are nonfunctional, whereas all the carrier proteins in the ordered domains are fully functional! This latter observation is quite curious, since reconstitution studies with a large variety of membrane enzymes and transport systems have almost uniformly demonstrated a requirement for fluid lipid for activity.[34] Moreover, Overath et al.[92] have recently presented evidence that, contrary to the original reports, dansyl galactosides are, in fact, transported by *E. coli* and that the fluorescence increase observed upon energization of cytoplasmic membrane vesicles is due, at least in part, to a nonspecific binding of dansyl galactoside to the membrane. Overath and co-workers thus maintain that since transport and nonspecific binding, as well as possible specific binding to the *lac* carrier proteins, can all induce changes in the fluorescence of dansyl galactoside, the studies of Shechter and co-workers could not really distinguish between changes in the number of *lac* carriers which are functional and the rate at which these carriers are transporting dansyl galactoside. Thus, the original partition hypothesis of Thilo et al. remains a viable one pending further clarification of the nature of the interaction of dansyl galactosides with the *E. coli* cytoplasmic membrane.

Teather et al.[93] have recently constructed an *E. coli* strain which, in addition to being auxotropic for unsaturated fatty acids, contains a multicopy plasmid coding for the Y gene product of the *lac* operon, the lactose permease protein. Transport rates in this lactose permease-overproducing strain are six to ten times higher than in normal cells. Using this strain, Wright et al.[94] recently demonstrated that substrate binding to the lactose carrier protein in the membrane is not affected by the membrane lipid phase transition. That is, the temperature dependence of the dissociation constant of lactose binding is linear, exhibiting no breaks in the region of the lipid phase transition, and the number of binding sites also remains constant over the physiological temperature range. These findings suggest that changes in the rate of lactose translocation across the membrane, and not in the number of functional lactose permease molecules, are responsible for the characteristic Arrhenius plot shapes observed for *E. coli* cells of varying fatty acid composition. Interestingly, however, no ''breaks'' or upward inflections in the slope of the Arrhenius plot of lactose transport at the lower boundary temperature of the lipid phase transition could be detected in this lactose carrier-enriched strain, in contrast to the earlier study using a ''normal'' unsaturated fatty acid auxotroph.[41] Instead, Arrhenius plots which are curved at the higher temperatures and become linear and more steeply sloping at the lower temperatures are observed (see Figure 9). This behavior is not that predicted by the partition hypothesis put forward by Thilo et al.[41] These results indicate that the lactose permease can be completely inactivated below the lipid phase transition lower boundary.

The dependence of the rate-temperature profile of the β-glucoside transport system of *E. coli* on the phase state of the membrane lipids has often been studied in parallel with the β-galactoside transport system by Fox and co-workers[72,74,76-79] and by Overath and co-workers.[41] In all cases the β-glucoside and the lactose transport systems exhibited almost identical behavior. Thus, the β-glucoside transport system responds to the order-disorder transition of the membrane lipids just as does the lactose transport system previously discussed.

It is instructive to compare the relative rates of β-galactoside and β-glucoside transport in *E. coli* unsaturated fatty acid auxotrophs at temperatures above the Arrhenius plot break temperatures, where the membrane lipid exists predominantly or exclusively in the liquid-

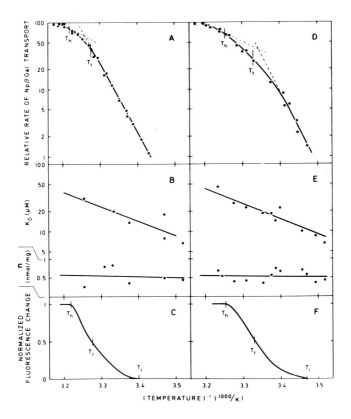

FIGURE 9. Arrhenius plot of the relative rate of β-galactoside transport, measured as indicated in Figure 8, vs. the transport assay temperature, for an unsaturated fatty acid auxotroph of *E. coli* which also overproduces the lactose permease. (Reprinted with permission from Reference 94. Copyright 1981 American Chemical Society.)

crystalline state. In this way the effects of changes in membrane lipid fluidity, rather than changes in membrane lipid phase state, on the rates of sugar transport in this organism can be determined. In certain *E. coli* auxotrophs, Overath and co-workers[73,87] have reported that the rate of β-galactoside transport is little affected by variations in fatty acid composition, so long as transport rates are measured between 30 and 40°C, where most or all of the membrane lipid is in the fluid state; similar results were also reported for optimal growth rates and maximum rates of oxygen consumption, as well as for rates of sugar efflux, in the same mutant. In several other *E. coli* unsaturated fatty acid auxotrophs, Fox and co-workers[72,74,76-78] have reported that β-galactoside and β-glucoside transport rates were almost comparable in cells enriched in oleic, linoleic, or dibromostearic acids, but that the transport rates in elaidic acid-enriched cells were reduced by 30 to 50%, when assayed at a temperature (about 40°C) above the elaidate-supplemented upper break temperature. However, even at 40°C appreciable amounts of gel-state lipid exist in cells enriched with elaidic acid. Thus, it may well be that the reduced rates of sugar transport sometimes seen in elaidic acid-enriched cells are due to a suboptimal mixture of gel and liquid-crystalline lipid rather than to a suboptimal fluidity in the fluid lipid domains. Taken together, these results indicate that sugar transport rates in *E. coli* are either not, or at most, weakly, affected by membrane lipid fluidity per se.

The transport of glucose in *E. coli* takes place via the phosphoenolpyruvate sugar phosphotransferase system, in contrast to the transport of β-galactosides and β-glucosides, which are driven by the electrochemical proton gradient across the cytoplasmic membrane.[16] Schechter et al.[95] have reported that the rate-temperature profiles of glucose transport, into cytoplasmic membrane vesicles prepared from an *E. coli* unsaturated fatty acid auxotroph, are not dependent on the fatty acid composition or on the phase state of the membrane lipids. The lack of a "break" in the region of the membrane lipid phase transition was also shown by Rottem et al. for α-methylglucoside uptake in a *M. mycoides* var. *capri* strain adapted to grow with low levels of cholesterol, although the apparent activation energy for α-methylglucoside uptake was much higher than in the native (cholesterol-rich) strain.[96] These workers also reported that the rates and apparent activation energies for α-methylglucoside phosphorylation were the same for isolated membranes of each strain. Since α-methylglucoside transport into *M. mycoides* var. *capri* also occurs via the phosphoenol pyruvate-dependent sugar phosphotransferase system,[97] these observations imply that vectorial, group-translocation transport systems may function without "mobile carrier" components which are sensitive to the phase state of the membrane lipid bilayer, in contrast to most "classical" active transport systems. Alternatively, the different types of energy-coupling mechanisms operating in the two types of transport systems may explain their different responses to the membrane lipid gel to liquid-crystalline phase transition. Interestingly, however, Arrhenius plots of α-methylglucoside efflux from *M. mycoides* var. *capri* showed breaks at temperatures corresponding to those of the lipid phase transitions. It is not clear whether α-methylglucoside efflux is a passive or protein-mediated process in this organism.

The effect of variations in the fatty acid composition and cholesterol content of the *A. laidlawii* B membrane on the rate-temperature profile of glucose uptake into intact cells has been studied by Read and McElhaney.[98] Glucose transport in this organism occurs via an electrochemical potential-driven, active transport process and not via the phosphoenolpyruvate sugar phosphotransferase system.[97] These workers reported that the rate of glucose uptake (at 37°C, for example) increases as the calorimetrically determined gel to liquid-crystalline membrane lipid phase transition temperature decreases. Moreover, the presence of cholesterol reduces the rate of glucose uptake for each fatty acid enrichment tested (see Figure 10). These results indicate that the absolute rate of glucose transport increases with the increasing fluidity of the membrane lipids and suggest that the glucose carrier protein(s) interacts intimately with the membrane lipids. In contrast to transport rates, the apparent activation energy for glucose uptake (above the phase transition midpoint temperature) was found to be independent of membrane lipid fatty acid composition and cholesterol content, suggesting that the apparent activation energy is determined by glucose binding to the carrier protein at the membrane surface, or at least by some process not influenced by the fluidity of the membrane lipids. Accurate estimates of glucose transport rates at temperatures below the lipid phase transition midpoint temperatures could not be obtained due to the mechanical fragility and leakiness of the cells under these conditions. This is one of the few studies in which active transport rates appear to be significantly affected by membrane lipid fluidity per se.

B. Amino Acid Transport

The influence of the fatty acid composition and phase state of the membrane lipids on the rates and temperature dependence of proline uptake by isolated membrane vesicles prepared from *E. coli* unsaturated fatty acid auxotrophs has been studied by two groups. Both groups utilized X-ray diffraction techniques to monitor the phase state of the membrane lipids. Esfahani et al.[99] reported single Arrhenius plot breaks at 26, 19, and 14°C for elaidic, oleic, and linolenic acid-enriched membrane vesicles, respectively (see Figure 11). The apparent activation energy for proline transport was found to be independent of fatty acid composition above the Arrhenius break temperature, but varied slightly below it. Shechter and co-workers[95] also reported Arrhenius plots with single breaks, but at temperatures of

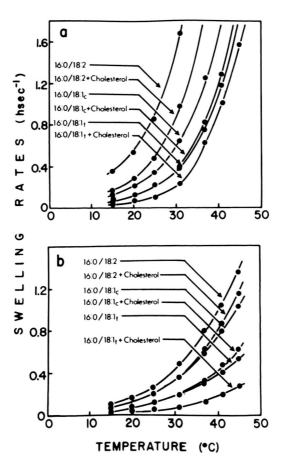

FIGURE 10. Swelling rates of *A. laidlawii* in erythritol and glucose. Swelling rates are shown as a function of temperature for cells grown in the presence of the lipid supplements indicated. (a) Erythritol; (b) glucose. Swelling rates are measures of the rate of the passive entry of erythritol and the protein-mediated entry of glucose into whole cells. (From Read, B. D. and McElhaney, R. N., *J. Bacteriol.*, 123, 47, 1975. With permission.)

38, 22, and 19°C for elaidic, oleic, and linolenic acid-enriched vesicles, respectively (see Figure 12). Moreover, the apparent activation energy of proline transport was reported to vary markedly both above and below the break temperature in this latter study. Thus, the agreement between these two studies of proline transport is only fair, as is the agreement between either study and investigations of the temperature dependence of the β-galactoside and β-glucoside transport systems reviewed earlier. However, in both studies the rate-temperature profile of the proline uptake system correlated at least qualitatively with the membrane lipid order-disorder transitions detected by X-ray diffraction.

Esfahani et al.[99] found that at temperatures above 26°C, the elaidate-enriched Arrhenius inflection point, the maximum rate of proline transport was about 25% higher in oleate-enriched than in elaidate- or linolenate-enriched *E. coli* cells, which were comparable.[99] On the other hand, Schechter et al.[95] reported that in vesicles prepared from *E. coli* auxotrophs the relative transport rates were about twofold higher in oleate-enriched than in linoleate- or linolenate-enriched systems, which were in turn about 2.5-fold higher than in the elaidate-

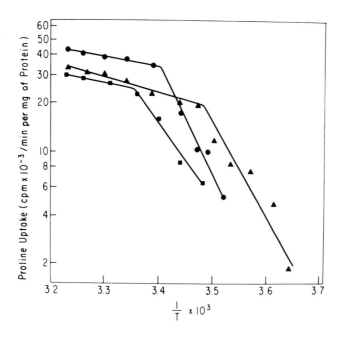

FIGURE 11.   Arrhenius plots of the rate of proline uptake by membrane vesicles prepared from an *E. coli* unsaturated fatty acid auxotroph grown in the presence of oleic (●), elaidic (■), or linolenic (▲) acids. (From Esfahani, M., Limbrick, A. R., Knutton, S., Oka, S., and Wakil, S. J., *Proc. Natl. Acad. Sci. U.S.A.*, 68, 3180, 1971. With permission.)

enriched vesicles. However, there was some evidence in this latter study that the vesicles prepared from elaidic acid-grown cells were not very transport competent, since the rates of glucose transport, which are not supposed to be affected by membrane lipid fluidity or phase state, were only about one tenth those found in vesicles enriched in the various *cis*-unsaturated fatty acids. Taken together, then, these studies suggest only a small dependence of proline transport rates on membrane lipid fatty acid composition and thus on membrane lipid fluidity in the liquid-crystalline state, with cells enriched in *cis*-monounsaturated fatty acids producing the greatest transport rates. Certainly these experiments do not indicate that transport rates simply increase regularly with increases in membrane lipid fluidity, since the phase transition temperatures were found to decrease and, presumably, the membrane lipid fluidities to increase, in the relative order elaidate- > oleate- > linoleate- > linolenate-enriched membranes. If anything, transport rates appeared to be highest in vesicles containing liquid-crystalline lipids of "intermediate" fluidity.

The rate-temperature profiles of arginine and glycine transport into cells of an *E. coli* unsaturated fatty acid auxotroph were also reported by Rosen and Hackette[100] to be influenced by the unsaturated fatty acid supplementation employed. Both transport systems produced Arrhenius plots with single breaks at about 30 and 13°C for elaidic and oleic acid-supplemented cells, respectively. These break temperatures are in reasonable agreement with those reported by some investigators for the lactose transport system, but only in fair agreement with the values for proline transport just discussed. One should note that the arginine transport system in *E. coli* is an osmotic shock-sensitive system which is dependent on the presence of a periplasmic binding protein for optimal function, while the glycine transport system is an osmotic shock-insensitive system which does not have a periplasmic binding protein component; there is some evidence that the mechanism of energization of these two types of systems is different, with the former being driven by ATP hydrolysis and the latter directly

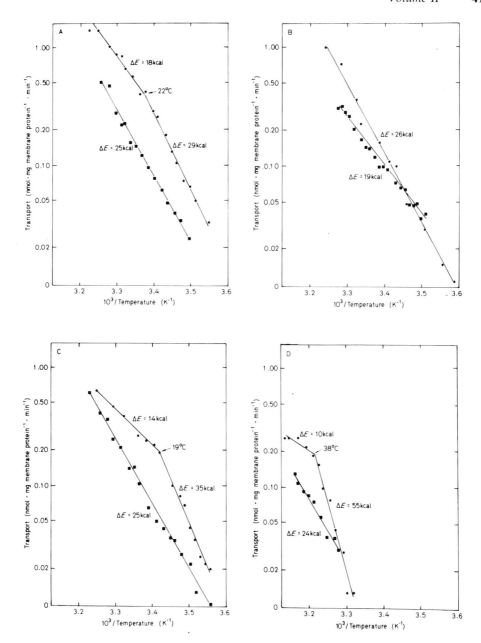

FIGURE 12. Arrhenius plots for proline (●) and glucose (■) transport across cytoplasmic membrane vesicles isolated from *E. coli* cells grown in the presence of oleic acid (A), linoleic acid (B), linolenic acid (C), and elaidic acid (D). (From Shechter, E., Letellier, L., and Gulik-Krzywicki, T., *Eur. J. Biochem.*, 49, 61, 1974. With permission.)

by the transmembrane electrochemical proton gradient.[16] Fatty acid compositions or phase transition temperatures were not reported in this study, so a more detailed comparison with other rate-temperature investigations is difficult. Moreover, transport rates for arginine and glycine were not gathered in the same temperature range, so an assessment of the effect of variations in membrane lipid fluidity on amino acid transport rates at temperatures above 30°C is not possible.

The temperature dependence of the osmotic shock-sensitive glutamine and osmotic shock-insensitive proline transport systems, and the response of these transport systems to variations in the fatty acid composition and phase state of the membrane lipids in intact cells of an *E. coli* unsaturated fatty acid auxotroph, have recently been studied by Eze.[101] The thermotropic phase behavior of the *E. coli* lipids in the cytoplasmic membrane fraction was determined by differential thermal analysis. Arrhenius plot breaks for glutamine uptake occur at 29, 23, 17, 14, and < 10°C for cells enriched in elaidic, palmitelaidic, palmitoleic, oleic, and linolenic acids, respectively. Quite similar behavior is observed for the proline transport system, except that in each case the break temperature is 1 to 3°C below that exhibited by the glutamine transport system in cells with the same fatty acid composition. Thus, the agreement with the proline uptake break temperatures reported by Esfahani et al.[99] and Shechter et al.[95] is reasonably good. With the exception of the palmitoleic acid value, the break temperatures are in qualitative agreement with the membrane lipid phase transition temperatures. However, the exact position of the break temperature with regard to the gel to liquid-crystalline phase transition varies considerably. In elaidic acid-enriched cells, the break temperatures fell near the lower boundary of the membrane lipid transition, while in palmitelaidic acid-enriched cells the break is observed between the lower boundary and midpoint temperature; in oleic acid-enriched cells the break occurs at the transition midpoint temperature, but in palmitoleic acid-enriched cells the break temperature occurs near the upper boundary of the lipid transition! A careful analysis of the rate-temperature profiles of some of the other sugar and amino acid transport systems discussed earlier also reveals a similar, although generally less pronounced, variation in the response of the system to the degree of completion of the lipid phase transition. A convincing molecular explanation for these observations is not yet at hand.

Eze[101] also studied the effects of alterations in membrane lipid fluidity in the liquid-crystalline state on amino acid transport rates. Above their respective break temperatures, the transport rates for glutamine and proline were generally two- to threefold higher for cells enriched in palmitelaidic, oleate, or palmitoleic acid than for cells enriched in either elaidic or linoleic acid. Again, no clear and simple correlation between membrane lipid fluidity and maximum transport rates was discernible, except that transport rates seemed to decrease in membranes enriched in either the highest-melting (elaidic) or lowest-melting (linoleic) fatty acids.

The effect of palmitate enrichment (and consequently of oleate depletion) on the relative rates of transport of glutamate, leucine, and proline was recently studied in yet another *E. coli* fatty acid auxotroph by Ingram et al.[102] Palmitate-enriched cells exhibited about a 50% reduction of glutamate transport rates when assayed at 37°C in comparison to control cells not starved for oleic acid. However, the maximum transport rates for leucine and proline were not affected by palmitic acid enrichment. Thus, the presence of a substantial amount of gel-state lipid appears to inhibit some but not all amino acid transport systems in *E. coli*.

The initial rates of uptake of a number of amino acids in an *E. coli* unsaturated fatty acid auxotroph enriched in various unsaturated fatty acids was studied by Holden et al.[103] These workers found that proline and threonine transport rates (measured at about 21°C) were much more depressed by enrichment with *trans*-unsaturated fatty acids (relative to *cis*-vaccenic acid enrichment) than were lysine and asparagine transport rates. Aspartic acid and leucine initial uptake rates exhibited intermediate degrees of inhibition by the *trans*-unsaturated fatty acid enrichment. Lipid enrichment with very low-melting fatty acids also reduced relative initial transport rates for alanine (markedly) and for arginine (moderately), but appeared not to affect, or even to slightly stimulate, asparagine uptake. Although differential changes in the relative number of functional transport systems in response to alterations in fatty acid composition might explain these results, the temperature-activity profiles of proline and lysine transport, in cells enriched in *cis*-vaccenic or palmitelaidic acids, suggested that nonuniform alterations in transport rates must also be involved. In general, it appeared that binding protein-dependent (osmotic shock-sensitive) amino acid transport systems were less sensitive to fatty acid compositional alterations than were binding protein-independent (osmotic shock-insensitive) transport systems. Again the heterogeneous response of various amino acid transport systems to membrane lipid compositional alterations was noted. The effects of membrane lipid phase state and fluidity cannot be easily separated in this study.

The effects of growth temperature on membrane lipid fatty acid composition, membrane lipid fluidity, and amino acid transport rates have also been studied in *Streptococcus faecalis* by Wilkins.[104] A shift in growth temperature from 37 to 10°C resulted in an increase in the proportion of unsaturated fatty acids in the membrane lipids. Electron spin resonance spectroscopy indicated that at 10°C cells grown at 37°C were considerably more ''fluid'' than those grown at 10°C, while at 37°C the opposite was true. Nevertheless, growth temperature had no significant effect on the rates of active transport of alanine or leucine; i.e., uptake rates at 10 or 37°C were the same in cells grown at either 10 or 37°C. Thus, the relative rates of amino acid transport in *S. faecalis* are apparently insensitive to the adaptive changes in membrane lipid fatty acid composition and fluidity which accompany alterations in the temperature of growth. Similar observations were reported previously for several other transport systems in other psychotropic and psychophilic bacteria by Wilkins[105] and by Herbert and Bell.[106]

C. Ion Transport

Little work has been done on the relationship between membrane lipid fluidity and phase state and ion transport in prokaryotes. Cho and Morowitz[107] studied the effect of growth temperature on the rates of K^+ influx and efflux, and on K^+ steady-state levels, in *A. laidlawii* B. These workers found that the rate of K^+ uptake (measured at 37°C) in cells grown at 37°C was 2.5-fold higher than for cells grown at 25°C, whereas the rate of K^+ efflux (at 15°C) and the steady-state K^+ levels (at 37°C) were independent of the temperature of growth. They thus concluded that K^+ influx in this organism is sensitive to the fluidity and phase state of the membrane lipids, whereas K^+ efflux and K^+ accumulation are not. However, these conclusions rest on the assumption that altering the growth temperature of this organism will also alter the fatty acid composition and thus the fluidity and phase state of its membrane lipids, as is true for most bacteria. However, most other studies have shown that the fatty acid composition and phase transition temperature of the membrane lipids of *A. laidlawii* B are essentially growth-temperature invariant.[9] Therefore, these studies do not appear to provide any information on the possible effects of membrane lipid physical state on K^+ translocation in this organism.

Le Grimellec and Leblanc[57] investigated the relationship between membrane lipid fatty acid composition and cholesterol content, and potassium transport rates and levels of accumulation in *M. mycoides* var. *capri*. These workers studied the growth characteristics, intracellular K^+ content, and the ability to extrude protons of native *M. mycoides* cells grown on medium supplemented with cholesterol and either palmitic plus oleic acids or with elaidic acid, and of a strain adapted to grow on low levels of cholesterol in the presence of elaidic acid; cholesterol accounts for 20 to 25% of the total membrane lipid in the native strain and less than 2% in the adapted strain. Native organisms grown on cholesterol-rich medium exhibited identical growth characteristics, intracellular K^+ contents and medium acidification properties, irrespective of fatty acid supplementation. In contrast, cholesterol-deficient organisms were unable to grow below pH 6.5 (instead of pH 5.2 as in the native strain), exhibited lowered intracellular K^+ levels, and a reduced ability to extrude protons. Moreover, K^+ passive permeability was drastically increased in the adapted strain, although K^+ remained in equilibrium with the (reduced) transmembrane potential, and the intracellular Na^+ content increased. Replenishing cholesterol in membranes of cholesterol-deficient cells

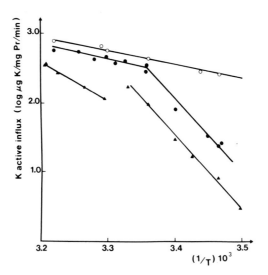

FIGURE 13.   Arrhenius plot of $^{42}K^+$ active influx by
M. mycoides cells as function of membrane lipid composition. O, Cells grown in cholesterol plus oleic and
palmitic acids; ▲, cells grown in cholesterol and elaidic
acid; and ●, cells grown without cholesterol but with
elaidic acid. (From Le Grimellec, C. and Leblanc, G.,
Biochim. Biophys. Acta, 599, 639, 1980. With
permission.)

resulted in a recovery of native growth characteristics, intracellular K^+ level, and acidification potential. These authors suggested that cholesterol depletion produces its characteristic
effects by inducing an increase in proton permeability, which, in turn, reduces the transmembrane electrochemical proton gradient that can be generated by this organism, thereby
reducing intracellular K^+ accumulation and limiting growth at lower pH values. The changes
observed in K^+ passive permeability did not appear to be involved in determining intracellular
K^+ levels. This study is an important one in that it demonstrates that lipid-dependent changes
in the energy state of a cell can affect transport processes.

Using the same organism, Le Grimellec and Leblanc[108] subsequently investigated the
temperature-activity relationship of K^+ active influx, Mg^{2+}-ATPase activity, transmembrane
potential, and membrane lipid composition. Arrhenius plots of the initial rates of K^+ exchange influx in the native strain enriched in palmitic plus oleic acids gave a linear relationship
(see Figure 13). On the other hand, the native strain enriched with elaidic acid produced a
biphasic linear Arrhenius plot with a discontinuity at about 28 to 30°C. Finally, the adapted
strain grown in the presence of elaidic acid exhibited a biphasic, linear Arrhenius plot with
a break at about 23°C. A broad endothermic lipid phase transition, occurring between 20
and 48°C, was observed by differential scanning calorimetry for membranes from the cholesterol-deficient strain, while no phase transitions could be detected with this technique in
membranes of the native strain, irrespective of fatty acid composition. However, diphenyl
hexatriene fluorescence polarization results suggested the presence of a "phase separation"
between 29 and 31°C in both palmitic plus oleic acid and in elaidic acid-containing native
strain membranes. Thus, the rates of active K^+ influx appear to be sensitive in a complex
manner to the phase state, and possibly to the fluidity, of the membrane lipids. However,
above the Arrhenius inflection points, the relative rates of K^+ influx do not correlate with
the relative membrane lipid fluidities. In contrast, Arrhenius plots of Mg^{2+}-ATPase activity

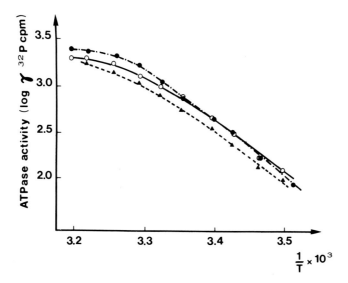

FIGURE 14.　Arrhenius plot of Mg^{2+}-ATPase in *M. mycoides* membranes isolated from cells grown with or without cholesterol in the presence of different fatty acids (symbols as in Figure 13). (From Le Grimellec, C. and Leblanc, G., *Biochim. Biophys. Acta,* 599, 639, 1980. With permission.)

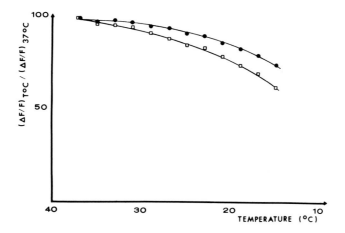

FIGURE 15.　Effect of temperature on the electrochemical transmembrane potential of *M. mycoides* cells grown in the presence of cholesterol and oleic and palmitic acids (□) or cholesterol and elaidic acid (●). Transmembrane potential was measured as the glucose-induced fluorescence intensity quenching of the dye merocyanine 540 at a given temperature $(\Delta F/F)_{t^\circ C}$ divided by the value obtained at 37°C $(\Delta F/F)_{37^\circ C}$. (From Le Grimellec, C. and Leblanc, G., *Biochim. Biophys. Acta,* 599, 639, 1980. With permission.)

and of transmembrane potential did not exhibit discontinuities or breaks (see Figures 14 and 15), and the activity-temperature profiles of the native and adapted strain were not significantly different. These workers thus concluded that the absolute Mg^{2+}-ATPase activity and its temperature dependence, as well as the temperature dependence of the transmembrane potential difference, were not affected by the order-disorder phase transition of the membrane lipids. Therefore, the observed alterations in the K^+ influx rates with membrane lipid composition must reflect the dependence of the K^+ carrier itself on the phase state of the membrane lipids.

D. Other Transport Systems

Dockter and Magnuson[109,110] and Dockter et al.[111] have studied chlortetracycline transport in *S. aureus* and in *Bacillus megaterium* and have related it to the lipid order-disorder phase transition of the cell membrane of these organisms. In *S. aureus*, the Arrhenius plot of initial rates of antibiotic transport was biphasic with a fairly pronounced break at a temperature of 27°C. Culturing *S. aureus* at 37°C in the presence of exogenous oleic acid, or at 10°C without exogenous fatty acid supplementation, increased the unsaturated fatty acid content of the cell membrane and reduced the Arrhenius plot break temperature by 8 to 9°C. Although the existence of membrane lipid phase transitions was not actually demonstrated by an independent physical technique, these workers suggested that the movement of chlortetracycline was facilitated by the more fluid lipid state existing above the presumed phase transition. In *B. megaterium*, Arrhenius plots of initial rates of antibiotic uptake were interpreted to be triphasic, with breaks occurring at 20 and 9.5°C. Correlative electron spin resonance and fluorescence probe experiments indicated that the break temperatures apparently observed for chlortetracycline transport corresponded to the upper and lower boundaries of the gel to liquid-crystalline ''lateral phase separation'' region. However, a careful examination of the Arrhenius plot of antibiotic transport reveals that, in fact, the experimental points fall on a single, slightly curving line. Thus, the assignment of discrete transport break temperatures seems quite arbitrary, and for this reason the significance of this study is unclear.

VII. CONCLUSIONS

The studies just reviewed appear to demonstrate that concomitant fatty acid and phospholipid biosynthesis are not necessary for the biosynthesis and assembly of functional transport systems in prokaryotic plasma membranes, provided that fatty acid and phospholipid starvation is not severe enough to affect cellular viability and membrane function generally. Whether or not the biosynthesis and assembly of functional transport systems can occur in plasma membranes whose lipids exist predominantly in gel state is not really known, since prokaryotic microorganisms whose membrane lipids are primarily in the solid state become leaky and rapidly lose viability. Definitive experiments to address this question are, therefore, difficult to design. It does seem clear, however, that there is no preferential, long-lived association between newly synthesized membrane lipids and newly synthesized transport proteins. There is, in fact, no good evidence for the preferential association of particular transport proteins with particular phospholipid molecular species in any prokaryotic plasma membrane, although such a possibility can certainly not be ruled out at the present time. Clearly, much additional experimental work, both with intact cells and reconstituted transport systems, will be necessary to resolve this question.

The results of many investigations support the idea that *most* membrane transport systems function optimally only when the majority of the membrane lipids exist in the liquid-crystalline state. However, there is a remarkably heterogeneous response to the existence of substantial amounts of gel state lipid in the membrane among the different transport systems examined, even in the same organism. Although a number of transport systems seem to begin to be inhibited when about half of the membrane lipids is converted to the solid state, other transport systems appear to be adversely affected by the formation of even small amounts of gel state lipid, while still others function normally until the fluid-to-solid lipid phase transition is almost complete. To further complicate the matter, even a single transport system in one particular organism may sometimes exhibit a differential response to lipid phase-state changes, depending upon the particular fatty acid compositional manipulations employed to alter the phase transition temperature. Convincing molecular explanations for this heterogeneous behavior are presently lacking, in part probably because the actual molecular mechanism of the translocation process, and of its coupling to the cellular energy transducing system, are still poorly understood. Again, much more work in this area remains for the future. In particular, the relationship between membrane lipid fluidity and phase state and membrane energization needs to be rigorously investigated.

Although it is often asserted that the activities of membrane transport systems increase with increases in lipid bilayer fluidity, a careful analysis of the published transport studies in prokaryotic microorganisms indicates that this is at best a crude generalization. Instead, the responses observed to changes in membrane lipid fluidity (within the liquid-crystalline state) are quite diverse. In a few systems, translocation rates do, indeed, seem to increase with increases in lipid fluidity, although this dependence in usually relatively small, with no more than a two- or threefold variation in transport rates being observed even with fairly marked alterations in the membrane lipid fatty acid composition or in the cholesterol content of the membrane. In contrast, some other transport systems seem to be almost insensitive to membrane lipid fluidity changes, provided that all or most of the lipid remains in the liquid-crystalline state. Moreover, a substantial number of transport systems actually exhibit apparently reduced rates of translocation in membranes whose lipids are far above their phase transition temperature, suggesting that "hyperfluid" lipid can actually inhibit the functioning of membrane transport systems. Since in most prokaryotes biochemical mechanisms exist to compensate for environmentally induced alterations in membrane lipid fluidity and phase state, it is presently unclear whether or not changes in membrane lipid fluidity are of general physiological importance in the regulation of membrane transport and other membrane-associated processes, as is often suggested. Again, the answers to this and many other important questions about biological membrane structural-function correlates must await the results of future studies.

ACKNOWLEDGMENT

The author's own work was generously supported by operating, personnel, and major equipment grants from the Medical Research Council of Canada and the Alberta Heritage Foundation for Medical Research.

REFERENCES

1. **Cronan, J. E. and Gelmann, E. P.,** Physical properties of membrane lipids: biological relevance and regulation, *Bacteriol. Rev.,* 39, 232, 1975.
2. **Cronan, J. E.,** Molecular biology of bacterial membrane lipids, *Ann. Rev. Biochem.,* 47, 163, 1978.
3. **Razin, S.,** The mycoplasma membrane, in *Organization of Procaryotic Cell Membranes,* Vol. 1, Ghosh, B. K., Ed., CRC Press, Boca Raton, Fla., 1982, 165.
4. **Neville, D. M.,** The preparation of cell surface membrane enriched fractions, in *Biochemical Analysis of Membranes,* Maddy, A. H., Ed., Chapman and Hall, London, 1976, chap. 2.
5. **Rohmer, M., Bovier, P., and Ourisson, G.,** Molecular evolution of biomembranes: structural equivalents and phylogenetic precursors of sterols, *Proc. Natl. Acad. Sci. U.S.A.,* 76, 847, 1979.
6. **Kannenberg, E., Blume, A., McElhaney, R. N., and Poralla, K.,** Monolayer and calorimetric studies of phosphatidylcholines containing branched-chain fatty acids and their interactions with cholesterol and with a bacterial hopanoid in model membranes, *Biochim. Biophys. Acta,* 733, 111, 1983.

7. **Raetz, C. R. H.,** Genetic control of phospholipid bilayer assembly, in *Phospholipids,* New Comprehensive Biochemistry, Vol. 4, Hawthorne, J. N. and Ansell, G. B., Eds., Elsevier, Amsterdam, 1982, 435.

8. **Archer, D. B., Rodwell, A. W., and Rodwell, E. S.,** The nature and location of *Acholeplasma laidlawii* membrane proteins investigated by two-dimensional gel electrophoresis, *Biochim. Biophys. Acta,* 513, 268, 1978.

9. **McElhaney, R. N.,** The structure and function of the *Acholeplasma laidlawii* plasma membrane, *Biochim. Biophys. Acta,* 779, 1, 1984.

10. **Rodwell, A. W. and Peterson, J. E.,** The effect of straight-chain saturated, monoenoic and branched-chain fatty acids on growth and fatty acid composition of *Mycoplasma* strain Y, *J. Gen. Microbiol.,* 68, 173, 1971.

11. **Silvius, J. R. and McElhaney, R. N.,** Lipid compositional manipulation in *Acholeplasma laidlawii* B. Effect of exogenous fatty acids on fatty acid composition and cell growth when endogenous fatty acid production is inhibited, *Can. J. Biochem.,* 56, 462, 1978.

12. **Eaton, L. C., Erdos, G. W., Vreeland, N. L., and Ingram, L. O.,** Failure of *Escherichia coli* to alter its fatty acid composition in response to cholesterol-induced changes in membrane lipid fluidity, *J. Bacteriol.,* 146, 1151, 1981.

13. **Rottem, S., Yashouv, J., Ne'eman, Z., and Razin, S.,** Cholesterol in *Mycoplasma* membranes. Composition, ultrastructure and biological properties of membranes from *Mycoplasma mycoides* var. *capri* cells adapted to grow with low cholesterol concentrations, *Biochim. Biophys. Acta,* 323, 495, 1973.

14. **Le Grimellec, C., Cardinal, J., Giocondi, M.-C., and Carriere, S.,** Control of membrane lipids in *Mycoplasma gallisepticum:* effect on lipid order, *J. Bacteriol.,* 146, 155, 1981.

15. **McElhaney, R. N.,** The use of differential scanning calorimetry and differential thermal analysis in studies of model and biological membranes, *Chem. Phys. Lipids,* 30, 229, 1982.

16. **Rosen, B. P.,** *Bacterial Transport,* Microbiology Series, Vol. 4, Marcel Dekker, New York, 1978.

17. **McElhaney, R. N.,** Effects of membrane lipids on transport and enzymic activities, in *Membrane Lipids of Prokaryotes,* Current Topics in Membranes and Transport, Vol. 17, Razin, S. and Rottem, S., Eds., Academic Press, New York, 1982, 317.

18. **Singer, S. J. and Nicholson, G. L.,** The fluid mosaic model of the structure of cell membranes, *Science,* 175, 720, 1972.

19. **Rothman, J. E. and Lenard, J.,** Membrane symmetry, *Science,* 175, 720, 1972.

20. **Vaz, W. L. C., Derzko, Z. I., and Jacobson, K. A.,** Photobleaching measurements of the lateral diffusion of lipids and proteins in artificial phospholipid bilayer membranes, in *Membrane Reconstitution,* Poste, G. and Nicolson, G. L., Eds., Elsevier, Amsterdam, 1982, 83.

21. **Cherry, R. J.,** Rotational and lateral diffusion of membrane proteins, *Biochim. Biophys. Acta,* 559, 289, 1979.

22. **Levine, Y. K., Birdsall, N. J. M., Lee, A. G., and Metcalfe, J. C.,** ^{13}C-nuclear magnetic resonance relaxation measurements of synthetic lecithins and the effect of spin-labelled lipids, *Biochemistry,* 11, 1416, 1972.

23. **Seelig, J. and Seelig, A.,** Lipid conformation in model membranes and biological membranes, *Q. Rev. Biophys.,* 13, 19, 1980.

24. **Davis, J. H.,** The description of membrane lipid conformation, order and dynamics by ^{2}H-NMR, *Biochim. Biophys. Acta,* 737, 117, 1983.

25. **Yeagle, P. L.,** ^{31}P nuclear magnetic resonance studies of the phospholipid-protein interface in cell membranes, *Biophys. J.,* 37, 227, 1982.

26. **Cornell, B. A., Hiller, R. G., Raison, J., Separovic, F., Smith, R., Vary, J. C., and Morris, C.,** Biological membranes are rich in low-frequency motion, *Biochim. Biophys. Acta,* 732, 473, 1983.

27. **Dettman, H. D., Weiner, J. H., and Sykes, B. D.,** ^{19}F-nuclear magnetic resonance studies of the coat protein of bacteriophage M13 in synthetic phospholipid vesicles and deoxycholate micelles, *Biophys. J.,* 37, 243, 1982.

28. **Kinsey, R. A., Kintanar, A., Tsai, M.-D., Smith, R. L., Janes, N., and Oldfield, E.,** First observation of amino acid side chain dynamics in membrane proteins using high field deuterium nuclear magnetic resonance spectroscopy, *J. Biol. Chem.,* 256, 4146, 1981.

29. **Jost, P. C., Griffith, O. H., Capaldi, R. A., and Vanderkooi, G.,** Evidence for boundary lipid in membranes, *Proc. Natl. Acad. Sci. U.S.A.,* 70, 480, 1973.

30. **Shinitzky, M. and Barenholtz, Y.,** Fluidity parameters of lipid regions determined by fluorescence polarization, *Biochim. Biophys. Acta,* 515, 367, 1978.

31. **Wolber, P. K. and Hudson, B. S.,** Bilayer acyl chain dynamics and lipid-protein interaction. The effect of the M13 bacteriophage coat protein on the decay of the fluorescence anisotropy of parinaric acid, *Biophys. J.,* 37, 253, 1982.

32. **Han, M. H.,** Non-linear Arrhenius plots in temperature-dependent kinetic studies of enzyme reactions. I. Single transition processes, *J. Theor. Biol.,* 35, 543, 1972.

33. **Gardiner, W. C.,** Temperature dependence of bimolecular gas reaction rates, *Acc. Chem. Res.,* 10, 326, 1977.

34. **Sandermann, H.,** Regulation of membrane enzymes by lipids, *Biochim. Biophys. Acta,* 515, 209, 1978.

35. **Raison, J. K.,** The influence of temperature-induced phase changes on the kinetics of respiratory and other membrane-associated enzyme systems, *Bioenergetics,* 4, 285, 1973.

36. **Kumamoto, J., Raison, J. K., and Lyons, J. M.,** Temperature "breaks" in Arrhenius plots: a thermodynamic consequence of a phase change, *J. Theor. Biol.,* 31, 47, 1971.

37. **Bagnall, D. J. and Wolfe, J.,** Arrhenius plots: information or noise?, *Cryo-Lett.,* 3, 7, 1982.

38. **Dixon, M. and Webb, E. C.,** *Enzymes,* 2nd ed., Academic Press, New York, 1964, 145.

39. **Wynn-Williams, A. T.,** An explanation of apparent sudden change in the activation energy of membrane enzymes, *Biochem. J.,* 157, 279, 1976.

40. **Jahnig, F. and Bramhall, J.,** The origin of a break in Arrhenius plots of membrane process, *Biochim. Biophys. Acta,* 690, 310, 1982.

41. **Thilo, L., Trauble, H., and Overath, P.,** Mechanistic interpretation of the influence of lipid phase transitions on transport functions, *Biochemistry,* 16, 1283, 1977.

42. **Silvius, J. R. and McElhaney, R. N.,** Membrane lipid physical state and modulation of the Na^+,Mg^{2+}-ATPase activity in *Acholeplasma laidlawii* B, *Proc. Natl. Acad. Sci. U.S.A.,* 77, 1255, 1980.

43. **Silvius, J. R. and McElhaney, R. N.,** Nonlinear Arrhenius plots and the analysis of reaction and motional rates in biological membranes, *J. Theor. Biol.,* 88, 135, 1980.

44. **Londesborough, J. and Varimo, K.,** The temperature dependence of adenylate cyclase from baker's yeast, *Biochem. J.,* 181, 539, 1979.

45. **Sprague, E. D., Larrabee, C. E., and Halsall, H. B.,** Statistical evaluation of alternative models: application to ligand-protein binding, *Anal. Biochem.,* 101, 175, 1980.

46. **Silvius, J. R., Read, B. D., and McElhaney, R. N.,** Membrane enzymes: artifacts in Arrhenius plots due to temperature dependence of substrate binding, *Science,* 199, 902, 1978.

47. **Keleti, T.,** Errors in the evaluation of Arrhenius and van't Hoff plots, *Biochem. J.,* 209, 277, 1983.

48. **McElhaney, R. N., De Gier, J., and van Deenen, L. L. M.,** The effect of alterations in fatty acid composition and cholesterol content on the permeability of *Mycoplasma laidlawii* B cells and derived liposomes, *Biochim. Biophys. Acta,* 219, 245, 1970.

49. **De Gier, J., Mandersloot, J. G., Hupkes, J. V., McElhaney, R. N., and van Beek, W. P.,** On the mechanism of nonelectrolyte permeation through lipid bilayers and through biomembranes, *Biochim. Biophys. Acta,* 233, 610, 1971.

50. **Romijn, J. C., van Golde, L. M. G., McElhaney, R. N., and van Deenen, L. L. M.,** Some studies on the fatty acid composition of total lipids and phosphatidylglycerol from *Acholeplasma laidlawii* B and their relation to the permeability of intact cells of this organism, *Biochim. Biophys. Acta,* 280, 22, 1972.

51. **De Kruijff, B., De Greef, W. J., van Eyk, R. V. W., Demel, R. A., and van Deenen, L. L. M.,** The effect of different fatty acid and sterol composition on the erythritol flux through the cell membrane of *Acholeplasma laidlawii,* *Biochim. Biophys. Acta,* 298, 479, 1973.

52. **McElhaney, R. N., De Gier, J., and van der Neut-Kok, E. C. M.,** The effect of alterations in fatty acid composition and cholesterol content on the nonelectrolyte permeability of *Acholeplasma laidlawii* B cells and derived liposomes, *Biochim. Biophys. Acta,* 298, 500, 1973.

53. **Silvius, J. R., Mak, N., and McElhaney, R. N.,** Why do prokaryotes regulate membrane fluidity?, in *Membrane Fluidity: Biophysical Techniques and Cellular Regulation,* Kates, M. and Kuksis, A., Eds., Humana Press, Clifton, N.J., 1980, 213.

54. **Eze, M. O. and McElhaney, R. N.,** The effect of alterations in the fluidity and phase state of the membrane lipids on the passive permeation and facilitated diffusion of glycerol in *Escherichia coli,* *J. Gen. Microbiol.,* 124, 299, 1981.

55. **Haest, C. W. M., De Gier, J., van Es, G. A., Verkleij, A. J., and van Deen, L. L. M.,** Fragility of the permeability barrier of *Escherichia coli,* *Biochim. Biophys. Acta,* 288, 43, 1972.

56. **Van Zoelen, E. J. J., van der Neut-Kok, E. C. M., De Gier, J., and van Deenen, L. L. M.,** Osmotic behavior of *Acholeplasma laidlawii* B cells with membrane lipids in liquid-crystalline and gel state, *Biochim. Biophys. Acta,* 394, 463, 1975.

57. **Le Grimellec, C. and Leblanc, G.,** Effect of membrane cholesterol on potassium transport in *Mycoplasma mycoides* var. *capri* (PG3), *Biochim. Biophys. Acta,* 514, 152, 1978.

58. **Chapman, D., Gomez-Fernandez, J. C., and Goni, F. M.,** Intrinsic protein-lipid interactions. Physical and biochemical evidence, *FEBS Lett.,* 98, 211, 1979.

59. **Eze, M. O. and McElhaney, R. N.,** Stopped-flow spectrophotometric assay of glycerol permeation in *Escherichia coli:* applicability and limitations, *J. Gen. Microbiol.,* 105, 233, 1978.

60. **Fox, C. F.,** A lipid requirement of induction of lactose transport in *Escherichia coli,* *Proc. Natl. Acad. Sci. U.S.A.,* 63, 850, 1969.

61. **Hsu, C. C. and Fox, C. F.,** Induction of the lactose transport system in a lipid-synthesis-defective mutant of *Escherichia coli,* *J. Bacteriol.,* 103, 410, 1970.

62. **Nunn, W. D. and Cronan, J. E.,** Unsaturated fatty acid synthesis is not required for induction of lactose transport in *Escherichia coli, J. Biol. Chem.,* 249, 724, 1974.

63. **Weisberg, L. J., Cronan, J. E., and Nunn, W. D.,** Induction of lactose transport in *Escherichia coli* during the absence of phospholipid synthesis, *J. Bacteriol.,* 123, 492, 1975.

64. **Overath, P., Hill, F. F., and Lamnek-Hirsch, I.,** Biogenesis of *E. coli* membrane: evidence for randomization of lipid phase, *Nature (London), New Biol.,* 234, 264, 1971.

65. **Robbins, A. R. and Rotman, B.,** Inhibition of methylgalactoside transport in *Escherichia coli* upon the cessation of unsaturated fatty acid biosynthesis, *Proc. Natl. Acad. Sci. U.S.A.,* 69, 2125, 1972.

66. **Mindich, L.,** Induction of *Staphylococcus aureus* lactose permease in the absence of glycerolipid synthesis, *Proc. Natl. Acad. Sci. U.S.A.,* 68, 420, 1971.

67. **Willecke, K. and Mindich, L.,** Induction of citrate transport in *Bacillus subtilis* during the absence of phospholipid synthesis, *J. Bacteriol.,* 106, 514, 1971.

68. **McElhaney, R. N.,** The relationship between membrane lipid fluidity and phase state and the ability of bacteria and mycoplasmas to grow and survive at various temperatures, in *Biomembranes,* Vol. 12, Kates, M. and Manson, L. A., Eds., Plenum Press, New York, 1984, 249.

69. **McIntyre, T. M. and Bell, R. M.,** Mutants of *Escherichia coli* defective in membrane phospholipid synthesis. Effect of cessation of net phospholipid synthesis on cytoplasmic and outer membranes, *J. Biol. Chem.,* 250, 9053, 1975.

70. **McIntyre, T. M., Chamberlain, B. K., Webster, R. E., and Bell, R. M.,** Mutants of *Escherichia coli* defective in membrane phospholipid synthesis. Effects of cessation and reinitiation of phospholipid synthesis on macromolecular synthesis and phospholipid turnover, *J. Biol. Chem.,* 252, 4487, 1977.

71. **Tsukagoshi, N. and Fox, C. F.,** Abortive assembly of the lactose transport system, *Biochemistry,* 12, 2816, 1973.

72. **Wilson, G. and Fox, C. F.,** Biogenesis of microbial transport systems: evidence for coupled incorporation of newly synthesized lipids and proteins into membrane, *J. Mol. Biol.,* 55, 49, 1971.

73. **Schairer, H. V. and Overath, P.,** Lipids containing transunsaturated fatty acids change the temperature characteristic of thiomethylgalactoside accumulation in *Escherichia coli, J. Mol. Biol.,* 44, 209, 1969.

74. **Wilson, G., Rose, S. P., and Fox, C. F.,** The effect of membrane lipid unsaturation on glycoside transport, *Biochem. Biophys. Res. Commun.,* 38, 617, 1970.

75. **Tsukagoshi, N. and Fox, C. F.,** Transport system assembly and the mobility of membrane lipids in *Escherichia coli, Biochemistry,* 12, 2822, 1973.

76. **Linden, C. D., Keith, A. D., and Fox, C. F.,** Correlations between fatty acid distribution in phospholipids and the temperature dependence of membrane physical state, *J. Supramol. Struct.,* 1, 523, 1973.

77. **Linden, C. D. and Fox, C. F.,** A comparison of characteristic temperatures for transport in two unsaturated fatty acid auxotrophs of *E. coli, J. Supramol. Struct.,* 1, 535, 1973.

78. **Linden, C. D., Wright, K. L., McConnell, H. M., and Fox, C. F.,** Lateral phase separations in membrane lipids and the mechanism of sugar transport in *Escherichia coli, Proc. Natl. Acad. Sci. U.S.A.,* 70, 2271, 1973.

79. **Davis, J. H., Nichol, C. P., Weeks, G., and Bloom, M.,** Study of the cytoplasmic and outer membranes of *Escherichia coli* by deuterium magnetic resonance, *Biochemistry,* 18, 2103, 1979.

80. **Gally, H. U., Pluschke, G., Overath, P., and Seelig, J.,** Structure of *Escherichia coli* membranes. Phospholipid conformation in model membranes and cells as studied by deuterium magnetic resonance, *Biochemistry,* 18, 5605, 1979.

81. **Kang, S. Y., Gutowsky, H. S., and Oldfield, E.,** Spectroscopic studies of specifically deuterium-labelled membrane systems. Nuclear magnetic resonance investigation of protein-lipid interactions in *Escherichia coli* membranes, *Biochemistry,* 18, 3268, 1979.

82. **Gally, H. U., Pluschke, G., Overath, P., and Seelig, J.,** Structure of *Escherichia coli* membranes. Fatty acyl chain order parameters in inner and outer membranes and derived liposomes, *Biochemistry,* 19, 1638, 1980.

83. **Nichol, C. P., Davis, J. H., Weeks, G., and Bloom, M.,** Quantitative study of the fluidity of *Escherichia coli* membranes using deuterium magnetic resonance, *Biochemistry,* 19, 451, 1980.

84. **Baroin, A., Bienvenue, A., and Devaux, P. F.,** Spin-label studies of protein-protein interactions in retinal rod outer segment membranes. Saturation transfer electron paramagnetic resonance spectroscopy, *Biochemistry,* 18, 1151, 1979.

85. **Kang, S. Y., Gutowsky, H. S., Hsung, J. C., Jacobs, R., King, T. E., Rice, D., and Oldfield, E.,** Nuclear magnetic resonance investigation of cytochrome oxidase-phospholipid interaction: a new model of boundary lipid, *Biochemistry,* 18, 3257, 1979.

86. **Tamm, L. K. and Seelig, J.,** Lipid solvation of cytochrome oxidase. Deuterium, nitrogen-14, and phosphorus-31 nuclear magnetic resonance studies on the phosphocholine head group and on cis-unsaturated fatty acyl chains, *Biochemistry,* 22, 1474, 1983.

87. **Overath, P., Schairer, H. U., and Stoffel, W.,** Correlation of *in vivo* and *in vitro* phase transitions of membrane lipids in *Escherichia coli, Proc. Natl. Acad. Sci. U.S.A.,* 67, 606, 1970.

88. **Overath, P. and Trauble, H.,** Phase transitions in cells, membranes and lipids of *Escherichia coli.* Detection by fluorescent probes, light scattering and dilatometry, *Biochemistry,* 12, 2625, 1973.

89. **Sackman, E., Trauble, H., Galla, H.-J., and Overath, P.,** Lateral diffusion, protein mobility, and phase transitions in *Escherichia coli* membranes. A spin-label study, *Biochemistry,* 12, 5360, 1973.

90. **Therisod, H., Letellier, L., Weil, R., and Shechter, E.,** Functional *lac* carrier proteins in cytoplasmic membrane vesicles isolated from *Escherichia coli.* I. Temperature dependence of dansyl galactoside binding and β-galactoside transport, *Biochemistry,* 16, 3772, 1977.

91. **Letellier, L., Weil, R., and Schechter, E.,** Functional *lac* carrier proteins in cytoplasmic membrane vesicles isolated from *Escherichia coli.* II. Experimental evidence for a segregation of the *lac* carrier proteins induced by a conformational transition of the membrane lipids, *Biochemistry,* 16, 3777, 1977.

92. **Overath, P., Teather, R. M., Simoni, R. D., Aichele, G., and Wilhelm, U.,** Lactose carrier protein of *Escherichia coli.* Transport and binding of 2'-(N-dansyl)aminoethyl β-D-thiogalactopyranoside and *p*-nitrophenyl α-D-galactopyranoside, *Biochemistry,* 18, 1, 1979.

93. **Teather, R. M., Bramhall, J., Riede, I., Wright, J. K., Furst, M., Aichele, G., Wilhelm, U., and Overath, P.,** Lactose carrier protein of *Escherichia coli.* Structure and expression of plasmids carrying the Y gene of the *lac* operon, *Eur. J. Biochem.,* 108, 223, 1980.

94. **Wright, J. K., Riede, I., and Overath, P.,** Lactose carrier protein of *Escherichia coli*: interaction with galactosides and protons, *Biochemistry,* 20, 6404, 1981.

95. **Shechter, E., Letellier, L., and Gulik-Krzywicki, T.,** Relations between structure and function in cytoplasmic membrane vesicles isolated from an *Escherichia coli* fatty acid auxotroph. High-angle X-ray diffraction, freeze-etch microscopy and transport studies, *Eur. J. Biochem.,* 49, 61, 1974.

96. **Rottem, S., Cirillo, V. P., de Kruyff, B., Shinitzky, M., and Razin, S.,** Cholesterol in mycoplasma membranes. Correlation of enzymic and transport activities with physical state of lipids in membranes of *Mycoplasma mycoides* var. *capri* adapted to grow with low cholesterol concentrations, *Biochim. Biophys. Acta,* 323, 509, 1973.

97. **Cirillo, V. P.,** Transport systems, in *The Mycoplasmas,* Vol. 1, Barile, M. F. and Razin, S., Eds., Academic Press, New York, 1979, 323.

98. **Read, B. D. and McElhaney, R. N.,** Glucose transport in *Acholeplasma laidlawii* B: dependence on the fluidity and physical state of the membrane lipids, *J. Bacteriol.,* 123, 47, 1975.

99. **Esfahani, M., Limbrick, A. R., Knutton, S., Oka, S., and Wakil, S. J.,** The molecular organization of lipids in the membrane of *Escherichia coli*: phase transitions, *Proc. Natl. Acad. Sci. U.S.A.,* 68, 3180, 1971.

100. **Rosen, B. P. and Hackette, S. L.,** Effects of fatty acid substitution on the release of enzymes by osmotic shock, *J. Bacteriol.,* 110, 1181, 1972.

101. **Eze, M. O.,** The Relationship between Membrane Lipid Fluidity and Phase State and Passive Permeation, Facilitated Diffusion and Active Transport Processes in *Escherichia coli,* Ph.D. thesis, University of Alberta, Edmonton, 1978.

102. **Ingram, L. O., Eaton, L. C., Erdos, G. W., Tedder, T. F., and Vreeland, N. L.,** Unsaturated fatty acid requirement in *Escherichia coli*: mechanism of palmitate-induced inhibition of growth of strain WN1, *J. Membrane Biol.,* 65, 31, 1982.

103. **Holden, J. T., Bolen, J., Easton, J. A., and de Groot, J.,** Heterogeneous amino acid transport rate changes in an *E. coli* unsaturated fatty acid auxotroph, *Biochem. Biophys. Res. Commun.,* 81, 588, 1978.

104. **Wilkins, P. O.,** Effects of growth temperature on transport and membrane viscosity in *Streptococcus faecalis, Arch. Microbiol.,* 132, 211, 1982.

105. **Wilkins, P. O.,** Psychotropic Gram-positive bacteria: temperature effects on growth and solute uptake, *Can. J. Microbiol.,* 19, 909, 1973.

106. **Herbert, R. A. and Bell, C. R.,** Growth characteristics of an obligately phycrophilic *Vibrio* sp., *Arch. Microbiol.,* 113, 215, 1977.

107. **Cho, H. W. and Morowitz, H. J.,** Characterization of the plasma membrane of *Mycoplasma laidlawii.* VIII. Effect of temperature shift and antimetabolites on K^+ transport, *Biochim. Biophys. Acta,* 274, 105, 1972.

108. **Le Grimellec, C. and Leblanc, G.,** Temperature-dependent relationship between K^+ influx, Mg^{2+}-ATPase activity, transmembrane potential and membrane lipid composition in mycoplasma, *Biochim. Biophys. Acta,* 599, 639, 1980.

109. **Dockter, M. E. and Magnuson, J. A.,** Characterization of the active transport of chlortetracycline in *Staphylococcus aureus* by a fluorescence technique, *J. Supramol. Struct.,* 2, 32, 1974.

110. **Dockter, M. E. and Magnuson, J. A.,** Membrane phase transitions and the transport of chlortetracycline, *Arch. Biochem. Biophys.,* 168, 81, 1975.

111. **Dockter, M. E., Trumble, W. R., and Magnuson, J. A.,** Membrane lateral phase separations and chlortetracycline transport by *Bacillus megaterium, Proc. Natl. Acad. Sci. U.S.A.,* 75, 1319, 1978.

Chapter 4

THE INFLUENCE OF MEMBRANE LIPIDS ON THE PERMEABILITY OF MEMBRANES TO Ca^{2+}

Ross P. Holmes

TABLE OF CONTENTS

I. INTRODUCTION

The permeability of membranes to Ca^{2+} is a property so crucial to cellular function that it should be regarded as central to the life process itself. The importance of Ca^{2+} permeability is intimately related to the role of Ca^{2+} as an intracellular messenger in living cells.[1] To regulate the flow of Ca^{2+} within the cell the endoplasmic reticulum (ER) and mitochondria sequester, store, and release Ca^{2+}, and the plasma membrane permits Ca^{2+} inflow from the extracellular fluid as well as actively extruding it from the cell.[2,3] These various mechanisms are illustrated in Figure 1. They must act in a concerted, closely regulated manner to enable Ca^{2+} to serve its messenger function.

Despite their critical importance, most of these processes are poorly defined and the proteins facilitating the Ca^{2+} movements have not been isolated and characterized. An offshoot of this lack of knowledge concerning these processes is that it has given credibility to claims that specific phospholipids rather than proteins can directly regulate Ca^{2+} movements across some membranes.[4-8] The best characterized processes are the uptake of Ca^{2+} by the sarcoplasmic reticulum (SR), which may be similar to that occurring in other types of ER, and the pumping of Ca^{2+} out of the cell across the plasma membrane, particularly the erythrocyte membrane. These activities are biochemically expressed as Ca^{2+}-ATPase activities in isolated membrane fragments. At the other extreme, relatively little is known about the plasma membrane channel permitting Ca^{2+} inflow into nonexcitable cells. It has not even been conclusively shown that such a channel is protein in nature, which is why in some cells phospholipids have been proposed to fulfill this function. It is possible that this channel is similar to that in excitable cells, which has been better characterized, and to that in the brush border membranes of kidney and intestinal tissues which absorb Ca^{2+} from extrahumoral fluids.

Lipid-protein interactions are central to processes controlling the Ca^{2+} permeability of membranes. Indeed, several lines of research indicate that the lipid composition of membranes can influence the properties of membrane-bound enzymes including those involved in their permeability to Ca^{2+}. In particular, under physiological conditions, membrane lipids appear to form an important link in modifying the properties of channels that permit Ca^{2+} to flow into cells down a strong electrochemical gradient. This may follow a hormone signal where Ca^{2+} is used as an intracellular messenger to trigger a cellular response, or in Ca^{2+} absorption from the intestine, stimulated by calcitriol (1,25-dihydroxyvitamin D_3). In this chapter, the relationship between membrane lipids and their permeability to Ca^{2+} will be explored in detail. While it is evident at least from Ca^{2+}-ATPase systems, that regulation of Ca^{2+} fluxes is complex involving regulatory proteins, cations, and phosphorylation and dephosphorylation steps, the role of lipids also has to be considered, not only in modifying the environment in which these reactions occur, but also as a direct modifier of enzyme properties. In addition, conditions under which permeation of Ca^{2+} through the lipid bilayer can occur are considered, mostly because of their relationship to the development of pathological states.

II. Ca^{2+} PERMEABILITY OF THE LIPID BILAYER

The observations that the negatively charged phospholipids, phosphatidic acid and cardiolipin, could translocate Ca^{2+} across an organic solvent barrier[5] led to speculation and apparent experimental evidence that phospholipids played a direct role in Ca^{2+}-gating.[9-12] Another line of experimentation revealed that certain phospholipids under appropriate conditions preferred to form nonbilayer phases in membranes rather than bilayers.[13-16] Cardiolipin was of particular interest because it was observed to form nonbilayer regions in artificial membranes when they were exposed to Ca^{2+}.[14,15] The proposed vesicular nature of some

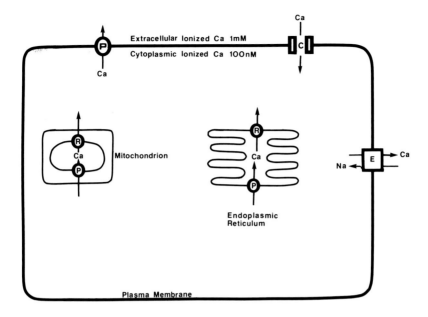

FIGURE 1.   Pathways for Ca^{2+} movements in a cell. C denotes a channel, E, an exchange process, P, a pumping process, and R, a release pathway, which may be a channel. The directions in which Ca^{2+} movements occur through these pathways are shown by arrows.

of these nonbilayer structures offered a possible explanation for how negatively charged phospholipids exerted their effects, both in membranes and in the organic solvent of a Pressman cell. Such a role for lipids in regulating membrane permeability would turn perception of the functional role of lipids in membranes through 180°. Instead of being inert matrices as they were depicted in early membrane models, or more recently as modifiers of protein conformation or positioning in the membrane, this new postulate suggests that they play a direct role in determining the permeability of membranes to some ions, notably Ca^{2+}. In view of its ramifications such a proposal and the evidence supporting it requires a detailed analysis.

A. Phospholipids as Ionophores

In a Pressman cell,[17,18] the assumption is made that a layer of organic solvent mimics the hydrocarbon milieu of membranes and provides a similar barrier to ion movements. The passage of a cation from one aqueous compartment (donor), across the organic solvent layer to another aqueous compartment (acceptor), is then proposed to provide a measure of membrane permeability. The ability of lipophilic solutes to translocate a cation between these compartments can be determined in this system. Ionophores, as might be predicted, show a high translocation rate.[5] A simpler model is to study the partitioning of cations between an aqueous phase and an organic phase containing a test lipophilic substance after efficiently mixing the two phases.[4,6,9] In both models cardiolipin and phosphatidic acid (PA) have been observed to transport Ca^{2+} into or across the organic layer in contrast to other phospholipids, although Green and colleagues[5,6] observed conflicting behavior of cardiolipin in the two different systems. A substance carrying a cation or anion into or through an organic solvent barrier may truly deserve to be termed an ionophore if this is, indeed, the definition of an ionophore. With reference to membranes, however, the definition usually has more narrow connotations in indicating that an ionophore can translocate ions across a membrane.

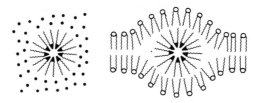

FIGURE 2.   Structure of inverted micelles formed by
phospholipids. The phospholipids believed to form such
structures usually have a cone-shaped head group as
depicted.[21,35] Their structures in either an organic solvent
(●) or in a phospholipid membrane are shown.

It is reasonable to propose that transfer of charged, lipid-insoluble compounds into an
organic phase containing a lipid solute occurs by formation of an inverted micelle as shown
in Figure 2. This is supported by ^{14}N NMR studies[19] and the disruption of Ca^{2+}-cardiolipin
partitioning in organic phases by phosphatidylcholine (PC),[20] a phospholipid which prefers
to form a bilayer. Whether the structural preference of cardiolipin is a factor in the latter
observation is not known, as the choline head group may disrupt calcium binding to the
cardiolipin head group.

The extrapolation of the formation of inverted vesicles in organic solvents and its role in
the transfer of cations to what may occur in a membrane is not simple. While there is strong
evidence that certain phospholipids prefer not to form bilayers, many of the observed properties of biological membranes depend on its bilayer nature. The stable existence of nonbilayer
phases as permanent structures in any biological membrane has not yet been established
with any certainty. Their transient existence may be important in membrane fusion.[13,21]
Microsomes derived from rat liver ER produce ^{31}P NMR spectra which suggest a large
percentage of the membrane phospholipids are undergoing isotropic motion.[22] To determine
whether nonbilayer phases do, in fact, occur in these membranes, other techniques should
be applied as phosphorylated proteins, vesicle tumbling, and the lateral diffusion of phospholipids may contribute to the ^{31}P spectra observed.[23] Hui et al.[24] demonstrated that in
phosphatidylethanolamine/phosphatidylcholine (PE/PC) mixtures the existence of hexagonal
H_{II} phases in membranes suggested by ^{31}P NMR spectra were not confirmed by X-ray
diffraction and freeze fracture electron microscopy. Their observations illustrate the caution
that must be exercised in relying on ^{31}P NMR spectra alone.

In postulating a role for nonbilayer phases in membrane permeability it would seem
essential that they be transient in nature to preserve the normal impermeability of membranes
to cations. They would only become permeable in response to a stimulus or under certain
conditions. Given the dynamic composition of biological membranes due to the presence of
regulatory enzymes that can rapidly modify their lipid composition, such a criterion can
feasibly be fulfilled. An example is the rapid turnover of phospholipids, mainly phosphoinositides, in certain cell types in response to a stimulus.[25] This will be discussed later in
more detail.

Phospholipids proposed to affect the Ca^{2+} permeability of biological membranes include
cardiolipin,[4] PA,[6,8-11] and lyso-PA.[26,27] It has been suggested that cardiolipin and PA promote
permeability through the formation of nonbilayer structures. Evidence supporting such a
role lies heavily on their ability to translocate Ca^{2+} in a Pressman cell or to facilitate its
partitioning into a hydrophobic environment. As I have outlined above, this evidence rests
on questionable grounds. There are other ways to approach this question and they all point
to the unlikelihood of phospholipids acting as ionophores in biological membranes.

The most direct approach to determine if these phospholipids can translocate Ca^{2+} across membranes is to study the effect of these lipids in liposomes. Encapsulating a Ca^{2+}-sensitive dye such as arsenazo III inside the liposomes offers a sensitive technique for examining the translocation of Ca^{2+} from outside the liposome to the inside.[28] Initial results from Weissmann's laboratory suggested that PA, when added exogenously to preformed liposomes or when incorporated directly into their membranes, increased their permeability to Ca^{2+}.[7,8] However, we were unable to replicate this observation using two commercial sources of PA, using two different techniques, or when synthesizing it *in situ* from PC using phospholipase D.[29] A third commercial source was active, but this was due to oxidation of its unsaturated fatty acid components. We increased the sensitivity of the assay by following Ca^{2+} permeability changes for much longer periods of time. Care was taken to correct for Ca^{2+}-induced aggregation of liposomes which apparently occurred during prolonged incubation, by using a spectral scan to detect the Ca^{2+}-arsenazo III peak which formed around 650 nm. Even after 48 hr of incubation at room temperature, the liposomes were remarkably impermeable to Ca^{2+} when containing up to 10 mol% PA. The reason for the different results in Weissmann's laboratory may relate to differences in technique or differences in the PAs used. For instance, it is possible that the PA they used was contaminated with a compound possessing potent ionophoretic properties, or that there was a subtle, but important difference in the fatty acid composition of the PAs. The latter would seem unlikely given the remarkable similarity between the composition of the PAs we found to be inactive and that of the parent egg yolk PC. If the fatty acid composition is important, one molecular species to test would be 1-stearoyl, 2-arachidonyl-PA given its preponderance in the phosphoinositides that yield PA during stimulus-response coupling in stimulated cells.[30] With regard to possible differences in technique, in our hands there are two fundamental properties which distinguish our use of this experimental system from that in Weissmann's laboratory. First, we have found that a 10-*M* excess of EGTA over Ca^{2+} can completely reverse Ca^{2+} uptake stimulated by the ionophore, A23187, in contrast to what was reported by Weissmann et al.[28] This reversal of uptake is compatible with their dissociation constants: $\sim 10^{-11}M$ for a Ca EGTA complex compared to $\sim 10^{-4}M$ for an arsenazo III · Ca complex,[31] although there are still uncertainties as to the nature of the complexes formed between Ca^{2+} and the dye. Weissmann et al.[28] also reported that in this experimental system, the ionophore, A23187, induced a much greater permeability towards Ca^{2+} than to Mg^{2+}. In contrast we found that A23187 induced similar permeabilities consistent with its observed behavior in other membrane systems.[32,33]

The next point to consider in proposing that PA acts as a Ca^{2+} ionophore by forming nonbilayer phases is the sequence of steps that is believed to occur during stimulus-response coupling. Phosphoinositides located on the inner leaflet of the plasma membrane are hydrolyzed by phospholipase C producing diacylglycerol which, in turn, is rapidly phosphorylated to yield PA.[25] Supporting this, phosphatidylinositol (PI) and presumably other phosphoinositides appear to be located primarily in this leaflet of the plasma membrane in many cell types.[34,38] For the PA produced to form an inverted vesicle or, perhaps, some other nonbilayer phase in the membrane, certain conditions must be met. First, the aggregation of PA into a PA-rich domain is required. This may occur if the parent phosphoinositides were localized in such a domain and after hydrolysis following stimulation, the PA produced remained in such a domain. Alternatively, it is possible that some factor triggers the aggregation of diffusely produced PA. In artificial membranes Ca^{2+} has been shown to fulfill such a role.[36] Whether this could occur in the inner leaflet of the plasma membrane of a stimulated cell given a more complex phospholipid composition, the presence of sterols and proteins, and the presence of other negatively charged phospholipids with a strong affinity for Ca^{2+}, including phosphatidylserine (PS) and PI, is not known. It is not known if sufficient Ca^{2+} is available to trigger such movements. The concentration in the cytoplasm is too low

(100 nM), but the inner surface of the plasma membrane may contain significant amounts of Ca^{2+} bound to phospholipids and proteins.[37,38] If the intracellular Ca^{2+} in the erythrocyte was evenly distributed in the cytoplasm, for instance, the concentration of Ca^{2+} would be 20 μM.[37]

The second question is if a PA-rich domain is produced could a nonbilayer structure form. It is not known whether similar concentrations of PA in apposing regions of bilayer leaflets are required. If this is so, the source of PA for the outer leaflet would be a major problem for this model. There is very little of any negatively charged phospholipids in the outer leaflet of plasma membranes examined to date.[34] It could possibly arise from transbilayer movement of some of the PA produced in the inner leaflet. As it has been proposed that nonbilayer phases may be important in transbilayer movements of phospholipids, this would create a cyclical argument.[4,13] It is possible that a nonbilayer-forming phospholipid such as PE could form nonbilayer structures which incorporate PA and shuttle it back and forth across the membrane. The limitation of transbilayer movement being a factor, regardless of the mechanism by which it occurs, is the time-frame of such movements. The half-time for the movement of PA across a membrane, when produced *in situ* by phospholipase D, is approximately 30 min.[39] Thus, during stimulus-response coupling only a small amount of the PA produced is likely to cross the membrane. More questions could be advanced as to how Ca^{2+} could move from the extracellular compartment to the cytoplasm via a nonbilayer structure, as an apparent collapse and reformation of these structures would be required.

An important conceptual feature of molecular models explaining how PA may facilitate Ca^{2+} movements across membranes is that this phospholipid can form nonbilayer structures in artificial membranes. However, the available evidence does not support this notion. Papahadjopoulos et al.[40] observed that at pH 6 and pH 8, Ca^{2+} did not induce nonbilayer phases in sonicated vesicles consisting of unsaturated PA. X-ray diffraction and freeze fracture analyses revealed a lamellar structure. Mg^{2+}, on the other hand, did cause nonlamellar structures to form at pH 6. Similarly, with disaturated PAs Liao and Prestegard[41] found that lamellar structures formed upon the addition of Ca^{2+}. However, dioleoyl PA will form a hexagonal phase at pH <5 with low levels of Ca^{2+},[158] and at pH 6 with excess Ca^{2+}.[159]

By contrast, there is sound evidence that cardiolipin forms nonbilayer phases when complexed to Ca^{2+}.[15,16] The nature of the nonbilayer phase formed depends on, amongst other things, the concentration of Ca^{2+}. At high Ca^{2+} concentrations (>1 mM) hexagonal H_{II} phases form. When cardiolipin liposomes are exposed to locally high concentrations of Ca^{2+} an intermediate, "isotropic" structure forms.[16] It is termed isotropic because of the ^{31}P NMR signal it generates, but its exact structural form is not known, although suggestions that it is an inverted vesicle seem compatible with the appearance of lipid particles in these membranes. Much lower concentrations of Ca^{2+} (100 μM) induced these structures in mixed dioleoylphosphatidylcholine (DOPC)/cardiolipin (1:1) liposomes.[16] At 5 mol% cardiolipin does not facilitate transfer of Ca^{2+} across membranes consisting of predominantly PC.[7,8] Also, at 25 mol% cardiolipin did not translocate Ca^{2+} or K^+ across black lipid membranes formed from PC and PE (Drzymala and Shamoo, unpublished results). Recent experiments in our laboratory indicate that liposomes containing high concentrations of cardiolipin (>30 mol%) are permeable to Ca^{2+}.[155] We are examining the conditions under which permeability occurs in detail, to determine if they are of any physiological significance. Gerritsen et al.[42] observed that large unilamellar vesicles and the outer layer of multilamellar vesicles consisting of 1:1, cardiolipin to DOPC, were permeable to Mn^{2+}.

B. Oxygenated Lipids

1. Modification of Permeability by Oxygenated Lipids

Lipids containing polar hydroxy or keto groups appear to be a class of compounds that

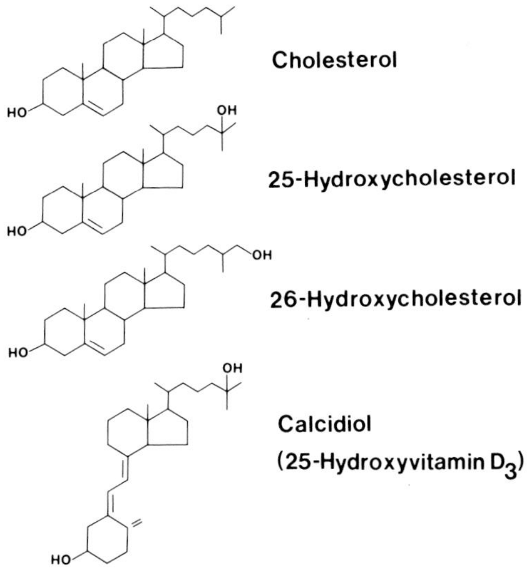

FIGURE 3. Structures of cholesterol, hydroxy derivatives, and the 25-hydroxy derivative of cholecalciferol (vitamin D₃).

affect the permeability of biological membranes to Ca^{2+} directly by perturbing the lipid bilayer. The sterols of importance are those that contain a polar group on the isoprenoid side chain of the steroid nucleus. The ones of most significance are 25-hydroxycholesterol, 26-hydroxycholesterol, and calcidiol, a seco-sterol. Their structures are shown in Figure 3. Fatty acid oxidation also introduces a polar group along a normally hydrophobic, hydrocarbon chain. I will consider the physiological relevance of these compounds, their observed relationships to changes in the Ca^{2+} permeability of tissue cells, and evidence that suggests their incorporation into membranes will affect the flow of Ca^{2+} and other cations through the lipid bilayer.

a. 25-Hydroxycholesterol

This sterol arises as an autooxidation product of cholesterol.[43,44] Thus, it is possible that significant amounts are ingested by humans when consuming cholesterol-containing foods, particularly those exposed to prooxidant conditions for extended periods of time. As this sterol and other oxidized sterols are quite toxic, concern has been expressed that significant amounts may be ingested in human foods and contribute to the onset of diseased states.[43] To verify whether significant amounts do occur in foods we have developed a method for HPLC analysis of 25-hydroxycholesterol in foods.[156] Our preliminary results suggest that in some of the most common cholesterol-containing foods, significant amounts of 25-hydroxycholesterol do not form. When this observation is coupled with evidence indicating that its intestinal absorption is low, the basis for concern over its ingestion appears unfounded. Analyses of serum and plasma using mass spectrometer detection indicate that levels in normal individuals are low. One study indicated an absence of detectable 25-hydroxycholesterol,[45] whereas the other indicated it present at 5 ng/mℓ in fresh plasma and 20 ng/mℓ in aged plasma.[46] There is some evidence that 25-hydroxycholesterol can be produced in the liver,[44] but to date this has not been shown to occur under physiological conditions.

25-Hydroxysterol has been extensively used as an in vitro tool for inhibiting the activity of 3-hydroxy-3-methylglutarate coenzyme A (HMG-CoA) reductase.[47] The method by which this inhibition occurs is not certain, but recent experiments suggest it affects both the rate of synthesis and the rate of degradation of the reductase.[48] The effect on reductase activity could be accounted for solely by the increased degradation rate, and the change in synthetic rate was not due to changed mRNA levels. 25-Hydroxycholesterol not only inhibits HMG-CoA reductase activity, but affects other cellular properties as well, including monovalent cation permeability, DNA synthesis, and cell morphology.[47] Attempts have been made to explain the pleiotropic effects of 25-hydroxycholesterol in terms of a reduced membrane sterol content resulting from inhibition of cholesterol biosynthesis.[47] This has been shown not to be true in several ways. First, the membrane cholesterol content of a mutant line of Chinese Hamster Ovary (CHO) cells can be manipulated by the amount of cholesterol that is included in the culture medium.[49] Using this model, a reduced membrane cholesterol level was shown not to influence DNA synthesis.[49] Second, when a variety of oxidized sterols were examined for their effects on cellular HMG-CoA reductase activity and DNA synthesis, discordant effects were observed.[50] Notably, DNA synthesis was affected without any apparent reduction in HMG-CoA reductase activity. Third, in an examination of the inhibition of DNA and cholesterol synthesis during the cell cycle, Astruc et al.[51] concluded that 25-hydroxycholesterol inhibited DNA synthesis under conditions where membrane cholesterol was most likely unaltered.

A plausible proposal to explain the effects of oxidized sterols on cellular properties is that because of their lipophilic properties they become inserted in cellular membranes, particularly the plasma membrane. To determine the effects of the incorporation of 25-hydroxycholesterol into membranes, we have studied its insertion in artificial membranes. We have found that levels as low as 0.5 mol% can affect the permeability of these vesicles to cations, including several divalent cations and Na^+.[52] We estimated that at this concentrations, the intravesicular concentration of Ca^{2+} was raised to 50 μM within 1 min when the external concentration of Ca^{2+} was 1 mM. We suggest that the pleiotropic effects observed with 25-hydroxycholesterol can be totally or at least partially explained by its insertion in membranes and the resulting disturbance of normal ionic gradients across these membranes. Ca^{2+} may be one of the most critical in view of its large concentration gradient, particularly across the plasma membrane. These ionic fluxes may tax membrane pumps and other regulatory systems that serve to maintain and control these gradients. Thus, there may be a net flow of Ca^{2+} and Na^+ into a cell exposed to 25-hydroxycholesterol, whereas K^+ and to a lesser extent Mg^{2+} may flow out. This would explain the observed effects of 25-hydroxycholesterol on the K^+ and Na^+ contents of cultured fibroblasts.[53] The authors interpreted their results as being due to a lowered membrane cholesterol content. However, the effects of 25-hydroxycholesterol were not reversed by mevalonate, and other experiments with ovarian cells indicate that variations in membrane cholesterol content do not perturb monovalent cation fluxes.[49]

The above hypothesis is speculative in that it remains to be shown that 25-hydroxycholesterol is incorporated into the membranes of cells under conditions where these pleiotropic effects are observed.

b. 26-Hydroxycholesterol

This hydroxylated sterol is produced in the liver as a precursor in the biosynthesis of bile acids.[54] Significant amounts have been identified in normal human serum, with levels ranging from 92 to 256 ng/mℓ.[45,46] It is carried by lipoproteins, two thirds of it esterified to one or two fatty acids. The binding to lipoproteins suggests that it would be taken up by cells by receptor-mediated endocytosis when bound to low-density lipoprotein. Either exchange between lipoproteins and membranes or breakdown of the particles in lysosomes may result in 26-hydroxycholesterol incorporation into membranes. However, it is likely that such membrane incorporation may be much less for lipoprotein-bound sterols than that occurring in in vitro studies and cultured cells where the oxygenated sterols may be in a variety of structural and bound forms.

Although it has not yet been tested, it is anticipated that incorporation of 26-hydroxy-cholesterol in membranes would produce similar effects to the incorporation of 25-hydrox-ycholesterol. 26-Hydroxycholesterol has been shown to inhibit HMG CoA reductase activity in cultured cells and is slightly more inhibitory than 25-hydroxycholesterol.[55]

c. Calcidiol

Calcidiol, the 25-hydroxy derivative of vitamin D, is the main circulating form of vitamin D. During hypervitaminosis D in experimental animals, serum levels increase substantially and can attain levels of almost 2 μg/mℓ.[56] At such levels death will occur within a few days. Serum Ca^{2+} may or may not increase concomitant with the calcidiol increase. We have proposed that it is the increased calcidiol levels in serum that precipitates tissue cal-cification.[56,57] If calcidiol is the calcinogenic agent, its in vivo toxic concentration is similar to the concentration of 25-hydroxycholesterol toxic to cultured smooth muscle cells, 1 to 5 μg/mℓ culture medium. We have observed that during hypervitaminosis D, high levels of both calcidiol and vitamin D are found in tissues that calcify, consistent with the hypothesis.[57] It seems clear that under these conditions of hypervitaminosis D, cells have an increased permeability to Ca^{2+}, but it remains to be shown that this is due to the incorporation of calcidiol into cellular membranes.

d. Oxygenated Fatty Acid Derivatives

The double bonds in polyunsaturated fatty acids are susceptible to oxidation under the right environmental conditions. They may also be oxidized in a controlled fashion by cellular enzymes, the lipoxygenase pathway being of particular importance.

Considering the lipoxygenase products, leukotriene B_4 [(5S), (12R)-dihydroxy-6,8,11,14-(*cis,trans,trans,cis*)-eicosatetraenoic acid] appears to be important in Ca^{2+} homeostasis. In neutrophils, for instance, it has been shown to increase the rate of calcium influx.[58] The mechanism by which it affects neutrophils is not known. Naccache et al.[58] suggest that because of a specific requirement for various functional groups on the leukotriene B_4, it acts through a membrane receptor. This is also compatible with its activity at low concentrations, 10^{-7} to 10^{-9} M. Using the dye encapsulation technique, Serhan et al.[7] presented evidence that it can translocate Ca^{2+} across artificial membranes, suggesting it acts directly on the lipid bilayer. Using the same experimental system, however, we have not observed any effect with exogenous concentrations of leukotriene B_4 up to 10^{-5} M (unpublished results).

The nonenzymatic oxidation of membrane fatty acids modified the functional properties of membranes, including an alteration in their permeability. This has been shown to include the permeability of membranes to Ca^{2+}.[8,29] The formation of hydroperoxy derivatives may perturb fatty acyl interactions in membranes making them leaky to cations. Alternatively, the formation of short-chain fatty acids and malonaldehyde may be important in modifying membrane permeability.

2. Perturbation of Membranes by Oxidized Sterols

We have probed the mechanisms by which 25-hydroxysterols perturb membranes.[59] Using fatty acid probes, with spin labels attached at various points along the carbon chain, we compared the effects of cholesterol and 25-hydroxycholesterol on the spectral properties of the probes when incorporated in PC liposomes. In contrast to cholesterol, 25-hydroxycho-lesterol was not able to order the acyl chains of the phospholipids. The effect appeared most pronounced when the spin label was deepest in the membrane bilayer, as 25-hydroxycho-lesterol had no effect at all on the spectral parameters. Verma et al.[60] have applied Raman spectroscopy to the study of the interactions between 25-hydroxysterols and phospholipids. They observed concentrations as low as 0.2 mol% calcidiol influenced the phase transition of dilaurylphosphatidylcholine (DLPC) vesicles. Sensitivity decreased with increasing fatty

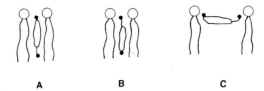

A   **B**   **C**

FIGURE 4.   Possible orientations of 25-hydroxy cholesterol in membranes. A is similar to cholesterol with
the isoprenoid side chain deep in the phospholipid leaflet; B is the reverse orientation with the 3β-OH group
deep in the leaflet; in C both hydroxyl groups are exposed to the aqueous environment.

acyl chain length, such that no effect was observed on distearoylphosphatidylcholine (DSPC) liposomes at concentrations up to 20 mol%.

When incorporated into membranes, 25-hydroxycholesterol could be orientated in three ways as shown in Figure 4. If the orientation is similar to cholesterol (A), the 25-hydroxyl group would perturb hydrocarbon interactions in the hydrophobic milieu of the membrane. The reverse orientation (B) would seem unlikely as the bulky steroid nucleus as well as the 3β-hydroxyl group could perturb acyl chain interactions, and it does not account for rather specific interactions that occur between the 3β-hydroxyl group and phospholipid head groups. An orientation with both hydroxyl groups at the surface (C) would also influence phospholipid interactions. However, it appears awkward as hydrophobic residues would be exposed to hydrophilic regions of the glycerol or choline moieties, or to the aqueous layer. Nevertheless, Gallay et al.[61] suggested that 22,R-hydroxycholesterol is orientated in this way based on its interfacial properties in monomolecular films. Using molecular models, Verma et al.[60] found that calcidiol could be folded in such a fashion, but not 25-hydroxycholesterol. Whatever the orientation of the sterols, the 25-hydroxyl groups apparently prevent the closer packing of phospholipid hydrocarbon chains that occurs with cholesterol. In this way the polarity of the hydroxyl group permits cation leak through the bilayer.

III. CALCIUM PUMPS

Calcium pumps appear to be ubiquitous components of eukaryotic cells and function to translocate Ca^{2+} across membranes against a Ca^{2+} gradient. This requires energy which is furnished by ATP hydrolysis and biochemically their activity is expressed as a Ca^{2+}-ATPase activity. It is also referred to as a $(Ca^{2+} + Mg^{2+})$ATPase activity, but there is some question as to whether such terminology is correct, as it is not certain whether free ATP, Mg ATP, or Ca ATP is the substrate.[62,63] In view of this uncertainty the activity is referred to as a Ca^{2+}-ATPase which makes no claims as to whether Mg^{2+} is a substrate cofactor or an effector. The best characterized enzyme systems are the SR Ca^{2+}-ATPase and the erythrocyte Ca-ATPase. The enzymes found in the ER and plasma membrane of the other tissues are similar, but important differences now appear to exist in their regulatory properties.

A. Sarcoplasmic Reticulum Ca^{2+}-ATPase

This transport system sequesters Ca^{2+} released in muscle cells to stimulate myofilament contraction. The catalytic protein has a molecular weight of 110,000;[64] 2 mol of Ca^{2+} are translocated for each mole of ATP hydrolyzed. Oligomeric complexes can form in membranes,[65] but recent evidence suggests that it is not required for optimal activity.[66] The enzyme appears to be associated with a proteolipid[67] and a 53,000-mol wt glycoprotein.[68] Phosphorylation appears to be an important step in regulating activity, particularly through a cyclic AMP-dependent protein kinase.[69] A calmodulin-dependent phosphorylation has been shown to also occur in cardiac SR.[70,71] A 22- to 24,000-dalton protein, phospholamban, has been identified as the protein phosphorylated in both instances. There is also evidence that calmodulin has a direct effect on the cardiac enzyme, but not that of skeletal muscle.[72]

The enzyme requires a hydrophobic environment to sustain activity, but there is apparently little specificity concerning this requirement. A fluid membrane is required for activity,[73,74] and protein rotation also appears to be essential.[74] A variety of nonionic detergents can substitute for phospholipids in a reconstituted system.[75] One of the best approaches to studying lipoprotein interactions in membranes is to study the purified protein reconstituted in liposomes of defined lipid composition. Using such an approach Johannsson et al.[76] showed that there was an optimal bilayer thickness of reconstituted membranes, using di-monounsaturated PCs. Maximal activity was obtained when there were 20 carbon atoms in the fatty acyl chains. Similarly, Navarro et al.[77] examined the nature of the head-group requirement. Their results indicated that lipids with a cone-shaped head group had the greatest coupling between the Ca^{2+} transported and the ATP hydrolyzed. This included DOPE and monogalactosyldiglyceride. Other phospholipids supported high rates of Ca^{2+}-dependent ATP hydrolysis but did not couple this hydrolysis to substantial Ca^{2+} uptake. This highlights the need to ensure that Ca^{2+} transport is accompanying ATP hydrolysis in reconstitution experiments, a requirement that has not been met in many studies. Early experiments suggested that cholesterol was excluded from the annulus of lipids surrounding the protein,[78] but recent more-refined experiments indicate that it is not completely excluded but associates with the protein to a lesser extent than do phospholipids.[79] This partial discrimination appears to reflect on the rigidity of the sterol nucleus. When considered with the preference of the enzyme for a fluid environment, these results suggest that lipids with some flexibility are required to interact with the protein surface.

Free fatty acids or their esters, amphiphiles that can be used to modify the fluidity of membranes,[80] affect the properties of the SR. The effects are complex. Both unsaturated and saturated fatty acids have been shown to inhibit Ca^{2+} uptake[81-83] at low concentrations, whereas high concentrations of palmitic acid (18 μM) stimulate uptake.[84] Under some conditions they inhibit Ca^{2+} release,[83] whereas under others they stimulate release.[82,84] Unsaturated acids were shown to stimulate Ca^{2+}-ATPase activity,[82] so this suggests that these fatty acids uncouple ATP hydrolysis from Ca^{2+} transport. Meddineo et al.[84] have developed a plausible model to explain some of these complex effects.

An indirect observation that the fatty acid composition of SR phospholipids may influence enzyme activity was obtained by subfractionating vesicles into two populations: a dense fraction believed to originate in terminal cisternae and a light fraction derived from the longitudinal reticulum.[85] Both fractions appeared to contain similar concentrations of the enzyme, but the activity of the heavy fraction was 43% higher than the light fraction. The phospholipid class distributions were similar, but the heavier fraction contained fewer saturated fatty acids and more unsaturated fatty acids, suggesting it may be more fluid and that this may account for the activity differences. Similar arguments relating the lipid composition of the SR from fast and slow twitch muscles, membrane fluidity, and the Ca^{2+}-ATPase activity of these membranes have been advanced.[86] In this case it was proposed that a greater cholesterol content and a greater sphingomyelin-to-PC ratio, conditions that will decrease membrane fluidity, were responsible for a decreased Ca^{2+}-ATPase activity in the slow twitch muscle.

B. Plasma Membrane Ca^{2+}-ATPase

The polypeptide carrying out this activity in the erythrocyte, the best characterized of these enzymes, spans the surface membrane, has a molecular weight of 140,000, and acts to pump Ca^{2+} out of the cell.[87-89] This enzyme maintains the intracellular Ca^{2+} level in erythrocytes at less than 1 μM. In the absence of the regulatory protein, calmodulin, it has a low affinity for Ca^{2+} ($K_{1/2} \sim 1 \times 10^{-5} M$). Calmodulin increases the affinity for Ca^{2+} ($K_{1/2} < 1 \times 10^{-6} M$).[87] Two Ca^{2+} ions are apparently transported across the membrane for each ATP that is hydrolyzed. It is not yet certain whether Ca ATP or Mg ATP serves as a substrate for this reaction. Mg^{2+} is required for activity and the monovalent cations, K^+ and Na^+, stimulate the reaction.[89] Cyclic AMP inhibits the activity of the enzyme, apparently by stimulating a cyclic AMP-dependent protein kinase which phosphorylates a 20,000-dalton protein which may modify the activity of the Ca^{2+}-ATPase.[90] This is in contrast to the SR where cyclic AMP stimulated Ca^{2+}-ATPase activity following phosphorylation of a regulatory protein of similar size, phospholamban.[70,71] Erythrocytes contain adenylate cyclase activity which is sensitive to hormone stimulation.[91] This suggests that a link exists between hormone fluctuations and the activity of the Ca^{2+} pump. Regulation of this enzyme, as with the SR enzyme, is complex. It is also evident that there is still much to be learned about the functioning of the pump in the erythrocyte and other cells. Further evidence of the complexity of the regulatory mechanisms and the role of protein phosphorylation in the inhibition of the enzyme was obtained in an examination of the effect of insulin on the activity of the enzyme in adipocyte plasma membranes.[92]

The lipid requirement for activity is broad with both mono- and diacylglycerophospholipids supporting activity, either zwitterionic or negatively charged.[93] Certain lipids can stimulate the activity in isolated membranes (inside-out vesicles). This includes PI, PS, lyso-PC and oleic acid. Hydrolysis of membrane lipids by phospholipase A is also stimulatory.[89] Recent studies on the interactions of the purified protein with phospholipids have confirmed the effect of lipids in modifying the properties of the enzyme.[87] When reconstituted in PC liposomes, calmodulin stimulates the enzyme's activity. However, in PS membranes the enzyme is fully stimulated in the absence of calmodulin and does not respond to the addition of calmodulin. As isolated erythrocyte membranes respond to calmodulin, this suggests that the enzyme is in a PC environment or one that is similarly responsive *in situ*. Furthermore, it raises the possibility that PS or other acidic phospholipids could serve as regulators of activity. Experiments with rat brain synaptosome plasma membranes support this possibility as low concentrations of PI 4,5-bisphosphate have been observed to stimulate the enzyme to a greater extent than saturating levels of calmodulin.[94] Furthermore, this identifies a possible connection between phosphoinositide turnover, to be discussed later, and Ca^{2+} efflux.

Two experimental approaches have indicated that the activity of the enzyme is sensitive to the fluidity of its lipid environment. In one approach, free fatty acids have been used as modifiers of membrane properties. Unsaturated fatty acids fluidize regions of the membrane, whereas saturated fatty acids and monounsaturated fatty acids with a *trans* double bond rigidify certain lipid domains in membranes.[80] The effects of free fatty acids on Ca^{2+}-ATPase activity are complex and the concentrations of free Ca^{2+}, fatty acid, and calmodulin are among the critical factors. Oleic acid is clearly stimulatory in membranes depleted of calmodulin and slightly stimulatory in membranes containing calmodulin.[95,96] These results suggest that changes in membrane fluidity could influence the activity of the enzyme in the presence and absence of the regulatory protein, calmodulin. This effect of fluidizing free fatty acids could partially be due to their negative charge. In support of this, fatty acids which rigidify membranes are also stimulatory. They evoked the same increase in Ca^{2+}-ATPase activity in calmodulin-deficient membranes, but the concentration of free Ca^{2+} inducing half-maximal activity, 3 μM, was double that in oleic acid-stimulated membranes.[95] The use of methylated fatty acids would distinguish between charge and fluidity effects.

The other approach used to examine the effects of membrane fluidity also suggests that this factor is an important determinant of enzyme activity. Enzymatic methylation of PE in erythrocyte membranes has been shown to increase membrane fluidity.[97] Concomitant with this increased fluidity was a 50% increase in Ca^{2+}-ATPase activity.[98] However, these membranes were apparently calmodulin deficient as they were washed in low ionic strength buffer and had properties similar to deficient membranes. Similarly, phospholipid methylation stimulated Ca^{2+}-ATPase activity in basolateral membranes from kidney cortex which were also calmodulin-deficient.[99] It would be of interest to determine the effect of phospholipid methylation in calmodulin replete membranes. While there may be questions as to the biological significance of phospholipid methylation as a physiological regulator of membrane fluidity, it may, nevertheless, be a useful technique for modifying membrane fluidity in vitro.

If the fluidity of membrane lipids does influence enzyme activity, it is possible that the saturation of dietary fatty acids may modify the enzyme's properties *in situ*. Galo et al.[100,101] have studied how dietary fats influence the activity of the enzyme, its cooperativity, and its stimulation by triiodothyronine. Weaknesses in the experimental design, the isolation of membranes, and the assay of activity mar the interpretation of their results, however.[102] The stimulation of the enzyme in calmodulin-deficient membranes by triiodothyronine has been confirmed by Davis and Blas.[103] The effect of triiodothyronine on calmodulin-replete membranes, and the effect of dietary lipids on this process are points warranting clarification.

We have studied how dietary lipids modify the properties of this enzyme in rat erythrocytes.[104] With 10% fat diets we observed an influence of the saturation of dietary fatty acids on the activity of the basal and calmodulin-stimulated activities. Erythrocytes from rats on the corn oil diet had an activity approximately 20% higher than those from rats on a lard diet. The dietary treatment produced small but significant changes in the fatty acid composition of the erythrocytes. With rats fed 15% fat diets by contrast, no differences in activities were observed.[157] The reasons for the difference between 10 and 15% fat diets are unclear, but they may be related to the slightly different fatty acid compositions that resulted from the dietary treatments. They do highlight the rather small influence of dietary lipids on the activity of this enzyme in the erythrocyte.

That this enzyme is an ubiquitous component of the plasma membrane of cells is becoming clear. Rather than listing the now large number of tissue cells in which this activity has been observed, it will suffice to comment that an observation that it does not exist would be highly significant. Some differences have been identified between the erythrocyte enzyme and that in other plasma membranes, and these have been recently reviewed.[94] These relate mainly to the requirement for Mg^{2+} and the activation by calmodulin. The liver enzyme, for instance, has been shown to be unresponsive to calmodulin, to respond to a different activator, and to also be influenced by an inhibitory protein.[105,106] The identification of a $(Ca^{2+} + Mg^{2+})$-ATPase activity on the surface membrane of pancreatic cells, which is an ecto-enzyme and is not apparently involved in Ca^{2+} transport, suggests that caution should be exercised in assigning this activity universally to a Ca^{2+} pump.[107] Another example appears to be a spectrin-dependent ATPase activity in human erythrocyte which is stimulated by Ca^{2+}.[108] Whether the plasma membrane enzymes in other tissues are identical to the erythrocyte enzyme is not yet clear, but recent experiments with the enzyme from heart sarcolemma indicate it has similar properties and shares immunological cross-reactivity with the erythrocyte enzyme.[109]

IV. CALCIUM CHANNEL

A Ca^{2+} channel is defined here as a protein which facilitates the selective movement of Ca^{2+} across a membrane down its concentration gradient. Convincing evidence in excitable cell membranes indicates that this occurs through an ion-selective pore or channel created by or in a protein rather than occurring by a uniport process. This latter is unlikely, as significant energy expenditure, which would be required to catalyze protein conformational changes, is not required for Ca^{2+} influx. Coanion influx or countercation efflux may be required to maintain a constant potential difference across the membrane, but it is not thought to be controlled by the same protein. A valid hypothesis to pursue is that such proteins are ubiquitous components of all plasma membranes and, perhaps, other types of membranes as well. Some of the evidence supporting such an hypothesis will be presented in this section. Such an hypothesis recognizes the relative impermeability of lipid bilayers to Ca^{2+}. We have reported that even after 48 hr of incubation in medium containing 1 mM Ca^{2+}, liposomes consisting predominantly of PC had not taken up significant amounts of Ca^{2+}, demonstrating the impermeability of bilayers to Ca^{2+}.[52] For Ca^{2+} to serve its messenger role within cells, the source of Ca^{2+} in some cells is derived partially from the extracellular fluid.[1] This may occur in nerve transmission, muscle contraction, and in hormone-stimulated cells. Bulk movement of Ca^{2+} in mammals also apparently requires a channel to regulate Ca^{2+} uptake from the intestine and the glomerular filtrate in the kidney, and Ca^{2+} deposition in bone cells. Ca^{2+} release-pathways from intracellular storage compartments, microsomes, and mitochondria may utilize channels as well.

A. The Channel in Excitable and Nonexcitable Cell Membranes

Ca^{2+} channels have been studied most extensively in excitable tissues. This is a consequence of their high concentration in the surface membrane of such tissue cells and the ability in some of these tissues to detect open channels as a Ca^{2+} current. The classification of cells as either excitable or unexcitable is somewhat arbitrary as discussed by Rasmussen.[1] In traditional physiological terms, cells have been referred to as excitable when they respond to a stimulus with an action potential, a property shared by such cells as the beta cells of Islets of Langerhans, which is usually regarded as a nonexcitable cell, and not by the cells of slow muscle, a tissue usually classified as excitable. There is some dispute as to the exact subcellular localization of channels in some tissues (usually identified by the high affinity binding of a radioactive Ca^{2+}-blocking agent such as nitredipine), but recent studies with smooth muscle show their exclusive localization in the plasma membrane.[110] The high rate of Ca^{2+} transfer in excitable membranes is one indication that the permeation occurs through a perm-selective pore. The modulation of these channels (whether they are open or closed) apparently depends on their phosphorylation by a cyclic AMP-dependent protein kinase.[111-113] A protein similar to phospholamban of the SR appears to be phosphorylated.[111,112] This creates an interesting link in the movement of Ca^{2+} in muscle cells if similar proteins in the plasma membrane and SR are stimulated by a cyclic AMP-dependent phosphorylation, but activate different Ca^{2+}-translocation processes.

It is probable that at least two types of Ca^{2+} channels exist: one that is potential dependent and another that is connected to receptor functions.[114] The potential-dependent channels are activated by membrane depolarization, whereas the receptor-linked channels are apparently insensitive to membrane potential.[115] A mobilization of intracellular Ca^{2+} stores may accompany activation of the receptor-linked channel.

The identification of a Ca^{2+} channel in the plasma membrane of many cells classified as nonexcitable has proved to be difficult because of the presence of the Ca^{2+} extrusion pump. In studying Ca^{2+} fluxes with the radioisotopic tracer, ^{45}Ca, this pumping mechanism apparently masks the influx due to the channel. Several approaches have been used to unmask the channel. In one, low temperature has been used to block the Ca^{2+}-ATPase in Ehrlich ascites cells[116] and liver cells[117] illustrating the presence of a channel. In another, vanadate has been utilized to inhibit the Ca^{2+}-ATPase in erythrocytes.[118] The influx observed in the presence and absence of vanadate suggested that a slow cycling of Ca^{2+} occurs across the erythrocyte membrane. In the absence of vanadate, Ca^{2+} uptake was barely detectable. The inhibition of Ca^{2+} uptake by Co^{2+} and La^{3+} in cultured kidney cells suggests that the Ca^{2+} channel shares similar characteristics to Ca^{2+} channels in excitable cells.[119] The interpretation

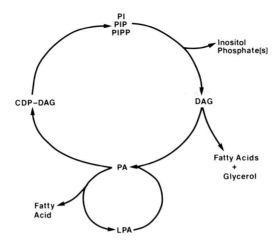

FIGURE 5. Possible cycles of phosphoinositide turnover during stimulus-response coupling in stimulated cells. PI = phosphatidylinositol, PIP = phosphatidylinositol 4-phosphate, PIPP = phosphatidylinositol 4,5 bisphosphate, DAG = diacylglycerol, PA = phosphatidic acid, LPA = lysophosphatidic acid, CDP-DAG = cytosine diphosphate-diacylglycerol.

of these experiments, however, warrants caution as erythrocytes, stored for up to 3 weeks in the cold, were used. Wiley et al.[120] have demonstrated that cold storage increases Ca^{2+} influx in erythrocytes using ATP or Mg^{2+} depletion to inhibit the Ca^{2+} pump. They suggested that storage may induce several changes that affect Ca^{2+} permeability, including proteolysis, membrane lipid changes, and membrane attachment of the third component of complement. It is not certain whether changes occurring during cold storage activate the Ca^{2+} channel or produce another Ca^{2+} influx pathway.

Ca^{2+}-blocking drugs such as verapamil and nifedipine inhibit Ca^{2+} influx in nonexcitable cells including erythrocytes,[118] platelets,[121-123] and cultured kidney cells.[119] Invariably, though, high concentrations of the inhibitors had to be used to achieve an effect. This suggests that either the characteristics of the Ca^{2+} channel are quite different in nonexcitable cells, or that the drugs are acting via some other mechanism to exert their effect. To date, no high affinity binding sites for these drugs have been identified in nonexcitable cells.

It seems probable that the release of Ca^{2+} from intracellular stores in the ER and in mitochondria may be mediated via a Ca^{2+} channel. As yet, little is known about these processes so it is premature to discuss what role membrane lipids may play.

B. Phosphoinositide Turnover and Ca^{2+} Influx in Stimulated Cells

Based on observations with a variety of stimulated cells, Michell[25] proposed that there was a close link between phosphoinositide turnover and Ca^{2+} influx. The polyphosphoinositides, in particular, PI 4,5-bisphosphate, appear to be more important than PI itself in this relationship.[124] The metabolism of phospholipids that may occur in such cells is outlined in Figure 5. Key events in this pathway for transmission of the stimulatory signal may be the synthesis of diacylglycerol, the release of inositol phosphates, an increase in intracellular cyclic GMP, and the release of arachidonic acid for eicosanoid synthesis.[125,126] Much attention has been directed at PA being a central intermediate particularly in acting as a Ca^{2+} ionophore.[7-10,12] As discussed above there appears to be no basis to such claims. Some experiments claiming support for PA in this role have been incorrectly interpreted. For example, in the study by Imai et al.[127] of thrombin stimulation of platelets, Ca^{2+} uptake from the medium increased for 20 sec before shutting off. This coincided exactly with the build-up of diacylglycerol which peaked at 20 sec. PA synthesis continued unabated for 5 min. Their experiments clearly show that diacylglycerol synthesis is closely coupled to Ca^{2+} influx, not PA as claimed. Diacylglycerol is apparently important through stimulating the activity of protein kinase C by increasing its affinity for Ca^{2+} and phospholipids, activators of the protein.[125] This provides a means of rapid transmission of the signal and allowing its dissipation with further metabolism of the diacylglycerol. A protein of molecular weight 40,000 is one apparent target for the kinase. The function of this protein remains unknown. An attractive but simplistic proposal would be that the 40,000-dalton protein, when phosphorylated, opens a Ca^{2+}-selective channel in the membrane. This is supported by evidence that the Ca^{2+} channel in excitable tissues is opened through phosphorylation.[113] Recent evidence using the Ca^{2+}-sensitive fluorescent probe, quin-2, suggests that this is not the case. A substantial rise in intracellular Ca^{2+} did not occur, but a cellular response was detected when neutrophils were stimulated with the chemotactic peptide, formylmethionylleucylphenylalanine,[128] and platelets with diacylglycerol or phorbol ester.[129] These stimulants are known to activate protein kinase C.[130,131] As discussed by Michell,[132] this suggests that there are two synergistic pathways coupling a natural stimulus to a cellular response. Both may need to be activated to elicit a full response. Thus, the relationship between the cycle of phosphoinositide degradation and synthesis depicted in Figure 5 and Ca^{2+} influx remains as nebulous as ever. In view of the purported association of phosphorylation of the Ca^{2+} channel in excitable tissues with its opening, the relationship between the phosphorylated 40,000-dalton protein and a Ca^{2+} channel seems worthy of attention.

C. Effect of Exogenous Phosphatidic Acids on Cells

Several experiments[10,11,133,134] have indicated that the addition of PA to cells simulates the effect of a stimulant and increases Ca^{2+} uptake. In these publications, the possible role of PA acting as a Ca^{2+} ionophore has been suggested. In light of the evidence presented above, that PA cannot translocate Ca^{2+} across membranes, such a proposal does not seem plausible, so an alternative explanation must be sought. A tacit assumption has been that the early events in the phosphoinositide turnover that normally accompany cellular stimulation are by-passed with PA addition. This is apparently not true as most of the reactions in the cycle illustrated in Figure 4 are reversible. This is evident in the studies of Pagano et al.[135] where the addition of PA to cultured cells, even at $2°C$, resulted in its rapid conversion to diacylglycerol. Hence, the exogenous addition of PA does not provide evidence that it plays a central role in the link between phosphoinositide turnover and Ca^{2+} influx in stimulated cells.

D. Calcitriol

The role of vitamin D in stimulating Ca^{2+} absorption in intestinal mucosal cells (enterocytes) is well recognized. The Ca^{2+} gradient between the intestinal fluid and the cytoplasm of the enterocyte, the impermeability of lipid bilayers to Ca^{2+}, the selectivity of Ca^{2+} uptake by the enterocyte, and its saturability are consistent with the existence of a perm-selective Ca^{2+} channel in the brush-border membranes of these cells. Evidence has recently been presented indicating that a change in the lipid composition of the brush-border membrane may be important in regulating the properties of such a channel.[136,137] The in vivo response of enterocytes to calcitriol (1,25-dihydroxyvitamin D_3) was correlated with an increase in the PC content of the brush-border membranes, and an increased incorporation of arachidonic acid relative to palmitic acid. Both these changes should increase the fluidity of the brush-border membrane, suggesting that fluidization of the lipid domain containing the Ca^{2+} channel may open the channel. The increase in PC synthesis most likely results from activation of CTP:phosphocholine cytidylyltransferase. This is a key regulatory enzyme in PC biosynthesis and it undergoes a cyclic AMP-dependent phosphorylation and a translocation from cytoplasm to the ER to modify its activity.[138] The importance of the increased polyunsaturated fatty acid content was confirmed by demonstrating that chicks deficient in essential fatty acids were unable to respond to calcitriol treatment.[139] Support for this reasoning was also obtained by exposing brush-border membranes to free fatty acids before assaying their Ca^{2+} uptake.[140] Exposure to the methyl ester of *cis*-vaccenic acid, which is known to fluidize membranes, simulated the effect of calcitriol, whereas the methyl ester of *trans*-vaccenic acid, which rigidifies membranes, had no effect. It is possible that calcitriol influences the Ca^{2+} channel in a wide variety of cells, through altering the membrane lipid composition. Skeletal muscle cells[141] and kidney cells[142] are two such examples.

V. SODIUM-CALCIUM EXCHANGE

The protein(s) involved in this process have not been identified and the contribution this exchange makes to regulating or maintaining cytoplasmic Ca^{2+} levels has not been precisely determined in any tissues. It has been identified primarily in excitable cells where it may play a role in dissipating a large transient increase in the cytoplasmic Ca^{2+} concentration. Compatible with this function it has a high capacity and relatively low affinity for Ca^{2+}. At resting cytoplasmic Ca^{2+} concentrations of 100 nM, it appears to contribute only marginally to the extrusion of Ca^{2+} from the cell.[143] The available evidence suggests this exchange process is not widely spread in nonexcitable tissues with evidence for its presence in lymphocyte,[144] kidney basolateral plasma membranes,[145] and spermatozoa plasma membranes,[146,147] whereas it is absent in neutrophils.[148]

Little is known concerning the influence of the membrane lipid environment on the functioning of this antiporter. Philipson and colleagues[149,150] have examined the effects of phospholipase treatments on Na^+-Ca^{2+} exchange in isolated cardiac sarcolemmal vesicles. Uptake was stimulated 10 to 70% by phospholipase C treatment.[149] This indicated that the lipids hydrolyzed, PC, PE, and sphingomyelin, were not essential for activity. The presence of the negatively charged phospholipids, PS and PI, appeared sufficient to support activity. Treatment with phospholipase D was even more stimulatory, increasing activity fourfold.[150] The apparent K_m for Ca^{2+} decreased from 18 to 6 μM, whereas the V_{max} almost doubled. These results suggest that PA produced by the phospholipase D digestion stimulates Na^+-Ca^{2+} exchange.

Clearly, more information is required concerning this exchange process in terms of its physical role, its tissue distribution, its regulation, and the role of membrane lipids.

VI. CONCLUSIONS

Much remains to be learned about the membrane processes facilitating Ca^{2+} movement across membranes. A greater understanding of the processes regulated by proteins is required as is clarification of the role of membrane lipids. The contribution that each process makes to intracellular homeostasis and the way Ca^{2+} signal is transmitted is also still not clear. With the Ca^{2+} pumping enzymes, the best-characterized class of proteins that facilitate Ca^{2+} movements across membranes, the physical properties of membrane lipids appear to be important in determining their level of activity. This is evident in an examination of the lipid composition of the SR in different types of muscle.[85,86] There is no strong evidence, however, that dynamic changes in the membrane lipid composition serve to regulate the activity. Regulatory proteins apparently serve this function, and protein phosphorylation is becoming evident as an important mechanism, not only with the pumping enzymes, but also with proteins that form Ca^{2+} channels in membranes. In contrast to the transporting ATPases, a dynamic change in membrane lipid composition appears to be central to modifying the properties of Ca^{2+} channels to stimulate Ca^{2+} influx. A change in membrane lipid composition in response to calcitriol has been linked to modifications of the Ca^{2+} channel in intestinal brush-border membranes. In a number of cell types stimulation by factors such as hormones precipitates a turnover of phosphoinositides concomitant with an increased Ca^{2+} inflow. The nature of the connection between the two events remains as shrouded as it was when Michell[25] reviewed the widespread nature of the correlation between the two in 1975. The clinical ramifications of alterations in these permeability pathways are obviously quite profound. Changes have been implicated in the cardiomyopathy accompanying diabetes,[151] atherosclerosis,[152] and aging[153,154] for instance. A greater understanding of membrane Ca^{2+} permeability may greatly enhance knowledge of how such pathological cellular changes progress.

REFERENCES

1. **Rasmussen, H.,** *Calcium and cAMP as Synarchic Messengers,* John Wiley & Sons, New York, 1981.
2. **Carafoli, E.,** The regulation of the cellular functions of Ca^{2+}, in *Disorders of Mineral Metabolism,* Vol. 2, *Calcium Physiology,* Bronner, F. and Coburn, J., Eds., Academic Press, New York, 1982, 1.
3. **Barritt, G. J.,** Calcium transport across cell membranes: progress toward molecular mechanisms, *Trends Biochem. Sci.,* 6, 322, 1981.
4. **Cullis, P. R., De Kruijff, B., Hope, M. J., Nayar, R., and Schmid, S. L.,** Phospholipids and membrane transport, *Can. J. Biochem.,* 58, 1091, 1980.
5. **Tyson, C. A., Vande Zande, H., and Green, D. E.,** Phospholipids as ionophores, *J. Biol. Chem.,* 251, 1326, 1976.
6. **Green, D. E., Fry, M., and Blondin, G. A.,** Phospholipids as the molecular instruments of ion and solute transport in biological membranes, *Proc. Natl. Acad. Sci. U.S.A.,* 77, 257, 1980.
7. **Serhan, C. N., Fridovich, J., Goetzl, E. J., Dunham, P. B., and Weissmann, G.,** Leukotriene B_4 and phosphatidic acid are calcium ionophores. Studies employing arsenazo III in liposomes, *J. Biol. Chem.,* 257, 4746, 1982.
8. **Serhan, C., Anderson, P., Goodman, E., Dunham, P., and Weissmann, G.,** Phosphatidate and oxidized fatty acids are calcium ionophores. Studies employing arsenazo III in liposomes, *J. Biol. Chem.,* 256, 2736, 1981.
9. **Putney, J. W., Jr., Weiss, S. J., Van De Walle, C. M., and Haddas, R. A.,** Is phosphatidic acid a calcium ionophore under neurohumoral control?, *Nature (London),* 284, 345, 1980.
10. **Salmon, D. M. and Honeyman, T. W.,** Proposed mechanism of cholinergic action in smooth muscle, *Nature (London),* 284, 344, 1980.
11. **Harris, R. A., Schmidt, J., Hitzemann, B. A., and Hitzemann, R. J.,** Phosphatidate as a molecular link between depolarization and neurotransmitter release in the brain, *Science,* 212, 1290, 1981.
12. **Putney, J. W., Jr.,** Recent hypotheses regarding the phosphatidylinositol effect, *Life Sci.,* 29, 1183, 1981.
13. **Cullis, P. R. and De Kruijff, B.,** Lipid polymorphism and the functional roles of lipids in biological membranes, *Biochim. Biophys. Acta,* 559, 399, 1979.
14. **Cullis, P. R. and De Kruijff, B.,** The polymorphic phase behaviour of phosphatidylethanolamines of natural and synthetic origin, A ^{31}P NMR study, *Biochim. Biophys. Acta,* 513, 31, 1978.
15. **Cullis, P. R., Verkleij, A. J., and Ververgaert, P. H. J. Th.,** Polymorphic phase behaviour of cardiolipin as detected by ^{31}P NMR and freeze-fracture techniques. Effects of calcium, dibucaine and chlorpromazine, *Biochim. Biophys. Acta,* 513, 11, 1978.
16. **De Kruijff, B., Verkleij, A. J., Leunissen-Bijvelt, J., Van Echteld, C. J. A., Hille, J., and Rijnbout, H.,** Further aspects of the Ca^{2+}-dependent polymorphism of bovine heart cardiolipin, *Biochim. Biophys. Acta,* 693, 1, 1982.
17. **Pressman, B. C.,** Ionophorus antibiotics as models for biological transport, *Fed. Proc., Fed. Am. Soc. Exp. Biol.,* 27, 1283, 1968.
18. **Pressman, B. C.,** Properties of ionophores with broad range cation selectivity, *Fed. Proc., Fed. Am. Soc. Exp. Biol.,* 32, 1698, 1973.

19. **Barrett-Bee, K., Radda, G. K., and Thomas, N. A.,** Interactions, perturbations and relaxations of membrane-bound molecules, in *Mitochondria: Biogenesis and Bioenergetics; Biomembranes: Molecular Arrangements and Transport Mechanisms,* Van den Bergh, S. G., Borst, P., Van Deenen, L. L. M., Riemersma, J. C., Slater, E. C., and Tager, J. M., Eds., Elsevier, Amsterdam, 1972, 231.

20. **Sokolove, P. M., Brenza, J. M., and Shamoo, A. E.,** Ca^{2+}-cardiolipin interaction in a model system. Selectivity and apparent high affinity, *Biochim. Biophys. Acta,* 732, 41, 1983.

21. **Verkleij, A. J.,** Lipidic intramembranous particles, *Biochim. Biophys. Acta,* 779, 43, 1984.

22. **De Kruijff, B., Van den Besselaar, A. M. H. P., Cullis, P. R., Van den Bosch, H., and Van Deenen, L. L. M.,** Evidence for isotropic motion of phospholipids in liver microsomal membranes. A ^{31}P NMR study, *Biochim. Biophys. Acta,* 514, 1, 1978.

23. **Burnell, E. E., Cullis, P. R., and De Kruijff, B.,** Effects of tumbling and lateral diffusion on phosphatidylcholine model membrane ^{31}P-NMR lineshapes, *Biochim. Biophys. Acta,* 603, 63, 1980.

24. **Hui, S. W., Stewart, T. P., Yeagle, P. L., and Albert, A. D.,** Bilayer to non-bilayer transition in mixtures of phosphatidylethanolamine and phosphatidylcholine: implications for membrane properties, *Arch. Biochem. Biophys.,* 207, 227, 1981.

25. **Michell, R. H.,** Inositol phospholipids and cell surface receptor function, *Biochim. Biophys. Acta,* 415, 81, 1975.

26. **Gerrard, J. M., Kindom, S. E., Perterson, D. A., Peller, J., Krantz, K. E., and White, J. G.,** Lysophosphatidic acids: influence on platelet aggregation and intracellular calcium flux, *Am. J. Pathol.,* 96, 423, 1979.

27. **Lapetina, E. G., Billah, M. M., and Cuatrecasas, P.,** Lysophosphatidic acid potentiates the thrombin-induced production of arachidonate metabolites in platelets, *J. Biol. Chem.,* 256, 11984, 1981.

28. **Weissmann, G., Anderson, P., Serhan, C., Samuelsson, E., and Goodman, E.,** A general method, employing arsenazo III in liposomes, for study of calcium ionophores: results with A23187 and prostaglandins, *Proc. Natl. Acad. Sci. U.S.A.,* 77, 1506, 1980.

29. **Holmes, R. P. and Yoss, N. L.,** Failure of phosphatidic acid to translocate Ca^{2+} across phosphatidylcholine membranes, *Nature (London),* 305, 637, 1983.

30. **Mahadevappa, V. G. and Holub, B. J.,** Degradation of different molecular species of phosphatidylinositol in thrombin-stimulated human platelets. Evidence for preferential degradation of 1-acyl 2-arachidonoyl species, *J. Biol. Chem.,* 258, 5337, 1983.

31. **Yingst, D. R. and Hoffman, J. F.,** Intracellular free Ca and Mg of human red blood cell ghosts measured with entrapped arsenazo III, *Anal. Biochem.,* 132, 431, 1983.

32. **Liu, C. and Hermann, T. E.,** Characterization of ionomycin as a calcium ionophore, *J. Biol. Chem.,* 253, 5892, 1978.

33. **Kauffman, R. F., Taylor, R. W., and Pfeiffer, D. R.,** Cation transport and specificity of ionomycin. Comparison with ionophore A 23187 in rat liver mitochondria, *J. Biol. Chem.,* 255, 2735, 1980.

34. **Op den Kamp, J. A. F.,** Lipid asymmetry in membranes, *Ann. Rev. Biochem.,* 48, 47, 1979.

35. **Benga, Gh. and Holmes, R. P.,** Molecular interactions between components in biological membranes and their implications for membrane function, *Prog. Biophys. Mol. Biol.,* 43, 195, 1984.

36. **Jacobson, K. and Papahadjopoulos, D.,** Phase transitions and phase separations in phospholipid membranes induced by changes in temperature, pH, and concentration of bivalent cations, *Biochemistry,* 14, 152, 1975.

37. **Harrison, D. G. and Long, C.,** The calcium content of human erythrocytes, *J. Physiol.,* 199, 367, 1968.

38. **Simson, J. A. V., Spicer, S. S., and Katsuyama, T.,** Cell membrane cation localization by pyroantimonate methods: correlation with cell function, in *Pathobiology of Cell Membranes,* Vol. 2, Trump, B. F. and Arstila, A. U., Eds., Academic Press, New York, 1980, 1.

39. **De Kruijff, B. and Baken, P.,** Rapid transbilayer movement of phospholipids induced by an asymmetrical perturbation of the bilayer, *Biochim. Biophys. Acta,* 507, 38, 1978.

40. **Papahadjopoulos, D., Vail, W. J., Pangborn, W. A., and Poste, G.,** Studies on membrane fusion. II. Induction of fusion in pure phospholipid membranes by calcium ions and other divalent metals, *Biochim. Biophys. Acta,* 448, 265, 1976.

41. **Liao, M. J. and Prestegard, J. H.,** Structural properties of a Ca^{2+}-phosphatidic acid complex. Small angle X-ray scattering and calorimetric results, *Biochim. Biophys. Acta,* 645, 149, 1981.

42. **Gerritsen, W. J., De Kruijff, B., Verkleij, A. J., De Gier, J., and Van Deenen, L. L. M.,** Ca^{2+}-induced isotropic motion and phosphatidylcholine flip-flop in phosphatidylcholine-cardiolipin bilayers, *Biochim. Biophys. Acta,* 598, 554, 1980.

43. **Taylor, C. B., Peng, S. K., Werthessen, N. T., Tham, P., and Lee, K. T.,** Spontaneously occurring angiotoxic derivatives of cholesterol, *Am. J. Clin. Nutr.,* 32, 40, 1979.

44. **Bjorkhem, I., Gustafsson, J., Johansson, G., and Persson, B.,** Biosynthesis of bile acids in man: hydroxylation of C_{27}-steroid side chain, *J. Clin. Invest.,* 55, 478, 1975.

45. **Javitt, N. B., Kok, E., Burstein, S., Cohen, B., and Kutscher, J.,** 26-Hydroxycholesterol. Identification and quantitation in human serum, *J. Biol. Chem.,* 256, 12644, 1981.

46. **Smith, L. L., Teng, J. I., Lin, Y. Y., Seitz, P. K., and McGehee, M. F.,** Sterol metabolism. XLVII. Oxidized cholesterol esters in human tissues, *J. Steroid Biochem.,* 14, 889, 1981.

47. **Kandutsch, A. A., Chen, H. W., and Heiniger, H. J.,** Biological activity of some oxygenated sterols, *Science,* 201, 498, 1978.

48. **Tanaka, R. D., Edwards, P. E., Lan, S. F., and Fogelman, A. M.,** Regulation of 3-hydroxy-3-methylglutaryl coenzyme A reductase activity in avian myeloblasts. Mode of action of 25-hydroxycholesterol, *J. Biol. Chem.,* 258, 13331, 1983.

49. **Bakker-Grunwald, T. and Sinensky, M.,** $^{86}Rb^+$ fluxes in Chinese hamster ovary cells as a function of membrane cholesterol content, *Biochim. Biophys. Acta,* 558, 296, 1979.

50. **Defay, R., Astruc, M. E., Roussillon, S., Descomps, B., and Crastes de Paulet, A.,** DNA synthesis and 3-hydroxy-3-methylglutaryl CoA reductase activity in PHA stimulated human lymphocytes: a comparative study of the inhibitory effects of some oxysterols with special reference to side chain hydroxylated derivatives, *Biochem. Biophys. Res. Commun.,* 106, 362, 1982.

51. **Astruc, M., Roussillon, S., Defay, R., Descomps, B., and Crastes de Paulet, A.,** DNA and cholesterol biosynthesis in synchronized embryonic rat fibroblasts. II. Effects of sterol biosynthesis inhibitors on cell division, *Biochim. Biophys. Acta,* 763, 11, 1983.

52. **Holmes, R. P. and Yoss, N. L.,** 25-Hydroxysterols increase the permeability of liposomes to Ca^{2+} and other cations, *Biochim. Biophys. Acta,* 770, 15, 1984.

53. **Chen, H. W., Heiniger, H. J., and Kandutsch, A. A.,** Alteration of $^{86}Rb^+$ influx and efflux following depletion of membrane sterol in L-cells, *J. Biol. Chem.,* 253, 3180, 1978.

54. **Kok, E., Burstein, S., Javitt, N. B., Gut, M., and Byon, C. Y.,** Bile acid synthesis. Metabolism of 3β-hydroxy-5-cholenoic acid in the hamster, *J. Biol. Chem.,* 256, 6155, 1981.

55. **Esterman, A. L., Baum, H., Javitt, N. B., and Darlington, G. J.,** 26-Hydroxycholesterol: regulation of hydroxymethyl glutaryl-CoA reductase activity in Chinese hamster ovary cell culture, *J. Lipid Res.,* 24, 1304, 1983.

56. **Holmes, R. P. and Kummerow, F. A.,** The relationship of adequate and excessive intake of vitamin D to health and disease, *J. Am. Coll. Nutr.,* 2, 173, 1983.

57. **Holmes, R. P.,** Tissue infiltration of vitamin D and 25-hydroxyvitamin D during hypervitaminosis D, in *Vitamin D: Chemical, Biochemical and Clinical Endocrinology of Calcium Metabolism,* Norman, A. W., Schaefer, K., Herrath, D., and Grigoleit, H.-G., Eds., Walter de Gruyter, New York, 1982, 567.

58. **Naccache, P. H., Molski, T. F. P., Becker, E. L., Borgeat, P., Picard, S., Vallerand, P., and Sha'afi, R. I.,** Specificity of the effect of lipoxygenase metabolites of arachidonic acid on calcium homeostasis in neutrophils. Correlation with functional activity, *J. Biol. Chem.,* 257, 8608, 1982.

59. **Benga, Gh., Hodarnau, A., Ionescu, M., Pop, V. I., Frangopol, P. T., Strujan, V., Holmes, R. P., and Kummerow, F. A.,** A comparison of the effects of cholesterol and 25-hydroxycholesterol on egg yolk lecithin liposomes: spin label studies, *Ann. N.Y. Acad. Sci.,* 414, 140, 1983.

60. **Verma, S. P., Philippot, J. R., and Wallach, D. F. H.,** Chain length dependent modification of lipid organization by low levels of 25-hydroxycholesterol and 25-hydroxycholecalciferol. A laser Raman study, *Biochemistry,* 22, 4587, 1983.

61. **Gallay, J., De Kruijff, B., and Demel, R. A.,** Sterol-phospholipid interactions in model membranes. Effect of polar group substitutions in the cholesterol side-chain at C_{20} and C_{22}, *Biochim. Biophys. Acta,* 769, 96, 1984.

62. **Graf, E. and Penniston, J. T.,** CaATP: the substrate, at low ATP concentrations, of Ca^{2+}-ATPase from human erythrocyte membranes, *J. Biol. Chem.,* 256, 1587, 1981.

63. **Enyedi, A., Sarkadi, B., and Gardos, G.,** On the substrate specificity of the red cell calcium pump, *Biochim. Biophys. Acta,* 687, 109, 1982.

64. **MacLennan, D.H. and Reithmeier, R. A. F.,** The structure of the Ca^{2+}/Mg^{2+} ATPase of sarcoplasmic reticulum, in *Membranes and Transport,* Vol. 1, Martonosi, A., Ed., Plenum Press, New York, 1982, 567.

65. **Yamamoto, T. and Tonomura, Y.,** Ca^{2+}/Mg^{2+}-dependent ATPase in sarcoplasmic reticulum. Kinetic properties in its monomeric and oligomeric forms, in *Membranes and Transport,* Vol. 1, Martonosi, A., Eds., Plenum Press, New York, 1982, 573.

66. **Andersen, J. P., Skriver, E., Mahrous, T. S., and Moller, J. V.,** Reconstitution of sarcoplasmic reticulum Ca^{2+}-ATPase with excess lipid dispersion of the pump units, *Biochim. Biophys. Acta,* 728, 1, 1983.

67. **Shoshan, V., MacLennan, D. H., and Wood, D. S.,** A proton gradient controls a calcium-release channel in sarcoplasmic reticulum, *Proc. Natl. Acad. Sci. U.S.A.,* 78, 4828, 1981.

68. **Campbell, K. P. and MacLennan, D. H.,** Purification and characterization of the 53,000-dalton glycoprotein from the sarcoplasmic reticulum, *J. Biol. Chem.,* 256, 4626, 1981.

69. **Tada, M., Ohmori, F., Yamada, M., and Abe, H.,** Mechanism of the stimulation of Ca^{2+}-dependent ATPase of cardiac sarcoplasmic reticulum by adenosine 3':5'-monophosphate-dependent protein kinase. Role of the 22,000-dalton protein, *J. Biol. Chem.,* 254, 319, 1979.

70. **Davis, B. A., Schwartz, A., Samaha, F. J., and Kranias, E. G.,** Regulation of cardiac sarcoplasmic reticulum calcium transport by calcium-calmodulin-dependent phosphorylation, *J. Biol. Chem.,* 258, 13587, 1983.

71. **Plank, B., Wyskovsky, W., Hellmann, G., and Suko, J.,** Calmodulin-dependent elevation of calcium transport associated with calmodulin-dependent phosphorylation in cardiac sarcoplasmic reticulum, *Biochim. Biophys. Acta,* 732, 99, 1983.

72. **Oliva, J. M., de Meis, L., and Inesi, G.,** Calmodulin stimulates both adenosine 5′-triphosphate hydrolysis and synthesis catalyzed by a cardiac calcium ion dependent adenosine triphosphatase, *Biochemistry,* 22, 5822, 1983.

73. **Warren, G. B., Toon, P. A., Birdsall, N. J. M., Lee, A. G., and Metcalfe, J. C.,** Reversible lipid titrations of the activity of pure adenosine triphosphatase-lipid complexes, *Biochemistry,* 13, 5501, 1974.

74. **Hidalgo, C., Thomas, D. D., and Ikemoto, N.,** Effect of the lipid environment on protein motion and enzymatic activity of the sarcoplasmic reticulum calcium ATPase, *J. Biol. Chem.,* 253, 6879, 1978.

75. **Dean, W. L. and Tanford, C.,** Properties of a delipidated, detergent-activated Ca^{2+}-ATPase, *Biochemistry,* 17, 1683, 1978.

76. **Johannsson, A., Keightley, C. A., Smith, G. A., Richards, C. D., Hesketh, T. R., and Metcalf, J. C.,** The effect of bilayer thickness and *n*-alkanes on the activity of the $(Ca^{2+} + Mg^{2+})$-dependent ATPase of sarcoplasmic reticulum, *J. Biol. Chem.,* 256, 1643, 1981.

77. **Navarro, J., Toivio-Kinnucan, M., and Racker, E.,** Effect of lipid composition on the calcium adenosine 5′-triphosphate coupling ratio of the Ca^{2+}-ATPase of sarcoplasmic reticulum, *Biochemistry,* 23, 130, 1984.

78. **Warren, G. B., Houslay, M. D., and Metcalfe, J. C., and Birdsall, N. J. M.,** Cholesterol is excluded from the phospholipid annulus surrounding an active calcium transport protein, *Nature (London),* 255, 684, 1975.

79. **Silvius, J. R., McMillen, D. A., Saley, N. D., Jost, P. C., and Griffith, O. H.,** Competition between cholesterol and phosphatidylcholine for the hydrophobic surface of sarcoplasmic reticulum Ca^{2+}-ATPase, *Biochemistry,* 23, 538, 1984.

80. **Karnovsky, M. J., Kleinfeld, A. M., Hoover, R. L., and Klausner, R. D.,** The concept of lipid domains in membranes, *J. Cell Biol.,* 94, 1, 1982.

81. **Adams, R. J., Cohen, D. W., and Gupte, S., Johnson, J. D., Wallick, E. T., Wang, T., and Schwartz, A.,** *In vitro* effects of palmitylcarnitine on cardiac plasma membrane, Na, K-ATPase, and sarcoplasmic reticulum Ca^{2+}-ATPase and Ca^{2+} transport, *J. Biol. Chem.,* 254, 12404, 1979.

82. **Cheah, A. M.,** Effect of long chain unsaturated fatty acids on the calcium transport of sarcoplasmic reticulum, *Biochim. Biophys. Acta,* 648, 113, 1981.

83. **Katz, A. M., Nash-Adler, P., Watras, J., Messineo, F., Takenaka, H., and Louis, C. F.,** Fatty acid effects on calcium influx and efflux in sarcoplasmic reticulum vesicles from rabbit skeletal muscle, *Biochim. Biophys. Acta,* 687, 17, 1982.

84. **Messineo, F. C., Rathier, M., Favreau, C., Watras, J., and Takenaka, H.,** Mechanisms of fatty acid effects on sarcoplasmic reticulum. III. The effects of palmitic and oleic acids on sarcoplasmic reticulum function — a model for fatty acid membrane interactions, *J. Biol. Chem.,* 259, 1336, 1984.

85. **Van Winkle, W. B., Bick, R. J., Tucker, D. E., Tate, C. A., and Entman, M. L.,** Evidence for membrane microheterogeneity in the sarcoplasmic reticulum of fast twitch skeletal muscle, *J. Biol. Chem.,* 257, 11689, 1982.

86. **Borchman, D., Simon, R., and Bicknell-Brown, E.,** Variation in the lipid composition of rabbit muscle sarcoplasmic reticulum membrane with muscle type, *J. Biol. Chem.,* 257, 14136, 1982.

87. **Niggli, V., Adunyah, E. S., Penniston, J. T., and Carafoli, E.,** Purified $(Ca^{2+}-Mg^{2+})$-ATPase of the erythrocyte membrane. Reconstitution and effect of calmodulin and phospholipids, *J. Biol. Chem.,* 256, 395, 1981.

88. **Carafoli, E., Zurini, M., Niggli, V., and Krebs, J.,** The calcium-transporting ATPase of erythrocytes, *Ann. N. Y. Acad. Sci.,* 402, 304, 1982.

89. **Sakardi, B.,** Active calcium transport in human red cells, *Biochim. Biophys. Acta,* 604, 159, 1980.

90. **Cunningham, E. B., Varghese, S., Lee, S., Brissette, R., and Swislocki, N. I.,** Modulation of the activity of the $(Ca^{2+} + Mg^{2+})$-dependent adenosine triphosphatase of the human erythrocyte, *Biochim. Biophys. Acta,* 731, 378, 1983.

91. **Rodan, S. B., Rodan, G. A., and Sha'afi, R. I.,** Demonstration of adenylate cyclase activity in human red blood cell ghosts, *Biochim. Biophys. Acta,* 428, 509, 1976.

92. **Pershadsingh, H. A. and McDonald, J. M.,** Direct addition of insulin inhibits a high affinity Ca^{2+}-ATPase in isolated adipocyte plasma membranes, *Nature (London),* 281, 495, 1979.

93. **Roelofsen, B.,** The lipid requirement of the Ca^{2+}/Mg^{2+}-ATPase in the erythrocyte membrane, in *Membranes and Transport,* Vol. 1, Plenum Press, New York, 1982, 607.

94. **Penniston, J. T.,** Plasma membrane Ca^{2+}-pumping ATPases, *Ann. N.Y. Acad. Sci.,* 402, 296, 1982.

95. **Schmalzing, G. and Kutschera, P.,** Modulation of ATPase activities of human erythrocyte membranes by free fatty acids or phospholipase A_2, *J. Membr. Biol.,* 69, 65, 1982.

96. **Wetzker, R., Klinger, R., and Frunder, H.,** Effects of fatty acids on activity and calmodulin binding of Ca^{2+}-ATPase of human erythrocyte membranes, *Biochim. Biophys. Acta,* 730, 196, 1983.

97. **Hirata, F. and Axelrod, J.,** Enzymatic methylation of phosphatidylethanolamine increases erythrocyte membrane fluidity, *Nature (London),* 275, 219, 1978.

98. **Strittmatter, W. J., Hirata, F., and Axelrod, J.,** Increased Ca^{2+}-ATPase activity associated with methylation of phospholipids in human erythrocytes, *Biochem. Biophys. Res. Commun.,* 88, 147, 1979.

99. **Chauhan, V. P. S. and Kalra, V. K.,** Effect of phospholipid methylation on calcium transport and (Ca^{2+} + Mg^{2+})-ATPase activity in kidney cortex basolateral membranes, *Biochim. Biophys. Acta,* 727, 185, 1983.

100. **Galo, M. G., Unates, L. E., and Farias, R. N.,** Effect of membrane fatty acid composition on the action of thyroid hormones on (Ca^{2+} + Mg^{2+})-adenosine triphosphatase from rat erythrocyte, *J. Biol. Chem.,* 256, 7113, 1981.

101. **Galo, M. G., Bloj, B., and Farias, R. N.,** Kinetic changes of the erythrocyte (Mg^{2+} + Ca^{2+})-adenosine triphosphatase of rats fed different fat-supplemented diets, *J. Biol. Chem.,* 250, 6204, 1975.

102. **Holmes, R. P. and Kummerow, F. A.,** The effect of dietary lipids on the composition and properties of biological membranes, in *Structure and Properties of Cell Membranes,* Vol. 3, Benga, Gh., Ed., CRC Press, Boca Raton, Fla., 1985, chap. 9.

103. **Davis, P. J. and Blas, S. D.,** In vitro stimulation of human red blood cell Ca^{2+}-ATPase by thyroid hormone, *Biochem. Biophys. Res. Commun.,* 99, 1073, 1981.

104. **Holmes, R. P., Mahfouz, M., Travis, B. D., Yoss, N. L., and Keenan, M. J.,** The effect of membrane lipid composition on the permeability of membranes to Ca^{2+}, *Ann. N.Y. Acad. Sci.,* 414, 44, 1983.

105. **Lotersztajn, S., Hanoune, J., and Pecker, F.,** A high affinity calcium-stimulated magnesium-dependent ATPase in rat liver plasma membranes. Dependence on an endogenous protein activator distinct from calmodulin, *J. Biol. Chem.,* 256, 11209, 1981.

106. **Lotersztajn, S. and Pecker, F.,** A membrane-bound protein inhibitor of the high affinity Ca ATPase in rat liver plasma membranes, *J. Biol. Chem.,* 257, 6638, 1982.

107. **Hamlyn, J. M. and Senior, A. E.,** Evidence that Mg^{2+}- or Ca^{2+}-activated adenosine triphosphatase in rat pancreas is a plasma-membrane ectoenzyme, *Biochem. J.,* 214, 59, 1983.

108. **Baskin, G. S. and Langdon, R. G.,** A spectrin-dependent ATPase of the human erythrocyte membrane, *J. Biol. Chem.,* 256, 5428, 1981.

109. **Caroni, P., Zurini, M., Clark, A., and Carafoli, E.,** Further characterization and reconstitution of the purified Ca^{2+}-pumping ATPase of heart sarcolemma, *J. Biol. Chem.,* 258, 7305, 1983.

110. **Grover, A. K., Kwan, C.-Y., Luchowski, E., Daniel, E. E., and Triggle, D. J.,** Subcellular distribution of [^{3}H]nitrendipine binding in smooth muscle, *J. Biol. Chem.,* 259, 2223, 1984.

111. **Rinaldi, M. L., LePeuch, C. J., and Demaille, J. G.,** The epinephrine-induced activation of the cardiac slow Ca^{2+} channel is mediated by the cAMP-dependent phosphorylation of calciductin, a 23,000 M_r sarcolemmal protein, *FEBS Lett.,* 129, 277, 1981.

112. **Rinaldi, M. L., Capony, J.-P., and Demaille, J. G.,** The cyclic AMP-dependent modulation of cardiac sarcolemmal slow calcium channels, *J. Mol. Cell Cardiol.,* 14, 279, 1982.

113. **Cachelin, A. B., de Peyer, J. E., Kokubun, S., and Reuter, H.,** Ca^{2+} channel modulation by 8-bromocyclic AMP in cultured heart cells, *Nature (London),* 304, 462, 1983.

114. **Triggle, D. J. and Swamy, V. C.,** Calcium antagonists. Some chemical-pharmacologic aspects, *Circ. Res.,* 52, I-17, 1983.

115. **Bolton, T. B.,** Mechanisms of action of transmitters and other substances on smooth muscle, *Physiol. Rev.,* 59, 606, 1979.

116. **Hinnen, R., Miyamoto, H., and Racker, E.,** Ca^{2+} translocation in Ehrlich ascites tumor cells, *J. Membr. Biol.,* 49, 309, 1979.

117. **Cittadini, A. and Van Rossum, G. D. V.,** Properties of the calcium-extruding mechanism of liver cells, *J. Physiol.,* 281, 29, 1978.

118. **Varecka, L. and Carafoli, E.,** Vanadate-induced movements of Ca^{2+} and K^+ in human red blood cells, *J. Biol. Chem.,* 257, 7414, 1982.

119. **Sandvig, J. and Olsnes, S.,** Entry of the toxic proteins abrin, modeccin, ricin, and diptheria toxin into cells. I. Requirement for calcium, *J. Biol. Chem.,* 257, 7495, 1982.

120. **Wiley, J. S., McCulloch, K. E., and Bowden, D. S.,** Increased calcium permeability of cold-stored erythrocytes, *Blood,* 60, 92, 1982.

121. **Ikeda, Y., Kikuchi, M., Toyama, K., Watanabe, K., and Ando, Y.,** Inhibition of human platelet functions by verapamil, *Thromb. Haemostas.,* 45, 158, 1981.

122. **Kiyomoto, A., Sasaki, Y., Odawara, A., and Morita, T.,** Inhibition of platelet aggregation by diltiazem. Comparison with verapamil and nifedipine and inhibitory potencies of diltiazem metabolites, *Circ. Res.,* 52, I-115, 1983.

123. **Han, P., Boatwright, C., and Ardlie, N. G.,** Effect of the calcium-entry blocking agent nifedipine on activation of human platelets and comparison with verapamil, *Thromb. Haemostas.,* 50, 513, 1983.

124. **Fisher, S. K., Van Rooijen, L. A. A., and Agranoff, B. W.,** Renewed interest in the polyphosphoinositides, *Trends Biochem. Sci.,* 9, 53, 1984.

125. **Nishizuka, Y.,** Phospholipid degradation and signal translation for protein phosphorylation, *Trends Biochem. Sci.,* 8, 13, 1983.

126. **Streb, H., Irvine, R. F., Berridge, M. J., and Schulz, I.,** Release of Ca^{2+} from a nonmitochondrial intracellular store in pancreatic acinar cells by inositol-1,4,5-triphosphate, *Nature (London),* 306, 67, 1983.

127. **Imai, A., Ishizuka, Y., Kawai, K., and Nozawa, Y.,** Evidence for coupling of phosphatidic acid formation and calcium influx in thrombin-activated human platelets, *Biochem. Biophys. Res. Commun.,* 108, 752, 1982.

128. **Pozzan, T., Lew, D. P., Wollheim, C. B., and Tsien, R. Y.,** Is cytosolic ionized calcium regulating neutrophil activation?, *Science,* 221, 1413, 1983.

129. **Rink, T. J., Sanchez, A., and Hallam, T. J.,** Diacylglycerol and phorbol ester stimulate secretion without raising cytoplasmic free calcium in human platelets, *Nature (London),* 305, 317, 1983.

130. **Kaibuchi, K., Takai, Y., Sawamura, M., Hoshijima, M., Fujikura, T., and Nishizuka, Y.,** Synergistic functions of protein phosphorylation and calcium mobilization in platelet activation, *J. Biol. Chem.,* 258, 6701, 1983.

131. **Castagna, M., Takai, Y., Kaibuchi, K., Sano, K., Kikkawa, U., and Nishizuka, Y.,** Direct activation of calcium-activated, phospholipid-dependent protein kinase by tumor-promoting phorbol esters, *J. Biol. Chem.,* 257, 7847, 1982.

132. **Michell, R.,** Ca^{2+} and protein kinase C: two synergistic cellular signals, *Trends Biochem. Sci.,* 8, 263, 1983.

133. **Ohsako, S. and Deguchi, T.,** Stimulation by phosphatidic acid of calcium influx and cyclic GMP synthesis in neuroblastoma cells, *J. Biol. Chem.,* 256, 10945, 1981.

134. **Barritt, G. J., Dalton, K. A., and Whiting, J. A.,** Evidence that phosphatidic acid stimulates the uptake of calcium by liver cells but not calcium release from mitochondria, *FEBS Lett.,* 125, 137, 1981.

135. **Pagano, R. E., Longmuir, K. J., Martin, O. C., and Struck, D. K.,** Metabolism and intracellular localization of a fluorescently labeled intermediate in lipid biosynthesis within cultured fibroblasts, *J. Cell Biol.,* 91, 872, 1981.

136. **Matsumoto, T., Fontaine, O., and Rasmussen, H.,** Effect of 1,25-dihydroxy-vitamin D_3 on phospholipid metabolism in chick duodenal mucosal cell. Relationship to its mechanism of action, *J. Biol. Chem.,* 256, 3354, 1981.

137. **O'Doherty, P. J. A.,** 1,25-Dihydroxyvitamin D_3 increases the activity of the intestinal phosphatidylcholine deacylation-reacylation cycle, *Lipids,* 14, 75, 1979.

138. **Vance, D. E. and Pelech, S. L.,** Enzyme translocation in the regulation of phosphatidylcholine biosynthesis, *Trends Biochem. Sci.,* 9, 17, 1984.

139. **Kreutter, D., Matsumoto, T., Peckham, R., Zawalich, K., Wen, W. H., Zolock, D. T., and Rasmussen, H.,** The effect of essential fatty acid deficiency on the stimulation of intestinal calcium transport by 1,25-dihydroxyvitamin D_3, *J. Biol. Chem.,* 258, 4977, 1983.

140. **Fontaine, O., Matsumoto, T., Goodman, D. B. P., and Rasmussen, H.,** Liponomic control of Ca^{2+} transport: relationship to mechanism of action of 1,25-dihydroxyvitamin D_3, *Proc. Natl. Acad. Sci. U.S.A.,* 78, 1751, 1981.

141. **De Boland, A. R., Gallego, S., and Boland, R.,** Effects of vitamin D_3 on phosphate and calcium transport across and composition of skeletal plasma cell membranes, *Biochim. Biophys. Acta,* 733, 264, 1983.

142. **Tsutsumi, M., Alvarez, U., Avioli, L. V., and Hruska, K. A.,** Stimulation of renal brush border membrane vesicle (BBMV) phospholipid (PL) metabolism by 1,25-dihydroxycholecalciferol (1,25(OH)$_2$D$_3$), *Fed. Proc., Fed. Am. Soc. Exp. Biol.,* 43, 631, 1984.

143. **DiPolo, R. and Beauge, L.,** The calcium pump and sodium-calcium exchange in squid axons, *Ann. Rev. Physiol.,* 45, 313, 1983.

144. **Ueda, T.,** Na^+-Ca^{2+} exchange activity in rabbit lymphocyte plasma membranes, *Biochim. Biophys. Acta,* 734, 342, 1983.

145. **Gmaj, P., Murer, H., and Kinne, R.,** Calcium ion transport across plasma membranes isolated from rat kidney cortex, *Biochem. J.,* 178, 549, 1979.

146. **Bradley, M. P. and Forrester, I. T.,** A sodium-calcium exchange mechanism in plasma membrane vesicles isolated from ram sperm flagella, *FEBS Lett.,* 121, 15, 1980.

147. **Rufo, G. A., Jr., Schoff, P. K., and Lardy, H. A.,** Regulation of calcium content in bovine spermatozoa, *J. Biol. Chem.,* 259, 2547, 1984.

148. **Volpi, M., Naccache, P. H., and Sha'afi, R. I.,** Calcium transport in inside-out membrane vesicles prepared from rabbit neutrophils, *J. Biol. Chem.,* 258, 4153, 1983.

149. **Philipson, K. D., Frank, J. S., and Nishimoto, A. Y.,** Effects of phospholipase C on the Na^+-Ca^{2+} exchange and Ca^{2+} permeability of cardiac sarcolemmal vesicles, *J. Biol. Chem.,* 258, 5905, 1983.

150. **Philipson, K. D. and Nishimoto, A. Y.,** Stimulation of Na^+-Ca^{2+} exchange in cardiac sarcolemmal vesicles by phospholipase D, *J. Biol. Chem.,* 259, 16, 1984.

151. **Pierce, G. N., Kutryk, M. J. B., and Dhalla, N. S.,** Alterations in Ca^{2+} binding by and composition of the cardiac sarcolemmal membrane in chronic diabetes, *Proc. Natl. Acad. Sci. U.S.A.,* 80, 5412, 1983.
152. **Kramsch, D. M., Aspen, A. J., and Rozler, L. J.,** Atherosclerosis: prevention by agents not affecting abnormal levels of blood lipids, *Science,* 213, 1511, 1981.
153. **Shapiro, B. L. and Lam, L. F.-H.,** Calcium and age in fibroblasts from control subjects and patients with cystic fibrosis, *Science,* 216, 417, 1982.
154. **Peterson, C. and Gibson, G. E.,** Aging and 3,4-diamino-pyridine alter synatosomal calcium uptake, *J. Biol. Chem.,* 258, 11482, 1983.
155. **Holmes, R. P. and Yoss, N. L.,** unpublished results.
156. **Kou, I. and Holmes, R. P.,** unpublished results.
157. **Travis, B. D. and Holmes, R. P.,** unpublished results.
158. **Farren, S. B., Hope, M. J., and Cullis, P. R.,** Polymorphic phase preference of phosphatidic acid: a ^{31}P and ^{2}H NMR study, *Biochem. Biophys. Res. Commun.,* 111, 675, 1983.
159. **Verkleij, A. J., DeMaagd, R., Leunissen-Bijvelt, J., and De Kruijff, B.,** Divalent cations and chlorpromazine can induce non-bilayer structures in phosphatidic acid-containing model membranes, *Biochim. Biophys. Acta,* 684, 255, 1982.

Chapter 5

MOLECULAR ASPECTS OF STRUCTURE-FUNCTION RELATIONSHIP IN MITOCHONDRIAL H$^+$-ATPase

Jan Kopecký, Josef Houštěk, and Zdeněk Drahota

TABLE OF CONTENTS

I. INTRODUCTION

Among various ATPases involved in active transport across biological membranes, the proton translocating ATPase (H^+-ATPase) of mitochondria, chloroplasts, and bacteria represents the most complicated enzyme complex. Due to the key role of this enzyme in energy transduction, the structure and function of H^+-ATPase have been extensively studied during the past two decades in many laboratories. In spite of such a tremendous effort, the molecular mechanism of coupling between H^+-translocation and ATP synthesis catalyzed by H^+-ATPase remains unclear.

In mitochondria which are phylogenetically younger than chloroplasts and bacteria, the structure of H^+-ATPase appears to be more complicated, similar to some other membrane-bound enzyme complexes, e.g., cytochrome *c* oxidase.[1,2] The higher complexity mirrors probably much broader and advanced systems of regulation as exist in eukaryotic cells. Therefore, for any generalization with respect to the structural arrangement and function of H^+-ATPase, it is important to investigate the degree of homology among H^+-ATPases from different sources.

Various aspects of the structure, function, and genetics of mitochondrial H^+-ATPase were summarized in several recent reviews.[3-12] It is the aim of this article to focus mainly on the subunit composition and structural organization of mitochondrial H^+-ATPase, particularly in the light of the knowledge of H^+-ATPases from other energy-transducing membranes where a pronounced progress has recently been reached. In the second part of the article attention will be paid to new data analyzing the primary event in enzyme function, i.e., proton translocation and the overall mechanism of coupling of proton transport with catalytic function of mitochondrial H^+-ATPase.

II. DEFINITION OF H^+-ATPase

According to chemiosmotic theory,[13] which is now generally accepted,[14,15] formation of electrochemical gradient of protons across the membrane $\Delta\mu H^+$ is the primary event of oxidative phosphorylation and photophosphorylation. In all types of energy-transducing membranes it is the H^+-translocating ATPase which couples the flow of protons down their electrochemical gradient with the synthesis of ATP and thus utilizes the energy of $\Delta\mu H^+$. As the reaction is reversible, the hydrolysis of ATP results in an opposite process, formation of proton gradient.

Physiologically, H^+-ATPase in mitochondria and chloroplasts operates only in the synthetic mode (oxidative phosphorylation, photophosphorylation). In anaerobic and facultatively anaerobic bacteria, the hydrolytic mode is also important. When ATP is hydrolyzed, created $\Delta\mu H^+$ is used as a driving force for coupled movements of various ions,[6,16] e.g., cations and amino acids.

H^+-ATPases can be easily distinguished from other transport ATPases by means of lipophilic compounds — oligomycin and/or DCCD.[17] Specific sensitivity to these inhibitors is in contrast with the sensitivity of other ATPases to vanadate.[18] H^+-ATPases found in different types of membranes, i.e., mitochondria, chloroplast, and bacteria, are both structurally and functionally similar but not entirely homologous.[10,14,16,19,20] They can be defined as typical ionic pumps[21] consisting of a highly selective proton channel which spans the membrane and of a catalytic part which gates the channel and represents the entire site of ATP synthesis or hydrolysis. The two functionally different multisubunit parts of the enzyme complex are also distinct in their structure. With respect to the membrane the overall organization of the enzyme is thus asymmetric.

As far as the established terminology originating from Racker's coupling factors[22] is concerned, the membrane-bound proton translocating moiety of H^+-ATPase is called F_0 and the catalytic moiety is called F_1. These symbols will be further specified in this article according to the types of membrane from which the enzyme originates (mitochondria — MF_1, MF_0; chloroplast — CF_0, CF_1; bacteria — BF_0, BF_1).

III. ISOLATION OF H^+-ATPase

In general, the isolation of H^+-ATPase consists in solubilization of the enzyme from membrane by treatment with detergents and subsequent purification. Ideally, a homogenous protein complex should be obtained which, when properly assembled into the phospholipid membrane, is capable of synthesis and hydrolysis of ATP coupled with the transmembrane H^+-movement. However, the methodology of reconstitution experiments is still difficult. Thus, DCCD or oligomycin-sensitive ATP hydrolysis, P_i-ATP exchange reaction, or ATP-dependent proton translocation is used to verify, whether the function of the isolated enzyme is preserved, no matter that these activities can serve only as partial criteria during the purification.[23,24]

The crude preparation of H^+-ATPase was obtained in 1966 for the first time using solubilization of mitochondria with cholate.[15] Later various procedures for the isolation of H^+-ATPase from bovine heart,[26-30] rat liver,[32-35] and yeast[36-38] mitochondria were described. In these procedures H^+-ATPase was solubilized by cholate,[25,27,29] deoxycholate,[31] Triton® X-100,[26,28,32,36,38] lysolecithin,[30] octylglucoside,[34] or by using zwitterionic detergent CHAPS.[35] Separation from other membrane constituents was achieved by decreasing the detergent concentration by dialysis[28,29] and by ammonium sulfate precipitation in the presence of cholate.[25,27,29,38] Density gradient centrifugation,[27,30,38] chromatography on agarose 5 M,[29] or affinity chromatography[26] are used for the final purification. Using specific antibodies against F_1 subunits it was possible to precipitate the whole H^+-ATPase from small amounts of Triton®-solubilized mitochondria.[33,37,39-41] The method developed in Hatefi's laboratory[29,42] for the isolation of H^+-ATPase, termed complex V, allows for the isolation of all components of respiratory chain in the form of four complexes (I to IV) from the same batch of mitochondria.

Various preparations of isolated mitochondrial H^+-ATPase differ in state of aggregation, content of phospholipids, homogeneity, and the ability to catalyze partial reactions of oxidative phosphorylation. Preparation of H^+-ATPase isolated from rat liver mitochondria is well dispersed[31] and its gross structure can be analyzed by electron microscopy technique. Typical tripartite structural arrangement of the enzyme was evaluated after negative staining (see Section V).

As far as homogeneity of isolated mitochondrial H^+-ATPase is concerned, the evidence used for soluble enzymes, such as sedimentation velocity, sedimentation equilibrium, gel filtration, or nondenaturating electrophoresis, is difficult to apply directly to membrane proteins which always require a detergent to keep them in solution. Instead, polyacrylamide gel electrophoresis in the presence of sodium dodecyl sulfate (SDS-PAGE) is the best method of evaluation of the number and apparent molecular weights of the enzyme protein subunits at present. The combination of the results of SDS-PAGE with functional criteria (see above) is then a usual but not optimal approach to show what are the true subunits of H^+-ATPase and what are contaminants.

In bacteria, preparation of H^+-ATPase of *Escherichia coli*[43,44] and thermophilic bacteria PS 3 are most extensively characterized. They both contain only eight polypeptide subunits, five of them being the components of F_1 and three of them subunits of F_0. All BF_0 subunits were verified[10,19,23,46] (see also Section IV.B) both functionally and, very importantly, genetically. There exists, in fact, a genetic evidence that a fourth polypeptide may be present in BF_0 or might be involved in assembly of the enzyme into the membrane (see Kanazawa and Futai[46]). In H^+-ATPase isolated from chloroplasts eight to nine types of protein subunits were detected.[47]

When the subunit composition of mitochondrial H^+-ATPase preparations is tested by one-dimensional SDS-PAGE, usually around 10 to 12 polypeptide bands are identified, while up to 18 bands can be clearly resolved using two-dimensional gels.[39-48] The resolving power of various electrophoretic systems in combination with different sensitivity of staining techniques clearly affects the number of bands detected.

Of the two most frequently used SDS-PAGE systems, i.e., the phosphate system according to Weber and Osborne[49] and Tris-glycine system of Laemmli,[50] the latter has a much higher resolving power. Good resolution of low molecular weight polypeptides (below 10K) is obtained only when urea is included.[51] Further improvement of electrophoretic resolution of the Laemmli's system can be reached by using long linear gradient gels with sucrose.[52] Thus, 18 polypeptide bands are resolved in the two commonly used H^+-ATPase preparations by one-dimensional SDS-PAGE.[52] Since the electrophoretic patterns obtained by different SDS-PAGE systems are difficult to compare, this adaptation of Laemmli's procedure might serve as a simple reference system for identification of H^+-ATPase protein subunits. The use of several systems in the same study has so far been rather exceptional.[29,35]

In recent preparations of H^+-ATPase the contamination by components of the respiratory chain is very low and does not explain the high complexity of the mitochondrial H^+-ATPase. Also, the enzyme immunoprecipitated by specific antibodies against F_1[33,37,39,41] has a polypeptide composition similar to that of the isolated complex V. The last modification of the isolation procedure for the complex V[29] yields probably the "purest" preparation of mitochondrial H^+-ATPase. As compared with preparations of Serrano et al.[27] and Berden and Voorn Brouwer[28] this enzyme lacks almost completely the 30K band, possibly the adenine nucleotide translocator.

All preparations of isolated H^+-ATPase from mitochondria catalyze oligomycin and DCCD-sensitive hydrolysis of ATP, which can be stimulated by the addition of phospholipids. Most preparations also catalyze the ATP-Pi exchange reaction sensitive to oligomycin, DCCD, and uncouplers.[27,30,31,34,35] The ATP-Pi exchange reaction requires the addition of phospholipids and is believed to be dependent on a vesicular structure, although in some preparations a high activity is found in the absence of phospholipids.[30,53,54] Because of hydrophobic F_0, H^+-ATPase tends to aggregate. Thus, under specific conditions, i.e., when carrying significant amounts of phospholipids, the formation of small vesicle-like aggregates composed only from H^+-ATPase molecules is not unlikely. In fact, such aggregates were shown by electron microscopy of rat liver H^+-ATPase.[35]

When H^+-ATPase isolated from yeast[55] and bovine heart[27] was incorporated into liposomes, the ATP-dependent H^+-pumping was observed directly by measuring pH of the medium with pH electrode,[27,55] or indirectly using fluorescent probes.[28] As stated above, the evidence that isolated H^+-ATPase can synthetize ATP is of utmost importance. With the exception of an early crude preparation[56,57] this fact was demonstrated only with mitochondrial H^+-ATPase purified from yeasts.[55] Therefore, at present there exists only one preparation of mitochondrial H^+-ATPase that can be regarded as fully functional.

IV. IDENTIFICATION AND FUNCTION OF H^+-ATPase SUBUNITS

Authenticity of individual subunits of mitochondrial H^+-ATPase is still a continuous point of contention, particularly in case of subunits of F_0. In contrast, in bacteria and chloroplasts enormous progress has been achieved during the past years and F_0 subunits are well defined with respect to their size, number, stoichiometry, and physicochemical properties. This holds partly even for their functional role and arrangement within F_0. There are several reasons for the existing discrepancy in the knowledge of F_0 in comparison with mitochondria. Structurally simpler bacterial H^+-ATPases are conclusively and reproducibly well purified at present; bacteria allow for genetic experiments which provide valuable evidence inde-

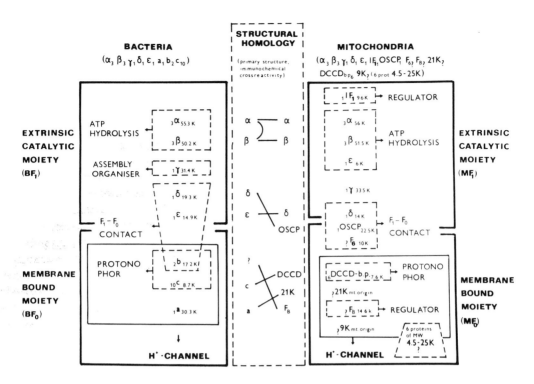

FIGURE 1. Structure and function relationships of H⁺-ATPase subunits. Molecular weights of bacterial enzyme subunits refer to DNA sequence (H⁺-ATPase operon) in *E. coli*.[46] Molecular weights of mitochondrial enzyme subunits refer to primary structure (DCCD-binding protein,[114] F$_B$,[102] IF$_1$[82]) or to mobility in SDS-PAGE.[33,52]

pendent of classical approaches; bacterial H⁺-ATPase and its subunits are more resistant to irreversible denaturation, which, particularly in thermophilic bacteria, permits successful fragmentation and reconstitution studies.

A. Subunits of F$_1$

When analyzed by SDS-PAGE, all preparations of mitochondrial H⁺-ATPase exhibit five polypeptides corresponding to subunits of isolated F$_1$ (for reviews on the isolation and properties of isolated F$_1$ see Pedersen[16]). They can be recognized by coelectrophoresis of the isolated H⁺-ATPase with isolated F$_1$. Also, specific physicochemical properties of individual F$_1$ subunits allow for the additional ways of their detection.

The identification of the three large subunits of F$_1$ in H⁺-ATPase preparations is particularly simple as their molecular weights 50 to 60K (α), 49 to 55K (β), and 32 to 33K (γ) are distinct from those of all remaining subunits of H⁺-ATPase (Figure 1). Subunits α and β can be detected using radioactive photoactivable derivatives of adenine nucleotides[58,60] and, in addition, the subunit β by its specific interaction with the fluorescent probe aurovertin,[60-62] by labeling with [¹⁴C]-DCCD,[63] and by labeling with phosphate analog, 4-azido-2-nitrophenylphosphate.[64] As compared with β and δ subunits, subunits α, γ, and ε of MF$_1$ contain SH-groups reacting with [¹⁴C]-*N*-ethylmaleimide.[65,66]

Individual subunits of MF$_1$ can also be detected with the aid of antibodies.[34,67] With respect to the origin of subunits (mitochondria, bacteria, and chloroplast) antibodies against δ and ε were found to be clearly tissue-specific. The β subunit cross-reacted in all possible combinations of the antigen and antibody. A minor cross-reactivity was also observed between α and between γ subunits in some combinations. In MF$_1$, subunits α and β and subunits β and γ contain common antigenic determinants.[67]

To detect the small F_1 subunits at the level of H^+-ATPase or rather distinguish them from F_0 subunits of similar molecular weights, the chemical modification with hydrophobic and hydrophilic reagents appears to be valuable. Thus, all MF_1 subunits, but only MF_0 subunits of molecular weight higher than 20K, were labeled with the hydrophilic modifying agent [^{35}S]-diazobenzensulfonate[39] and none of the MF_1 subunits reacted with photoreactive radioactive phospholipids.[52]

As preparations of F_1 isolated from bacteria and chloroplasts contain five protein subunits similar to the MF_1, the equivalence of all of the F_1 subunits in mitochondria, bacteria, and chloroplasts is often assumed. However, it was shown that the small subunits δ and ϵ of the BF_1 are not related with those in the mitochondrial enzyme.[20] The bacterial ϵ is the counterpart of mitochondrial δ and the mitochondrial OSCP, the subunit involved in binding of F_1 to F_0 (see Section IV.B), is equivalent to bacterial δ. In bacteria, both δ and ϵ are required for the binding of F_1.[10,21,23] Also, in chloroplasts[68,69] and mitochondria[70] the requirement of the δ subunit for F_1 binding is well demonstrated. As far as the two large subunits of F_1 are concerned, a high degree of homology in some regions of the β subunit from *E. coli* F_1 and bovine heart F_1 was found.[46] Some similarities between the sequence of bacterial β and α subunits were also observed.[46] Therefore, it can be concluded that only subunits α and β (and possibly also γ) of F_1 from various types of coupling membranes are really homologous. Thus, the present nomenclature of small F_1 subunits from various sources is misleading, as the same symbols are used to designate nonhomologous subunits, at least in mitochondria and bacteria. Moreover, the subunit ϵ of MF_1 has not any counterpart in BF_1 (Figure 1).

Authenticity and function of three large F_1 subunits α, β, and γ (more than 90% mass of the molecule) were best demonstrated with the bacterial and chloroplast enzymes. The fragments of bacterial[71] and chloroplast[72] F_1 obtained by trypsin digestion and containing only α and β subunits hydrolyzed ATP, similarly to two subunits "enzyme" (α, β) isolated from *Micrococcus lysodeikticus* by butanol extraction.[73] The simplest active fragments of MF_1 contained α, β, and ϵ subunits.[74] In reconstitution experiments, where complexes active in ATP hydrolysis were reformed from isolated bacterial subunits,[10,46,75] α, β, and γ subunits were required for activity. The most active was the complex with $\alpha_3 \beta_3 \gamma_1$ in accordance with the most probably subunit stoichiometry of F_1 ($\alpha_3 \beta_3 \gamma_1 \delta_1 \epsilon_1$). Thus, γ subunit is needed for proper reassociation of α and β subunits rather than for the entire activity and might serve as template for these subunits. The hybrid reconstitution of the active α, β, γ complex from subunits of thermophilic and mesophilic bacteria[76] supports the view that the role of these subunits is identical in all types of H^+-ATPases. Subunit γ is further involved in gating of the H^+-flux as shown in chloroplasts[77] and bacteria.[78,78] It is also required for manifestation of the ϵ subunit activity in bacteria and chloroplasts.[72,80]

All these results indicate that α and β subunits participate in the formation of the catalytic domain of the enzyme. Chemical modifications of F_1 subunits (for review see Cross[8]) as well as modification by photoactivable derivatives of adenine nucleotides[58-60] specify further that β subunit bears the catalytic site, whereas α subunit has a regulatory function in catalysis.

In mitochondria F_1 is freely associated with the additional sixth subunit, a heat-stable protein of molecular weight 10K, ATPase inhibitor protein (IF_1).[81] IF_1 was purified to homogeneity[81] and recently its sequence has been determined.[82] IF_1 inhibits ATP hydrolysis and the initial, but not steady-state rate of ATP synthesis.[83,84] Upon the establishment of $\Delta\mu H^+$ IF_1 is released from H^+-ATPase.[84] IF_1 can be detected on SDS-PAGE using coelectrophoresis of the isolated H^+-ATPase and IF_1.[85] IF_1 is cross-reactive; for example, IF_1 of bovine heart is active in yeasts and liver,[86,87] and yeast or candida IF_1 in bovine heart.[86-88] In bacteria and chloroplast a similar role might be played by subunit ϵ.[80,89]

B. Subunits of F_0

As with F_0 subunits, we understand all subunits of the H^+-ATPase complex that remain associated with the membrane after the removal of F_1. Hence, F_0 is isolated by removing F_1 from the purified H^+-ATPase (see References 9 and 48).

In MF_0, at least two subunits are involved in the interaction with F_1.[90,92] These include basic proteins of molecular weight 18 to 24K and extremely heat-stable protein of molecular weight 8 to 10K. They are called oligomycin sensitivity conferring protein (OSCP) and F_6, respectively. Both subunits are hydrophilic and are not required for the catalytic activity of F_1. Their association with F_0 is weaker than that of other F_0 subunits and both OSCP and F_6 can be released and purified.[93,94] Under specific conditions of F_1 extraction by chloroform at 0°C and alkaline pH, a protein resembling OSCP by its molecular weight and peptide fragments remains associated with F_1.[92] Also, the bacterial subunit δ, that is homologous with the mitochondrial OSCP with respect to both primary structure and function in H^+-ATPase (see above), is isolated with F_1. It appears that F_6 and OSCP constitute separate links between MF_1 and MF_0.[91,92] However, OSCP increases the affinity of the binding and is essential for expression of the inhibitory effect of oligomycin and DCCD on the catalytic activities of F_1.[90,91] Furthermore, OSCP appears to be required for the gate function of F_1.[91] F_6 is also involved in Pi-ATP exchange reaction.[92]

Molecular weights of the other proteins that are, or appear to be, subunits of the MF_0 are within the range of 4.5 to 22K.[9,52] Of them only the subunit, which binds DCCD, and to a certain extent the subunit called factor B, are well characterized structurally and functionally. Some others can be defined and distinguished by their physicochemical properties, site of their synthesis, or possible functional role. For example, 21 and 9K proteins are coded for by mitochondrial DNA[32,33,95,96,224,225] and/or are sensitive to trypsin.[91] The remaining "subunits" are known only due to their presence in various H^+-ATPase preparations, which is more or less reproducibly demonstrated. As the composition of F_0 in chloroplasts and bacteria is quite clear, the subunits of the MF_0 will be discussed in the light of the composition of F_0 from bacteria and chloroplasts.

Both mesophilic (*E. coli*)[43] and thermophilic (PS 3)[45] bacteria have F_0 composed of three subunits with molecular weights (SDS-PAGE) of 24, 19, and 8.4 and 19, 13.5, and 5.4K, respectively. Analogous subunits of the CF_0[47] have molecular weights of 15, 12.5, and 8K. These three subunits are designated a, b, and c in bacteria and I, II, and III in chloroplasts. The clearest evidence for their authenticity comes from genetic experiments. In *E. coli*, where the organization of genes for H^+-ATPase is well mapped (see References 23, 46), a λ transducing phage carrying this segment of DNA was constructed. The induction of this gene resulted in the overproduction of H^+-ATPase which, when isolated, exhibited an unaltered stoichiometry ($a_1 b_2 c_{10}$) of subunits with respect to control cells.[97] Other evidence comes from the construction of mutants of *E. coli* which were defective in one or more of F_0 subunits, as determined genetically or immunologically.[19] None of the strains which contained only one type or two types of subunits in any combination exhibited H^+-conductance. Whereas these three subunits of BF_0 are well demonstrated both structurally and functionally, the existence of the additional fourth protein subunit remains unresolved. An open frame capable of coding for 130 amino acids was found before the gene for the largest BF_0 subunit.[46] A protein of a corresponding molecular weight of about 14K is mostly absent in isolated bacterial H^+-ATPase, in spite of the fact that the gene belongs most likely to the H^+-ATPase, as the promotor was found in the proceeding region of the gene. The "missing" 14K protein might function as an organizer of the H^+-ATPase assembly into the membrane without being itself incorporated.

The presently available information about physicochemical properties of individual F_0 subunits in prokaryocytes is rather extensive. Their primary structure[19,46] was determined either indirectly from DNA sequence analysis or directly by sequencing the isolated protein. The two approaches, when used for subunit c, yielded results that were essentially in perfect agreement. In case of the subunits a and b, the N-terminal sequences were determined recently using isolated subunits.[19]

The subunit a, the largest BF_0 subunit, is highly hydrophobic (seven hydrophobic sequences).[19,46] Subunit of MF_0 with molecular weight close to 21K which is synthesized on mitochondrial ribosomes[32,33,95,96] exerts partial sequence homology with bacterial subunit a.[19,98] In mitochondria this subunit appears to be required for oligomycin sensitivity.[98] In yeasts, there is good evidence that panthothenic acid might be attached to this subunit.[99]

The medium size BF_0 subunit (b) in *E. coli* is mostly hydrophilic.[19,46] In contrast to both subunits a and c, subunit b is fully accessible to various proteases.[19] A partial cleavage of the subunit abolished the H^+-translocation but did not affect H^+-conductivity and F_1 binding. Since only relatively a small part of the subunit was cleaved, this fact is not in contradiction with the proposed role of the subunit b in binding of BF_1. Upon purification the b subunit of PS 3 binds BF_1[100] and its requirement for BF_1 binding is further inferred by experiments with mutants.[101] Only the b subunit of FB_0 is labeled with lipophilic, membrane-permeating photolabels, the label being associated with a relatively small N-terminal segment of the subunit.[19] A conclusive evidence for a homologous subunit in mitochondria has not yet been obtained.

A likely subunit of MF_0 is factor B (F_B). It was purified and characterized by Sanadi and co-workers.[102] F_B is an SH-protein (2 Cys) with molecular weight of 14.6K (amino acid composition), which aggregates when purified. It is stained poorly but can be detected with, e.g., $^{115}Cd^{2+}$, a dithiol group reagent, which also blocks the F_B activity. Monofunctional SH-reagents are not effective. Using rabbit antibodies against bovine heart F_B, a cross-reactivity was detected in chloroplasts and bacteria. It has not been determined which of the BF_0 subunits was cross-reacting. The function of F_B is somewhat obscure. It is not needed for oligomycin-sensitive recoupling of MF_1 to MF_0, it does neither affect significantly the ATPase activity, nor does it affect the passive H^+-translocation in submitochondrial particles. However, F_B converts an energy-uncoupled H^+-ATPase to a coupled H^+-ATPase capable of energy transduction, as indicated by stimulation of all ATP-dependent reactions, including Pi-ATP exchange. The preparation of H^+-ATPase enriched in the content of F_B, which is highly active in Pi-ATP exchange reaction, was isolated from bovine heart mitochondria.[30]

A small protein directly involved in H^+-translocation which is found in F_0 of all types of energy-transducing membranes is clearly the best-characterized subunit of H^+-ATPase. This holds true both for its function and particularly for its structure. As this subunit is the site of action of DCCD,[103-105] in mitochondria it is usually called the DCCD-binding protein, whereas in bacteria the protein is designated subunit c[23] and in chloroplasts subunit III.[47]

The DCCD-binding protein is exceptionally hydrophobic, which together with its reactivity to DCCD mainly helped in its characterization. When analyzed on SDS-PAGE it does not stain well[107] but is easily detected using radioactive DCCD.[103,104] Multiple forms observed on SDS-PAGE refer either to multiple aggregations (isolated protein) or functional states (see Section VI.B) of the protein.[9] The isolation is based on solubility of the protein in organic solvents[104,105] which in chloroplasts (butanol) allows for one-step purification.[106] In mitochondria and bacteria the solubilization (chloroform/methanol) is less specific and it has to be followed by a further purification — precipitation with diethyl ether, adsorption,[104,108-110] or thin-layer[111,112] or high-pressure liquid[113] chromatography. Isolated protein is composed from 76 to 81 amino acids in various H^+-ATPases and its molecular weight ranges from 6.5K in bacteria to 8K in chloroplasts. Amino acid composition of all preparations analyzed up to now shows a very low percentage of polar residues (16 to 25%). Tryptophan and histidine are generally absent; cysteine, serine, and lysine are missing in some of the preparations. At present, amino acid sequence of the DCCD-binding protein has been established in nine different organisms.[19,114] Numerous hydrophobic sequences are localized in two segments, each containing 20 to 25 amino acids, which are separated by a short polar chain. Most of the remaining polar residues are accumulated at the N-terminus;

very few are found in the hydrophobic segments. Various prediction methods for the secondary structure revealed β-turn in the central polar loop, and for hydrophobic segments a high tendency for α-helix formation in the lipid bilayer is predicted. The predicted secondary structure is in agreement with CD spectra of the yeast protein embedded in liposomes,[115] chemical modification data,[19,116,117] and genetic evidence concerning the structure of the DCCD-binding domain of the protein in bacteria.[118,119] All of these results indicate a hairpin-like arrangement of the protein in the membrane. While the two hydrophobic α-helical segments would transverse the hydrophobic part of the membrane, the central polar loop would face F_1 and the N and C termini would protrude on the other side of the membrane.

The DCCD-binding protein is the main constituent of H^+-channel through F_0. The purified protein from mitochondria and chloroplasts translocates protons when incorporated into liposomes (see Section VI.B.2). However, regarding the genetic evidence in the native membrane, the other F_0 subunits are likely to be involved in H^+-translocation, also.[19,120,121] The most convincing are the experiments in *E. coli* showing that none of the subunits of F_0 alone (a, b, c) and no combination of two subunits in the membrane lead to the expression of H^+-conductance.[19] There are several copies of the protein per F_0 which create a functional oligomer.[7,9,10,23,47] Their number is higher than of any F_0 subunit. In thermophilic bacteria PS 3, chloroplasts, and mitochondria six copies are likely, while in *E. coli* the estimated number[97] was as high as ten.

At least some of the polar amino acid residues of the protein are essential for H^+-translocation (for review see References 9, 12, and 122). This is best defined in the case of Glu/Asp at position 65 (numbering[107]) localized in the second hydrophobic domain, which represents the only reactive DCCD residue in the molecule.[19,114] Also, the modifications of tyrosine and arginine residues of DCCD-binding proteins of bacteria[123] and chloroplasts[110] were shown to inhibit their protonophoric function. In MF_0 various polar amino acids are involved in H^+-translocation and its regulation (see References 12 and 122). The definitive proof that these amino acids belong to DCCD-binding protein is, however, still missing. The involvement of the arginine present in DCCD-binding protein of mitochondria at position 45 is very likely.[122]

A hydrophobic segment around Glu65 is of particular importance for H^+-transport because it is also the site of action of the other inhibitors — oligomycin[124] and venturicidin.[54] The DCCD-binding protein is also the site of the inhibitory action of etidium bromide,[125] the reactive part of the protein remaining unknown.

As far as the biogenesis is concerned DCCD-binding protein of MF_0 can be of both nuclear[107,126] (fungi) and mitochondrial[96] (yeast) origin. In animal cells DCCD-binding protein is most probably synthesized outside mitochondria.[33,127]

When summarizing the question of authenticity of H^+-ATPase subunits in mitochondria there is a reasonable evidence, although not fully conclusive, that besides six subunits of F_1, other six proteins — DCCD-binding protein, OSCP, F_6, F_B, and 9 and 21K proteins — of mitochondrial origin are actual subunits of the enzyme. The two other proteins often discussed to belong to the enzyme, translocator of adenine nucleotides and uncoupler binding protein, were not shown to be required for the enzyme activity.[52,128] Therefore, of 18 polypeptides which are reproducibly found by SDS-PAGE in several preparations of the mitochondrial H^+-ATPase, still some six components of F_0 remain questionable. As some of them might also be proteolytic fragments of the other subunits, systematic analysis with the aid of antibodies and all available techniques of individual subunits detection would be of great value and might help to solve this particular problem.

V. ULTRASTRUCTURE OF H^+-ATPase

Several aspects of H^+-ATPase ultrastructure are well understood at present. The whole enzyme complex has molecular weight close to 500K and the major part of the enzyme mass is formed by membrane extrinsic protein — hydrophilic catalytic unit F_1 (360 to 380K).

The principal information on the gross structure of H^+-ATPase results from electron microscopy using the negative staining technique.[31,129,130] Relatively large F_1 is visualized in preparation of the inner mitochondrial membranes,[129] isolated H^+-ATPase,[31,130] or isolated F_1,[129,130] as globular spheres — head pieces of a diameter equal to 90 Å. They face the matrix side of the inner membrane. On electron micrographs of isolated H^+-ATPase,[31] head pieces are attached to base pieces which are equivalent to F_0. The thickness of the base piece is around 60 Å, which agrees with the view that F_0 spans the inner membrane. The spheres of F_1 are not in a direct contact with the base pieces — the connecting part forms a thin, longitudinal structure (40 to 50 Å long) appearing as a stalk. It should represent the structural and functional link[132] between F_0 and F_1. The stalk is clearly the smallest of the three parts of enzyme. Thus, the three parts create the image of a knob or mushroom which led to postulation of tripartite structure of H^+-ATPase.[31]

The data of electron microscopy are not fully consistent with other studies concerned with the gross organization of the enzyme. The most discussed is the "stalk" part which implies that F_1 is almost totally exposed to water phase and only a very small part is involved in connection with F_0. The accessibility of the F_1 surface to the hydrophilic modifying agent trinitrobenzosulfonate indicates that a relatively large part of the surface, more than $1/3$, is immersed in the hydrophobic membrane environment.[131] Thus, on cross-section the area of contact should appear much thicker than the stalk.

In mitochondria, at least two F_0 subunits (OSCP and F_6) and one F_1 subunit (δ) are functionally involved in the F_1-F_0 interaction.[70,90-92] The former two subunits are usually believed to form the stalk.[132] However, reconstitutions of the functional complex indicate[91,92] that F_6 and OSCP constitute separate links between F_1 and F_0 and, importantly, F_1 is attached to F_0 in the absence of either OSCP or F_6. While one copy of each OSCP[133] and δ[6-8,134] is present in the enzyme, the stoichiometry of F_6 is unknown.

The quaternary structure of MF_1 is mainly deduced from studies with isolated F_1, which was studied either in soluble form or in crystalline state. The experiments with bifunctional agents — cross-linking studies[60,135-139] — were originally the main source of information. These studies suggest that α subunits are in proximity of β subunits and also of other α subunits. The nucleotide binding site on β subunit (catalytic centrum) was located close to α subunit[60] and is supposed to be shielded[140] by α subunit. Subunit γ was found to be situated within 12 Å with respect to β and ϵ subunits. Subunit δ probably participates in contact[70] between F_1 and F_0 as well as γ subunit which was cross-linked with OSCP.[141] Subunit γ is protected from digestion by trypsin when F_1 is bound to the membrane[142] as well as against chemical modification by water-soluble trinitrobenzosulfonate.[131] Using the methylazidobenzimidate derivate of IF_1 as a photoaffinity labeling reagent, IF_1 binding site was localized on β subunit of MF_1.[143]

To specify further, the F_1 quaternary structure, an independent information which comes recently from computerized image reconstitution of electron micrographs and from X-ray diffraction studies, is particularly promising. The molecules of isolated BF_1 were crystallized for the first time[144] using enzyme from thermophilic bacteria PS 3. By means of computerized image reconstitution pseudohexagonal symmetry with central hollow region was observed[144] in accordance with CD spectra and small-angle neutron or X-ray scattering experiments[145,146] with soluble F_1. Small crystals (115 × 115 × 70 Å) were prepared later from MF_1.[147,148] "Three-dimensional crystallization" of the enzyme (large crystals) which allowed for X-ray analysis at 9 Å resolution[149] has so far been achieved only with MF_1 isolated from rat liver (dimensions 120 × 120 × 80 Å). The molecule seems to be formed by two equivalent halves, each of them being composed from three regions. When assuming that the central part of the molecule is devoid of the protein and that the most likely stoichiometry of F_1

subunits[134,150,151] is $\alpha_3 \beta_3 \gamma \delta \epsilon$, the authors concluded that the structure of F_1 is not perfectly dimeric and that α and β subunits do not possess equivalent positions in the complex. Data in favor of $\alpha_3\beta_3$ arrangement were also obtained with MF_1 using electron microscopic techniques.[153] In contrast to X-ray analysis, symmetric orientation of α and β in cyclohexane-type array was suggested, where the subunits β were oriented towards the central part of the molecule and the three small subunits of F_1 were localized in the center of the molecule.

Therefore, all of the above studies suggest that the main part of F_1 is hexagonal or pseudohexagonal, being formed by three dimers made up from α and β subunits, where the dimers are arranged symmetrically or slightly asymmetrically. α and β subunits do not have equivalent positions in the complex. β subunits of each dimer are supposed to be oriented towards the central hole in the molecule. Three small subunits, γ, δ, and ϵ, each in one copy per molecule, are thus probably situated close to the membrane-bound part of the enzyme, whereas IF_1 is attached to one of three β subunits.

It is still too early to draw any definitive conclusion concerning the quaternary structure of F_0 because only very few data are available at present. This is certainly due to a more difficult isolation of F_0 with respect to that of F_1 and more importantly due to technical problems with the analysis of hydrophobic intrinsic membrane proteins. Moreover, F_0 has apparently changed in the structure during phylogenesis from a relatively simple three-type subunit system in bacteria to a much more complex system in mitochondria. Thus, the information from one type of energy-transducing membrane can be applied to the other one only with reservations.

In all types of energy-transducing membranes, a relatively large part of F_0 (about 50% of its mass) is probably formed by an oligomer consisting of six to ten copies of the DCCD-binding protein (see Section IV.B). The structural arrangement of the oligomer is not known at present, but nonequivalence of the oligomer subunits with respect to interaction with spin-labeled analog of DCCD, NCCD,[154] and inhibition of H^+-translocation[9,41,155] might reflect the oligomer arrangement (see Section VI). It is very likely that the oligomer spans a substantial part of the membrane thickness in such a way that polypeptide chains are folded like hairpins with the central loops facing F_1 (see Section IV.B). In mitochondria, physical contacts among the individual polypeptide chains of the oligomer were demonstrated by cross-linking.[137]

In BF_0, the arrangement of the other two types of subunits, a and b, was demonstrated[19] in *E. coli*. The strongly hydrophobic subunit a might be totally buried in the membrane while two amphiphilic subunits b appear to be in contact with F_1. Regarding the cross-linking data, the subunit a is localized closely to subunit(s) c (DCCD-binding subunit) and two subunits b might form a dimer. With respect to stoichiometry[97] $a_1b_2c_{10}$ in BF_0, more numerous contacts between subunits are to be expected, e.g., between a and c. Scarcity of the studies as well as limited choice of hydrophobic cross-linkers is a possible reason for the fact that they have not been noted.

Subunit a of BF_0 probably permeates the membrane[19] which could also hold for a partly homologous 21K subunit of MF_0 synthesized in mitochondria. Interestingly, in the mitochondrial H^+-ATPase isolated from yeasts,[137] the DCCD-binding protein was cross-linked to subunit 6 and to subunit 4 (γ subunit of F_1).

With respect to the contact of F_0 subunits with phospholipids, significant labeling of b, marginal of c, and no labeling of a was observed in H^+-ATPase of *E. coli* using photoreactive lipid analogs.[52] In parallel experiments with mitochondrial H^+-ATPase numerous polypeptides of molecular weights 4.5 to 25K were labeled, all of them being presumably components of F_0.

Therefore, it is clear that approaches used to assess subunit interaction can be successfully applied to F_0 and thus more detailed and accurate information is to be expected in the near future. Also, the use of physical techniques so far applied only to F_1 is promising here as the crystallization of hydrophobic membrane proteins can be performed.[156]

VI. FUNCTION OF H^+-ATPase

Given that ATP synthesis is coupled to respiration through cyclic flow of protons between redox enzymes and H^+-ATPase,[13,15] the most crucial experiments to show the function of H^+-ATPase were to measure the net synthesis of ATP driven by proton gradient in liposomes with reconstituted mitochondrial H^+-ATPase and cytochrome oxidase[56] or bacteriorhodopsin.[57] Similar experiments were repeated later with very pure preparation of bacterial H^+-ATPase.[157] With isolated bacterial enzyme incorporated in liposomes,[158] submitochondrial particles,[159,160] and isolated rat liver mitochondria,[160] the net synthesis of ATP driven by external electric field was also demonstrated.

Various recent studies indicate localized energy transfer between the two major components of oxidative phosphorylation, redox enzymes of respiratory chain, and H^+-ATPase (see References 161 and 162). It is, therefore, discussed whether protons translocated by redox enzymes are utilized for synthesis of ATP before being equilibrated into bulk water phase, or whether localized electric field, in addition to bulk $\Delta\mu H^+$, might be involved in energy coupling.

It was suggested recently that anionic head groups of the membrane phospholipids have the capacity to bind and conduct protons along the surface.[162] In this context it should be stressed that inner mitochondrial membrane is very rich in anionic cardiolipin and that with isolated H^+-ATPase specific stimulatory effect of this phospholipid on oligomycin sensitivity has been demonstrated.[163]

The concept of coupling through local electric field predicts that the ratio between H^+-ATPase and respiratory chain components in the membrane must influence the threshold $\Delta\mu H^+$ value (bulk phase) of phosphorylation.[161] In most types of mitochondria (e.g., bovine heart) the above ratio is close to 1 while it is around 0.1 in brown adipose tissue mitochondria.[164-166] Hence, to test the above prediction the comparison of these two types of mitochondria was suggested.[161]

The molecular mechanism by which protons translocated via F_0 drive ATP synthesis is still a matter of discussion.[167-169] The evaluation of the exact number of protons translocated via F_0 per ATP synthesized (H^+/ATP) is very important in this respect. Only the H^+/ATP ratio of 2 is consistent with the "direct mechanism" of coupling between H^+-translocation and ATP synthesis.[168] The recent measurement of H^+/ATP ratio in mitochondria indicates, however, the value of 3 rather than 2 which was estimated in previous studies (see References 7, and 14). The former value is in favor of the "indirect mechanism"[167] of coupling (see below). Nevertheless, it should be kept in mind that the measurements of exact H^+/ATP value are still accompanied by technical problems[170] and thus definitive conclusion is difficult to draw.

The "indirect model" of coupling, where protons are supposed to drive the phosphorylation of ATP through conformational changes of F_1 subunits, was favored by most workers recently.[16] Conformational changes should cause changes in reactant binding, and the catalytic properties of F_1 are explained by the occurrence of both negative cooperativity of binding and positive cooperativity of multiple catalytic sites on F_1 molecule. In the most recent version of alternating site cooperativity model Boyer and co-workers[167] extended the originally suggested two-site mechanism to the three-site mechanism of catalysis. The latter model is thus more consistent with the presence of three catalytic domains[150,151,171] on F_1 molecule, which are presumably formed by three pairs of α and β subunits.[59,60]

As pointed out by Post (in discussion to Reference 167), the three domains which are supposed to bind differently ligands involved in ATP synthesis might be linked to conformational changes of F_1 molecule in a way that they would functionally rotate similarly as a stator of motor of Wankel type.

During ATP-hydrolysis there is a direct in-line transfer of phosphoric residue between

ATP and water which indicates that there is no phosphoenzyme intermediate.[172] The catalysis of P_i-ATP exchange reaction by isolated F_1 supports the view that ATP synthesis is simply the reverse of ATP hydrolysis.[173,174] It is likely, however, that the enzyme is so constructed that under appropriate conditions it could be effective as a one-way catalyst.[174] It was postulated[175] some time ago that ATPase activity of F_1 can differ due to conformational change. In a recent study direct evidence that F_1 is a "two-state" enzyme was obtained using nucleotide analogs.[176] Under physiological conditions H^+-ATPase hydrolytic activity is greatly suppressed by the regulatory subunit (IF_1; see Section IV.A).

It can be concluded that H^+-translocation represents the primary event in H^+-ATPase function. Therefore, in the following part of the review we shall focus on the molecular mechanism of H^+-transport through H^+-ATPase and on the overall mechanism of coupling between H^+-transport and catalytic functions of F_1. The results of the two types of experimental approaches are summarized here: (1) the studies of H^+-translocation coupled to the catalytic activity in H^+-ATPase and (2) the studies of the passive H^+-transport through H^+-ATPase at various degrees of resolution of the enzyme. The experiments on protonophoric activity of isolated DCCD-protein are also discussed.

A. Interaction of Intact H^+-ATPase with Inhibitors of H^+-Translocation

A significant progress in understanding the mechanism of H^+-translocation and overall mechanism of H^+-ATPase function has been reached using agents like DCCD and oligomycin which influence the catalytic activity of H^+-ATPase as a consequence of the inhibition of H^+-translocation through F_0.[9,12,17] It should be recalled that the modes of action of DCCD and oligomycin on F_0 are different.[17] DCCD is known to modify covalently a specific carboxyl group on DCCD-binding protein (see Section IV.B). The formation of the first reaction product, O-acylisourea, might be potentiated[107] by the high local concentration of protons around this carboxyl residue which is essential for H^+-translocation. In fact, local accumulation of protons might be partly responsible for a highly specific effect of DCCD on the protonophoric H^+-ATPase subunit (see below). At higher concentrations DCCD is known to modify a number of other membrane proteins, some of which are also involved in H^+-translocation (see References 177—179). The inhibitory effect of DCCD on F_1 by the covalent modification of β subunit was also noted.[63] This effect is, however, potentiated by acidic pH and requires higher concentrations of the inhibitor. Thus, in membrane-bound H^+-ATPase DCCD reacts preferentially with the F_0-component.[180]

The second-order rate constant of the DCCD-ATPase interaction in mitochondria is inversely related to the concentration of membranes indicating that DCCD reaches the inhibitory site by concentrating in the hydrophobic phospholipid environment.[180] Due to the high rate of DCCD degradation in the aqueous medium, which is further accelerated by the presence of mitochondrial phospholipids,[181] the concentration of the membranes during incubation with DCCD should be kept constant in inhibitory studies.[23,181]

Also, in the case of oligomycin the dependence of the kinetic constant of oligomycin-ATPase interaction on membrane concentration was observed.[182,183] It is likely that both oligomycin and DCCD as well as venturicidin are concentrated in a common hydrophobic domain in the membrane[54,124,184,185] before reaching the inhibitory site. It is probable that oligomycin disturbs the protonation of DCCD-reactive glutamic residue[186] which is essential for proton translocation. The binding sites of all three inhibitors are localized within the same hydrophobic segment of DCCD-binding protein (see Section IV.B). However, the binding sites for oligomycin and DCCD are not identical and their effects are not additive.[187] In contrast to DCCD the effect of oligomycin is reversible[188] and it is influenced by the conformational changes of F_0 subunits,[182,183,187] possibly in relation to the oligomeric state of the DCCD-binding protein.[7] The involvement of SH group in oligomycin sensitivity was suggested.[190] The SH group may be localized on F_0 subunit different from DCCD-binding protein and from OSCP.[191] Trypsin-sensitive site on MF_0 which is localized near the outer side of the inner membrane was also shown to be involved in the oligomycin sensitivity.[192]

The use of radioactive DCCD allows to correlate the inhibitory effect with the inhibitor binding and the titrations with radioactive DCCD thus provided a powerful approach to analyze the mechanism of H^+-translocation (see below). The observation that less than one third of the total amount of DCCD-reactive protein of F_0 in the membrane of *E. coli* was modified by DCCD under conditions which maximally inhibited H^+-ATPase hydrolytic activity[108] suggested, for the first time, that DCCD-binding protein is present in the membrane as an oligomer (see Section IV.B). It was shown later that only one to two subunits of DCCD-binding protein are modified under similar conditions in various types of coupling membranes (see Reference 9). Similarly, 1 mol oligomycin per 1 mol H^+-ATPase was shown to eliminate the H^+-ATPase hydrolytic activity in mitochondria.[187] While the exact titration of the inhibition of synthetic function of mitochondrial H^+-ATPase by radioactive DCCD is still missing, the inhibition of P_i-ATP exchange reaction in phosphorylating sub-mitochondrial particles was parallel with the inhibition of H^+-ATPase hydrolytic activity.[193] This is in contrast to a finding with the isolated enzyme where P_i-ATP exchange reaction was several times more sensitive.[27,54] The reason for this discrepancy is not clear. It is evident, however, that solubilization of the enzyme and replacement of the naturally occurring phospholipids with a detergent can change the properties of the enzyme substantially and can also influence the sensitivity to inhibitors.[194]

The view that 1 mol DCCD per mole H^+-ATPase eliminates enzyme activity was primarily based on, but not fully proved by, activity-binding data. As the amount of radioactive DCCD bound to specific (see below) inhibitory site was substoichimetric[41,180] (0.5 to 0.6 mol/mol enzyme), the true number of reacting molecules was most probably underestimated due to the condensing reaction of the DCCD-activated carboxyl group. Therefore, the DCCD reacting sites were assessed by an independent approach — kinetics of DCCD-ATPase interaction[54,180,189] which were calculated from inhibitory data of both ATP synthesis and hydrolysis. These experiments finally and clearly proved that 1 mol of DCCD per mole H^+-ATPase assembled in the mitochondrial membrane eliminates enzyme activity while attacking the inhibitory site without cooperativity.[180]

In mammalian mitochondria the inhibition of H^+-ATPase activity by radioactive DCCD is associated with predominant labeling of three membrane components (33, 16, and 8K) which are separated only in SDS-PAGE system of Laemmli.[28,33,41,107,184,185,195] While the 33K DCCD-binding protein does not belong to H^+-ATPase, the other two, 16 and 8K proteins, represent probably the dimeric and monomeric forms[28,33,184] of DCCD-binding protein of MF_0. Only the labeling of 8K component by radioactive DCCD correlates with the inhibition of H^+-ATPase hydrolytic activity, and saturation of this binding capacity coincides with the full inhibition.[41,84] The binding capacity of the 16K form is two to three times higher and labeling does not correlate with the inhibition of ATP hydrolysis. It was, therefore, suggested that the appearance of the two forms of DCCD-binding protein of F_0 on polyacrylamide gels reflects the asymmetrical structural arrangement of DCCD-binding oligomer in the native membrane.[41] Only those DCCD-binding proteins which are more accessible to DCCD and which are involved in the inhibition of ATP hydrolysis are easily dissociated by SDS as single polypeptide chains. Both 8 and 16K forms can be detected in chloroform-methanol extract of DCCD-labeled bovine heart mitochondria[41,184] as well as in isolated mammalian H^+-ATPase in the presence and absence of DCCD.[184] Thus, their existence is clearly not due to derivatization of the protein by DCCD. In fact, only a very small conformational change of the yeast DCCD-binding protein induced by the interaction with DCCD was detected by CD spectra.[115] It was also shown that solubilization by SDS does not markedly influence the protein conformation.[115] The experiments with spin-label analog of DCCD, NCCD, in chloroplasts were also in favor of structural asymmetry of

DCCD-binding oligomer.[154] It was shown that at least some monomers are localized at a maximal distance 15 to 20 Å from each other and that when only one third of DCCD-binding protein present was blocked by DCCD, the spin-spin interaction was abolished.

In summary, the full effect of DCCD (and oligomycin) on the catalytic activity of membrane-bound H^+-ATPase is caused by 1 mol inhibitor bound per 1 mol enzyme. The higher affinity towards the inhibitor of only one subunit of the oligomer is proposed to be an inherent property of the inhibitor binding site. These results have a direct impact to the elucidation of the process of H^+-translocation via F_0 and are helpful in clarifying some aspects of the mechanism of coupling between H^+-transport through F_0 and the catalytic turnover of membrane-bound F_1.

Concerning the H^+/ATP ratio higher than 1, several protons required for synthesis (hydrolysis) of 1 mol ATP might be either transported subsequently through one proton channel or in parallel through several (equivalent or nonequivalent) channels within each F_0. In the latter case the inhibition of H^+-transport through only one of the several channels would still eliminate the activity of the enzyme.

B. Passive H^+-Translocation via F_0

In intact H^+-ATPase, transport of protons through F_0 is under physiological conditions always coupled to ATP synthesis or hydrolysis. When F_1 is released from F_0 the protonophoric function of F_0 becomes fully expressed and the effective proton conductance of the membrane[196] (C_MH^+) in which F_0 is embedded substantially increases. The increased C_MH^+ can be lowered back by oligomycin, DCCD or F_1 indicating which part of C_MH^+ measured is related to the protonophoric function of F_0.

1. Techniques for Measurement

There are two ways how to measure C_MH^+ (and passive H^+-transport via F_0):

1. By estimating the magnitude of $\Delta\mu H^+$ at steady-state rate of respiration (mitochondria; submitochondrial particles; reconstituted system with respiratory proton pump; see Reference 14). Using this approach it was deduced that C_MH^+ in mitochondrial and chloroplast membrane is variable and that it increases above a certain threshold value of $\Delta\mu H^+$.[197-199] The error inherent to this method is the uncertainty concerning the exact value of H^+/e^- stoichiometry in respiratory chain. Qualitatively, the changes of C_MH^+ can be followed through the rate of steady-state respiration, e.g., the rate of NADH-oxidase activity.[187,193] In fact, these parameters and the reciprocal value of the half time of passive proton backflow correlate very well in oxygen pulse experiments.

2. The magnitude of C_MH^+ and, very importantly, the mechanism of passive H^+-translocation can be studied by analyzing the kinetics of the process.[12,23] In these experiments transmembrane Δ pH or $\Delta\Psi$ can be used as a driving force for the passive proton translocation. Both valinomycin and potassium ions are usually present — in the former case to prevent the formation of any $\Delta\Psi$ (which would inhibit the proton movement) while in the latter case to create the driving force of the process.

When Δ pH is the driving force it can be established by the addition of acid (or base) to the suspension of vesicles and the kinetics of the equilibration of H^+-gradient can be followed. The inverse value of the half time was shown to be proportional to C_MH^+.[196] Alternatively, Δ pH can be established by adding small amounts of oxygen to the anaerobic suspension of mitochondrial or submitochondrial particles and the anaerobic relaxation of the proton gradient can be recorded.[12] The onset of the process is not disturbed by problems with additions and mixing because anaerobiosis is spontaneously reached when oxygen is consumed. Therefore, the kinetics of the anaerobic proton decay can be accurately analyzed (see Section VI.B.3).

When $\Delta\Psi$ is the driving force it is usually generated by potassium diffusion potential in the presence of valinomycin. This method was used with F_0 derived from bacterial,[200,202] chloroplast,[203] and mitochondrial[12,91,204] H^+-ATPase when reconstituted in liposomes. Very recently this approach was also used to induce H^+-transport through F_0 in bovine heart submitochondrial particles.[12]

In order to follow the small change of H^+ concentration outside the vesicles glass pH electrode is usually used,[205] while spectrophotometric and fluorometric methods[206,207] with acid-base indicators can be applied to record the pH changes of the internal aqueous phase. The direct electrometrical measurement[205] using glass electrode can be used in continuously stirred suspensions with resolutions of 0.001 pH and overall response time (10 to 90% change) less than 0.5 sec. The electrometrical measurement is, therefore, quite suitable for the analysis of the kinetics of H^+-translocation through F_0. Spectrophotometric and fluorometric measurements might be faster than the electrometric recording, but their disadvantage is the nonlinear response of the probes used. The main difficulty is usually the binding of the probe to the membrane and uncertainty as to the distribution of the dye between the external and internal aqueous phase. Up to now only fluoresceine isothiocyanate-labeled dextran[193] trapped inside vesicles (submitochondrial particles) offers the possibility to record fluorometrically very fast pH changes inside the vesicles without any uncertainties concerning the probe binding to the membrane and its redistribution between internal and external compartments.

2. Reconstituted Systems

In BF_0,[200-202] CF_0,[203] and MF_0[12,91,204] reconstituted to liposomes, the existence of H^+-channel was definitely demonstrated. In BF_0 it was shown that H^+ rather than OH^- was the ion translocated.[200] Translocation of protons was inhibited by F_1 binding and it was also sensitive to DCCD but not to oligomycin. H^+-conductivity of BF_0 was blocked when only one third of DCCD-binding subunits was derivatized[123] by DCCD or tetranitromethan. The kinetics of H^+-translocation via BF_0 (after imposing $\Delta\Psi$) fit the kinetics predicted by the Goldman-flux equation,[201] indicating a single type of H^+-translocation via BF_0 under the experimental conditions used. Liposomes with MF_0[12,91,204] appeared to be less stable than those with BF_0.[200-202] The passive H^+-translocation in liposomes with MF_0 was blocked by both DCCD and oligomycin. H^+-transport in reconstituted systems with BF_0 and MF_0 thus shows the same specificity for inhibitors as H^+-ATPase in the original intact membranes.

In various preparations of DCCD-binding protein from mitochondria[115,208] and chloroplasts,[110,209-211] the stimulation of C_MH^+ in either liposomes or lipid-impregnated Millipore® filters was demonstrated. Increased C_MH^+ was specifically inhibited by DCCD and/or oligomycin. Recently, yeast DCCD-binding protein was also incorporated into artificial lipid bilayers[212] and single channel conductance (at pH 2.0), highly specific for protons, was recorded. The H^+-channel formation was related to dynamic aggregation of the DCCD-binding protein indicating that only oligomeric forms of DCCD-binding protein are capable of proton transport. These experiments substantiate the view that the DCCD-binding protein represents the main constituent of H^+-channel in F_0. In the native membrane, however, also other F_0 subunits might be involved in channel formation and regulation of its activity (see Section IV.B). This might be reflected by the fact that up to now a successful reconstitution of H^+-channel from isolated subunits of bacterial F_0 has not been reported.

3. Submitochondrial Particles

Measurements of H^+-translocation through F_0 embedded in the native membrane[12,213] (F_1-depleted submitochondrial particles) exclude possible artifacts related to the solubilization of H^+-ATPase, e.g., disturbance of protein-lipid and protein-protein interactions important for function of H^+-channel and for its regulation. In this system it is also possible to compare the sensitivity of passive H^+-translocation to specific inhibitors (DCCD, oligomycin) with that of intact H^+-ATPase. The approach is the use of submitochondrial particles which differ in degree of F_1 removal.

In oxygen pulse experiments it was shown that in partially F_1-depleted EDTA-submitochondrial particles (ESMP) passive H^+-transport during anaerobiosis follows biphasic kinetics.[12,213] In fully F_1-depleted urea-submitochondrial particles (USMP) the rate constant of monophasic anaerobic H^+-transport is similar to that of the slow phase in ESMP.[12,213] Low concentrations of oligomycin, DCCD, or binding of F_1 increase the aerobic $\Delta\mu H^+$ and concomitantly all these factors induce the fast phase of H^+-translocation. These observations suggest[213] that high transmembrane Δ pH enhances H^+-conductivity of F_0 in accordance with the experiment where the extent of aerobic $\Delta\mu H^+$ was decreased by progressive inhibition of succinate respiration by malonate.[155] As a result the fast phase of H^+-translocation was specifically depressed while the slow phase remained unaffected.[155]

At high concentrations of DCCD or oligomycin, both phases of H^+-translocation in ESMP as well as in USMP are inhibited.[155] As it emerges from titrations of the fast and the slow phase of H^+-translocation by radioactive DCCD,[155] maximal depression of the slow phase occurs around 1.0 mol DCCD bound per mole of F_0.[155] The depression is related to the inhibition of ATP hydrolysis and to the binding of DCCD to the 8K form of DCCD-binding protein. The fast phase is significantly less sensitive to DCCD (maximal inhibition at approximately 2 mol DCCD bound per mole F_0) which is consistent with the overall sensitivity of $C_M H^+$ to DCCD in submitochondrial particles.[187,193] The inhibition of the fast phase of H^+-translocation coincides with the DCCD binding to the 16K form of DCCD-binding protein.[155]

In contrast to the effect of DCCD, maximal inhibition of $C_M H^+$ in F_1-depleted submitochondrial particles by oligomycin was observed at 1 mol of oligomycin added per mole of F_0,[187] which is consistent with the sensitivity of hydrolytic activity of H^+-ATPase to oligomycin.[187]

C. Overall Mechanism of H^+-Translocation and the Two Modes of Enzyme Catalysis

Neither the molecular mechanism of H^+-translocation through F_0, nor its coupling to the catalytic function of F_1 can be well defined at present. However, recent experimental data, some of which were discussed in preceding paragraphs, allow for at least a partial characterization of the process of H^+-transport with respect to enzyme function (Table 1).

During ATP hydrolysis MF_0 seems to translocate protons as a single channel and translocation is fully inhibited by 1 mol of either DCCD or oligomycin per mole H^+-ATPase. According to this assumption, several protons translocated per each molecule of ATP hydrolyzed would enter the channel subsequently. H^+-conductivity state of MF_0, which is related to the hydrolytic mode of action of H^+-ATPase, seems to be preserved, also, in F_1-depleted submitochondrial particles under conditions when passive H^+-translocation is driven by $\Delta\mu H^+$ of low magnitude (see Section VI.B.3), or solely by its $\Delta\Psi$ component.[12] Also, with BF_0 in liposomes the kinetics of passive H^+-translocation driven by $\Delta\Psi$ and the kinetics of its inhibition by DCCD are consistent with one type of H^+-channel.[23,157,214]

The way the H^+-channel operates during ATP synthesis, i.e., under conditions when high value of $\Delta\mu H^+$ exists, may be different. When passive H^+-translocation was driven in submitochondrial particles by transmembrane Δ pH of large magnitude, the high conductivity state of MF_0 was observed (see Section VI.B.3). It is possible that such a high conductivity state of F_0 might reflect the synthetic mode of H^+-ATPase function. Interestingly, $\Delta\mu H^+$-dependent changes of H^+-conductivity of CF_0 were also described.[215,216]

The high conductivity state of F_0 may represent a new conformation of a single channel in F_0 as well as it could reflect unmasking of an additional H^+-channel. If the former possibility were true the H^+-channel in high conductivity state would become less sensitive

Table 1
STOICHIOMETRY OF DCCD AND OLIGOMYCIN INTERACTION WITH MITOCHONDRIAL H^+-ATPase

		Maximal effect (mol inhibitor per mol H^+-ATPase)	
Reaction	Preparation	DCCD (bound)	Oligomycin (added)
Catalytic activity			
ATP hydrolysis	Mitochondria, submito-chondrial particles	1 (155, 180[a], 187, 193, 223)	1[b] (182, 187)
	Isolated H^+-ATPase	>1 (54)	
ATP synthesis (state-3 respiration)	Mitochondria	1 (180[a])	1 (182)
P_i-ATP exchange	Mitochondria, submito-chondrial particles	<1 (54)	1 (182)
	Isolated H^+-ATPase		
Passive H^+-translocation driven by Δ pH			
Overall sensitivity	Submitochondrial particles	2—3 (155, 187, 193)	1 (187, 213)
	DCCD-binding protein in liposomes		1[c] (7, 208)
Low conducitivity state	Submitochondrial particles	1 (155)	≤1[d]
High conductivity state	Submitochondrial particles	2—3 (155)	≤1[d]

[a] Based on kinetics.
[b] Stoichiometry for rat liver mitochondria[182] is based on the assumption that the content of H^+-ATPase is about two times lower in rat liver in comparison with bovine heart mitochondria.[166,180]
[c] Related to the amount of DCCD-binding protein incorporated to the membrane.
[d] Based on the sensitivity of overall H^+-conductivity to oligomycin[187,213] and on the inhibition of both fast and slow phase of H^+-translocation.[213]

to DCCD while the sensitivity to oligomycin would remain unchanged. If two H^+-channels existed in F_0 at high values of transmembrane Δ pH, the overall sensitivity of C_MH^+ to oligomycin and DCCD would indicate that oligomycin rather than DCCD induced the cooperativity between the two channels, resulting in the full inhibition of H^+-transport at 1 mol inhibitor per mole F_0 (see Section VI.B.3).

The existence of transition from the low to the high conductivity state of F_0 suggests that conformational changes of F_0 subunits occur in H^+-ATPase, which is also in accordance with the kinetics of oligomycin-induced inhibition of H^+-ATPase activity,[184,187,193] the labeling of CF_0 subunits in chloroplasts with membrane nonpenetrating agents,[217-219] and with the temperature dependence[200] of H^+-translocation via BF_0.

Conformational changes of F_1 are likely to be directly involved in the catalytic mechanism (see Section VI) and the effect[213,220] of various ligands of F_1 on passive H^+-transport through F_0, when F_1-F_0 interaction is partially disturbed, can be explained by conformational changes induced in F_0 by F_1. It would not be surprising if the mechanism of coupling between F_1 and F_0 was mediated through both H^+-translocation and direct interaction between subunits of F_1 and F_0.

With respect to some other recent data, it is noteworthy that: (1) on the basis of experiments with binding of trypsin-treated F_1 to mitochondrial membrane, it was suggested by Pedersen[221] that conformational coupling between F_1 and F_0, or two H^+-channels F_0 exist in mitochondrial H^+-ATPase; and (2) in experiments with chloroplast H^+-ATPase in liposomes the dependence of the steady-state kinetics parameters of ATP synthesis (as well as hydrolysis) on Δ pH gradient was consistent with the mechanism where transport of three protons is sequential rather than concerted.[222]

In summary, it is tempting to suggest that during both ATP synthesis and ATP hydrolysis protons are translocated subsequently through the F_0 moiety of H^+-ATPase, and the H^+-conductivity is regulated directly at the level of F_0 by the magnitude of transmembrane Δ pH. The dependence of H^+-conductivity of Δ pH value may be, in fact, one of the mechanisms which favor the ATP-synthetic mode at the expense of ATP hydrolysis in mitochondrial H^+-ATPase.

ABBREVIATIONS

H^+-ATPase, H^+-translocating adenosine triphosphatase; F_1, extrinsic catalytic moiety of H^+-ATPase; F_0, membrane-bound moiety of H^+-ATPase; $\Delta\Psi$, membrane potential; Δ pH, difference in pH across membrane; $\Delta\mu H^+$, electrochemical potential difference of H^+ across the membrane ($\Delta\mu H^+ = \Delta\Psi - Z \Delta$ pH; $Z = 2.3$ RT/F $= 59$ at $25°C$ when expressed in millivolts); $C_M H^+$, effective proton conductance of membrane; DCCD, dicyclohexylcarbodiimide; SDS-PAGE, sodium dodecylsulfate polyacrylamide gel electrophoresis.

REFERENCES

1. **Capaldi, R. A.**, Structure of cytochrome c oxidase, in *Membrane Proteins in Energy Transduction*, Capaldi, R. A., Ed., Marcel Dekker, New York, 1979, 201.
2. **Ludwig, B. and Schatz, G.**, A two-subunit cytochrome c oxidase (cytochrome aa₃) from *Paracoccus denitrificans, Proc. Natl. Acad. Sci. U.S.A.*, 77, 196, 1980.
3. **Pedersen, P. L.**, Mitochondrial adenosine triphosphatase, *Bioenergetics*, 6, 243, 1975.
4. **Kozlov, I. A. and Skulachev, V. P.**, H^+-adenosine triphosphatase and membrane energy coupling, *Biochim. Biophys. Acta*, 463, 29, 1977.
5. **Penefsky, H. S.**, Mitochondrial ATPase, *Adv. Enzymol.*, 49, 223, 1979.
6. **Senior, A. E.**, The mitochondrial ATPase, in *Membrane Proteins in Energy Transduction*, Capaldi, R. A., Ed., Marcel Dekker, New York, 1979, 233.
7. **Criddle, R. A., Johnston, R. F., and Stack, R. J.**, Mitochondrial ATPases, *Curr. Top. Bioenerg.*, 89, 1979.
8. **Cross, R. L.**, The mechanism and regulation of ATP synthesis by F_1-ATPases, *Ann. Rev. Biochem.*, 50, 681, 1981.
9. **Houštěk, J., Kopecký, J., Svoboda, P., and Drahota, Z.**, Structure and function of the membrane-integral components of the mitochondrial H^+-ATPase, *J. Bioenerg. Biomembr.* 14, 1, 1982.
10. **Kagawa, Y.**, Structure and function of H^+-ATPase, in *Transport and Bioenergetics in Biomembranes*, Sato, R. and Kagawa, Y., Eds., Japan Scientific Society Press, 1982, 37.
11. **Tzagoloff, A.**, The mitochondrial adenosine triphosphatase, in *Mitochondria*, Siekavitz, P., Ed., Plenum Press, New York, 1982, 157.
12. **Papa, S., Guerrieri, F., Zanotti, F., and Scarfo, R.**, Flow and interactions of protons in the H^+-ATPase of mitochondria, in *Biological Membranes. Information and Energy Transduction in Biological Membranes*, Alan R. Liss, New York, in press.
13. **Mitchell, P.**, Coupling of phosphorylation to electron and hydrogen transfer by a chemi-osmotic type of mechanism, *Nature (London)*, 191, 144, 1961.
14. **Fillingame, R. H.**, The proton-translocating pumps of oxidative phosphorylation, *Ann. Rev. Biochem.*, 49, 1079, 1980.

15. **Boyer, P. D., Chance, B., Ernster, L., Mitchell, P., Racker, E., and Slater, E. C.,** Oxidative phosphorylation and photophosphorylation, *Ann. Rev. Biochem.,* 46, 955, 1977.

16. **Pedersen, P. L.,** H^+-ATPase in biological systems: an overview of their function, structure, mechanism, and regulatory properties, *Ann. N.Y. Acad. Sci.,* 402, 1, 1982.

17. **Linnett, P. E. and Beechey, R. B.,** Inhibitors of the ATP synthetase system, *Methods Enzymol.,* 55, 472, 1979.

18. **Simons, T. J. B.,** Vanadate — a new tool for biologists, *Nature (London),* 281, 337, 1979.

19. **Sebald, W., Friedl, P., Schairer, H. U., and Hoppe, J.,** Structure and genetics of the H^+-conducting F_0 portion of the ATPase, *Ann. N.Y. Acad. Sci.,* 402, 28, 1982.

20. **Walker, J. E., Runswick, M. J., and Saraste, M.,** Subunit equivalence in *Escherichia coli* and bovine heart mitochondrial F_1F_0 ATPase, *FEBS Lett.,* 146, 393, 1982.

21. **Kagawa, Y.,** Reconstitution of energy transformer, gate and channel subunit reassembly, crystalline ATPase and ATP synthesis, *Biochim. Biophys. Acta,* 505, 45, 1978.

22. **Racker, E. and Conover, T. E.,** Multiple coupling factors in oxidative phosphorylation, *Fed. Proc., Fed, Am. Soc. Exp. Biol.,* 22, 1088, 1963.

23. **Fillingame, R. H.,** Biochemistry and genetics of bacterial H^+-translocating ATPases, *Curr. Top. Bioenerg.,* 11, 35, 1981.

24. **Bossard, M. J. and Schuster, S. M.,** Catalysis of partial reactions of ATP synthesis by beef heart mitochondrial adenosine triphosphatase, *J. Biol. Chem.,* 256, 1518, 1981.

25. **Kagawa, Y. and Racker, E.,** Partial resolution of the enzyme catalyzing oxidative phosphorylation. IX. Reconstruction of oligomycin-sensitive adenosine triphosphatase, *J. Biol. Chem.,* 241, 2467, 1966.

26. **Swanljung, P., Frigeri, L., Ohlson, K., and Ernster, E.,** Studies on the activation of purified mitochondrial ATPase by phospholipids, *Biochim. Biophys. Acta,* 305, 519, 1973.

27. **Serrano, R., Kanner, B. I., and Racker, E.,** Purification and properties of the proton-translocating adenosine triphosphatase complex of bovine heart mitochondria, *J. Biol. Chem.,* 251, 2453, 1976.

28. **Berden, J. A. and Voorn-Brouwer, M. M.,** Studies on the ATPase complex from beef-heart mitochondria. I. Isolation and characterization of an oligomycin-sensitive and oligomycin-insensitive ATPase complex from beef heart mitochondria, *Biochim. Biophys. Acta,* 501, 424, 1978.

29. **Galante, Y. M., Wong, S. Y., and Hatefi, Y.,** Composition of complex V of the mitochondrial oxidative phosphorylation system, *J. Biol. Chem.,* 254, 12372, 1979.

30. **Hughes, J., Joshi, S., Torelz, K., and Sanadi, D. R.,** Isolation of a highly active H^+-ATPase from beef heart mitochondria, *J. Bioenerg. Biomembr.,* 14, 87, 1982.

31. **Soper, J. W., Decker, G. L., and Pedersen, P. L.,** Mitochondrial ATPase complex. A dispersed, cytochrome-deficient, oligomycin-sensitive preparation from rat liver containing molecules with tripartite structural arrangement, *J. Biol. Chem.,* 254, 11170, 1979.

32. **De Jong, L., Holtrop, M., and Kroon, A. M.,** The biogenesis of rat liver mitochondrial ATPase subunit composition of the normal ATPase complex and of the deficient complex formed when mitochondrial protein synthesis is blocked, *Biochim. Biophys. Acta,* 548, 48, 1979.

33. **Kužela, Š., Luciaková, K., and Lakota, J.,** Amino acid incorporation by isolated rat liver mitochondria into two protein components of mitochondrial ATPase complex, *FEBS Lett.,* 114, 197, 1980.

34. **Rott, R. and Nelson, N.,** Purification and immunological properties of proton-ATPase complexes from yeast and rat liver mitochondria, *J. Biol. Chem.,* 256, 9224, 1981.

35. **Mc Enery, M. W., Buhle, E. R. Ir., Aebi, U., and Pedersen, P. L.,** Proton ATPase of rat liver mitochondria — preparation of highly ordered, functional complex using the novel zwitterionic detergent 3-[(3-cholamidopropyl)-di-methyl-ammoniol]-1-propane-sulfonate (CHAPS), ASBC Meeting, San Francisco, 1983.

36. **Ryrie, I. J. and Gallagher, A.,** The yeast mitochondrial ATPase complex subunit composition and evidence for a latent protease contaminant, *Biochim. Biophys. Acta,* 545, 11, 1979.

37. **Sebald, W. and Wild, G.,** Mitochondrial ATPase complex from *Neurospora crassa, Methods Enzymol.,* 55, 344, 1979.

38. **Tzagoloff, A.,** Oligomycin-sensitive ATPase of *Saccharomyces cerevisiae, Methods Enzymol.,* 55, 351, 1979.

39. **Ludwig, B., Prochaska, L., and Capaldi, R. A.,** Arrangement of oligomycin-sensitive adenosine triphosphatase in the mitochondrial inner membrane, *Biochemistry,* 19, 1516, 1980.

40. **Robbins, B. A., Wong, S. Y., Hatefi, Y., and Galante, Y. M.,** Studies on the immunological properties of complex V, *Arch. Biochem. Biophys.,* 210, 489, 1981.

41. **Houštěk, J., Svoboda, P., Kopecký, J., Kužela, Š., and Drahota, Z.,** Differentiation of dicyclohexylcarbodiimide reactive sites of the ATPase complex in bovine heart mitochondria, *Biochim. Biophys. Acta,* 634, 331, 1981.

42. **Hatefi, Y., Galante, Y. M., Stiggall, D. L., and Ragan, C. F.,** Proteins, polypeptides, prosthetic groups and enzymic properties of complexes I. II, III, IV, and V of the mitochondrial oxidative phosphorylation system, *Methods Enzymol.,* 56, 577, 1979.

43. **Foster, D. L. and Fillingame, R. H.,** Energy-transducing H⁺-ATPase of *Escherichia coli*. Purification, reconstitution and subunit composition, *J. Biol. Chem.*, 254, 8230, 1979.

44. **Friedl, P., Friedl, C. H., and Schairer, H. U.,** The ATP synthetase of *Escherichia coli* K12: purification of the enzyme and reconstitution of energy-transducing activities, *Eur. J. Biochem.*, 100, 175, 1979.

45. **Yoshida, M., Sone, N., Hirata, H., and Kagawa, Y.,** A highly stable adenosine triphosphatase from a thermophilic bacterium, *J. Biol. Chem.*, 250, 7910, 1975.

46. **Kanazawa, H. and Futai, M.,** Structure and function of H⁺-ATPase: what we have learned from *Escherichia coli* H⁺-ATPase, *Ann. N.Y. Acad. Sci.*, 402, 45, 1982.

47. **Nelson, N.,** Proton-ATPase of chloroplast, *Curr. Top. Bioenerg.*, 2, 1, 1981.

48. **Galante, Y. M., Wong, S.-Y., and Hatefi, Y.,** Resolution and reconstitution of complex V of the mitochondrial oxidative phosphorylation system: properties and composition of the membrane sector, *Arch. Biochem. Biophys.*, 211, 643, 1981.

49. **Weber, K. and Osborn, M.,** The reliability of molecular weight determinations by dodecyl sulfate-polyacrylamide gel electrophoresis, *J. Biol. Chem.*, 244, 4406, 1969.

50. **Laemli, U. L.,** Cleavage of structural proteins during assembly of the head of bacteriophage T4, *Nature (London)*, 227, 680, 1970.

51. **Swank, R. T. and Munkers, K. D.,** Molecular weight analysis of oligopeptides by electrophoresis in polyacrylamide gel with SDS, *Anal. Biochem.*, 39, 462, 1971.

52. **Montecucco, C., Dabbeni-Sala, F., Friedl, P., and Galante, Y. M.,** Membrane topology of ATP synthase from bovine heart mitochondria and *Escherichia coli*, *Eur. J. Biochem.*, 132, 189, 1983.

53. **Stiggall, D. L., Galante, Y. M., and Hatefi, Y.,** Preparation and properties of an ATP-Pi exchange complex (complex V) from bovine heart mitochondria, *J. Biol. Chem.*, 253, 956, 1978.

54. **Kiehl, R. and Hatefi, Y.,** Interaction of [¹⁴C]dicyclohexylcarbodiimide with complex V, *Biochemistry*, 19, 541, 1980.

55. **Ryrie, I. J. and Blackmore, P. F.,** Energy-linked activities in reconstituted yeast adenosine triphosphatase proteoliposomes. ATP formation coupled with electron flow between ascorbate and ferricyanide, *Arch. Biochem. Biophys.*, 176, 127, 1976.

56. **Racker, E. and Kandrach, A.,** Reconstitution of the third site of oxidative phosphorylation, *J. Biol. Chem.*, 246, 7069, 1971.

57. **Racker, E. and Stockenius, W.,** Reconstitution of purple membrane vesicles catalyzing light-driven proton uptake and adenosine triphosphatase formation, *J. Biol. Chem.*, 249, 662, 1974.

58. **Vignais, P. V., Dianoux, A.-C., Klein, G., Lanquin, G. J. M., Lunardi, J., Pougeois, R., and Satre, M.,** Chemical approach to the structure and functioning of the H⁺-linked ATPases. Exploration of binding sites for natural ligands on the F₁-ATPases by photoaffinity labeling, in *Cell Function and Differentiation, Part B*, Alan R. Liss, New York, 1982, 439.

59. **Williams, N. and Coleman, P. S.,** Exploring the adenine nucleotide binding sites on mitochondria F₁-ATPase with a new photoaffinity probe, 3′-0-(4-benzoyl)benzoyl adenosine 5′-triphosphate, *J. Biol. Chem.*, 257, 2834, 1982.

60. **Schäfer, H. J., Mainka, L., Rathgeber, G., and Zimmer, G.,** Photoaffinity cross-linking of oligomycin-sensitive ATPase from beef heart mitochondria by 3′-arylazido-8-azids ATP, *Biochem. Biophys. Res. Commun.*, 111, 732, 1983.

61. **Müller, J. L. M., Rosing, J., and Slater, E. C.,** The binding of aurovertin to isolated F₁ (mitochondrial ATPase), *Biochim. Biophys. Acta*, 462, 422, 1977.

62. **Douglas, M. G., Koh, Y., Dockter, M. E., and Schatz, G.,** Aurovertin binds to the β subunit of yeast mitochondrial ATPase, *J. Biol. Chem.*, 252, 8333, 1977.

63. **Pougeois, R., Satre, M., and Vignais, P. V.,** Characterization of dicyclohexylcarbodiimide binding site on coupling factor 1 of mitochondrial and bacterial membrane-bound ATPases, *FEBS Lett.*, 117, 344, 1980.

64. **Pougeois, R., Languin, G. J.-M., and Vignais, P. V.,** Evidence that 4-azido-2-nitrophenylphosphate binds to the phosphate site on β-subunit of *Escherichia coli* BF₁-ATPase, *FEBS Lett.*, 153, 65, 1983.

65. **Senior, A. E.,** Mitochondrial adenosine triphosphatase. Localization of sulphydryl groups and disulfide bonds in the soluble enzyme from beef heart, *Biochemistry*, 14, 660, 1975.

66. **Satre, M., Klein, G., and Vignais, P. V.,** Structure of beef heart mitochondrial F₁-ATPase. Arrangement of subunits as disclosed by cross-linking reagents and selective labeling by radioactive ligands, *Biochim. Biophys. Acta*, 453, 111, 1976.

67. **Moradi, M., Grande, J., Godinot, C., Gautheron, D. C., and Huppert, J.,** Preparation of monoclonal antibodies reacting with different subunits of mitochondrial F₁-ATPase, *EBEC Rep.*, 1, 73, 1982.

68. **Nelson, N. and Karny, O.,** The role of δ subunit in the coupling activity of chloroplast CF₁, *FEBS Lett.*, 70, 249, 1976.

69. **Younis, H. M., Winget, G. D., and Racker, R.,** Requirement of the δ subunit of chloroplast coupling factor 1 for photophosphorylation, *J. Biol. Chem.*, 252, 1814, 1977.

70. **Gómez-Puyou, M. T., Gómez-Puyou, A., Nordebrand, K., and Ernster, L.,** Interaction of soluble ATPase (F₁) and its subunits with F₁-depleted submitochondrial particles, in *Function and Molecular Aspects of Biomembrane Transport,* Quagliariello, E., Ed., Elsevier/North-Holland, Amsterdam, 1979, 119.

71. **Nelson, N., Kanner, B. I., and Gutrick, O. L.,** Purification and properties of Mg^{2+}, Ca^{2+}-adenosine-triphosphatase from *Escherichia coli, Proc. Natl. Acad. Sci. U.S.A.,* 71, 2720, 1974.

72. **Deters, D. W., Racker, E., Nelson, N., and Nelson, H.,** Partial resolution of the enzymes catalyzing phosphorylation. XV. Approaches to the active site of coupling factor 1, *J. Biol. Chem.,* 250, 1041, 1975.

73. **Salton, M. N. J. and Schor, M. T.,** Subunit structure and properties of adenosine triphosphatase released from *Micrococcus lysodeikticus* membranes, *Biochim. Biophys. Res. Commun.,* 49, 350, 1972.

74. **Kozlov, I. A. and Mikelsaar, H. N.,** On the subunit structure of soluble mitochondrial ATPase, *FEBS Lett.,* 43, 212, 1974.

75. **Vogel, G. and Steinhart, R.,** ATPase of *Escherichia coli:* purification, dissociation, and reconstitution of the active complex from the isolated subunits, *Biochemistry,* 15, 208, 1976.

76. **Futai, M., Kanazawa, H., Takeda, K., and Kagawa, Y.,** Reconstitution of ATPase from the isolated subunits of coupling factor F₁'s of *Escherichia coli* and thermophilic bacterium PS3, *Biochim. Biophys. Res. Commun.,* 96, 227, 1980.

77. **McCarty, R. E.,** Regulation of proton flux through the H^+-ATPase of chloroplasts, *Ann. N. Y. Acad. Sci.,* 402, 84, 1982.

78. **Weiss, M. A. and McCarty, R. E.,** Cross-linking within a subunit of coupling factor 1 increases the proton permeability of spinach chloroplast thylakoids, *J. Biol. Chem.,* 252, 8007, 1977.

79. **Yoshida, M., Okamoto, H., Sone, N., Hirata, H., and Kagawa, Y.,** Reconstitution of thermostable ATPase capable of energy coupling from its purified subunits, *Proc. Natl. Acad. Sci. U.S.A.,* 74, 936, 1977.

80. **Smith, J. B. and Sternweiss, P. C.,** Purification of membrane attachment and inhibitory subunit of the proton translocating adenosine triphosphatase from *Escherichia coli, Biochemistry,* 16, 306, 1977.

81. **Pullmann, M. E. and Monroy, G. C.,** A naturally occuring inhibitor of mitochondrial adenosine triphosphatase, *J. Biol. Chem.,* 238, 3762, 1963.

82. **Frangione, B., Rosenwasser, E., Penefsky, H. S., and Pullmann, M. E.,** Amino acid sequence of the protein inhibitor of mitochondrial adenosine triphosphatase, *Proc. Natl. Acad. Sci. U.S.A.,* 78, 7403, 1981.

83. **Harris, D. A., Tscharner, V. von, and Radda, G. K.,** The ATPase inhibitor protein in oxidative phosphorylation. The rate-limiting factor to phosphorylation in submitochondrial particles, *Biochim. Biophys. Acta,* 548, 72, 1979.

84. **Gómez-Puyou, A., Sánchez-Bustamante, V. J., Darsson, A., and Gómez-Puyou, M. T.,** Interaction of mitochondrial ATPase with its inhibitor protein, *Ann. N. Y. Acad. Sci.,* 402, 164, 1982.

85. **Brooks,. J. C. and Senior, A. E.,** Studies on the mitochondrial oligomycin-insensitive ATPAse. II. The relationship of the specific protein inhibitor to the ATPase, *Arch. Biochem. Biophys.,* 147, 467, 1971.

86. **Chan, S. H. P. and Barbour, R. L.,** Purification and properties of ATPase inhibitor from rat liver mitochondria, *Biochim. Biophys. Acta,* 430, 426, 1976.

87. **Satre, M., de Jesphanian, M. B., Huet, J., and Vignais, P. V.,** ATPase inhibitor from yeast mitochondria purification and properties, *Biochim. Biophys. Acta,* 387, 241, 1975.

88. **Satre, M., Klein, G., and Vignais, P. V.,** ATPase inhibitor from *Candida utilis* mitochondria, *Methods Enzymol.,* 55, 421, 1979.

89. **Nelson, N., Nelson, H., and Racker, E.,** Partial resolution of the enzymes catalyzing phosphorylation. XII. Purification and properties of an inhibitor isolated from chloroplast coupling factor 1, *J. Biol. Chem.,* 247, 7657, 1972.

90. **Vàdinaenu, A., Berden, J. A., and Slater, E. C.,** Proteins required for the binding of mitochondrial ATPase to the mitochondrial inner membrane, *Biochim. Biophys. Acta,* 449, 468, 1976.

91. **Glaser, E., Norling, B. and Ernster, L.,** Reconstitution of mitochondrial oligomycin and DCCD-sensitive ATPase, *Eur. J. Biochem.,* 170, 225, 1980.

92. **Liang, A. M. and Fischer, R. J.,** Subunit interaction in the mitochondrial H^+-translocating ATPase. The role of oligomycin sensitivity conferral protein and coupling factor G in ATPase binding and P_i-ATP exchange in mitochondrial membranes, *J. Biol. Chem.,* 258, 4784, 1983.

93. **Racker, E.,** Preparation of coupling factor 6 (F₆), *Methods Enzymol.,* 55, 398, 1979.

94. **Senior, A. E.,** Oligomycin-sensitivity-conferring protein, *Methods Enzymol.,* 55, 391, 1979.

95. **Sebald, W.,** Biogenesis of mitochondrial ATPase, *Biochim. Biophys. Acta,* 463, 1, 1977.

96. **Trembath, M. K., Monk, B. C., Kellerman, G. M., and Linnane, A. W.,** Biogenesis of mitochondria 40. Phenotypic suppression of a mitochondrial mutation by a nuclear gene in *Saccharomyces cerevisiae, Mol. Gen. Genet.,* 140, 333, 1975.

97. **Foster, D. and Filingame, R. H.,** Stoichiometry of subunits in the H^+-ATPase complex of *Escherichia coli, J. Biol. Chem.,* 257, 2009, 1982.

98. **Macino, G. and Tzagoloff, A.,** Assembly of the mitochondrial membrane system: sequence analysis of a yeast mitochondrial ATPase gene containing the oli-2 and oli-4 loci, *Cell,* 20, 507, 1980.

99. **Griddle, R. S., Packer, L., and Sheih, P.,** Oligomycin-dependent ionophoric protein subunit of mitochondrial adenosinetriphosphatase, *Proc. Natl. Acad. Sci. U.S.A.,* 74, 4306, 1977.

100. **Sone, N., Yoshida, M., Hirata, H., and Kagawa, Y.,** Resolution of the membrane moiety of the H$^+$-ATPase complex into two kinds of subunits, *Proc. Natl. Acad. Sci. U.S.A.,* 75, 4219, 1978.

101. **Downie, J. A., Senior, A. E., Cox, G. B., and Gibson, F.,** A fifth gene (uni E) in operon concerned with oxidative phosphorylation in *Escherichia coli, J. Bacteriol.,* 137, 711, 1979.

102. **Sanadi, D. R.,** Mitochondrial coupling factor B properties and role in ATP synthesis, *Biochim. Biophys. Acta,* 683, 39, 1982.

103. **Knight, I. G., Holloway, C. T., Roberton, A. M., and Beechey, R. B.,** The chemical nature of the site of action of dicyclohexylcarbodi-imide in the mitochondria, *Biochem. J.,* 109, 27P, 1968.

104. **Cattel, K. J., Lindop, C. R., Knight, I. G., and Beechey, R. B.,** The identification of the site of action of N′, N′-dicyclohexylcarbodi-imide as proteolipid in mitochondrial membranes, *Biochem. J.,* 125, 169, 1971.

105. **Stekhoven, F. S., Waitkus, R. F., and van Moerkerk, H. Th. B.,** Identification of the dicyclohexyl-carbodiimide-binding protein in the oligomycin sensitive adenosine triphosphatase from bovine heart mitochondria, *Biochemistry,* 11, 1144, 1972.

106. **Sigrist, H., Sigrist-Nelson, K., and Gitler, C.,** Single-phase butanol extraction: a new tool for proteolipid isolation, *Biochem. Biophys. Res. Commun.,* 74, 178, 1977.

107. **Seblad, W., Graf, Th., and Lukins, H. B.,** The dicyclohexylcarbodiimide-binding protein of the mitochondrial ATPase complex from *Neurospora crassa* and *Saccharomyces cerevisiae, Eur. J. Biochem.,* 93, 587, 1979.

108. **Fillingame, R. H.,** Purification of the carbodiimide-reactive protein component of the ATP energy-transducing system of *Escherichia coli, J. Biol. Chem.,* 251, 6630, 1976.

109. **Graf, Th. and Seblad, W.,** The dicyclohexylcarbodiimide-binding protein of the mitochondrial ATPase complex from beef heart, *FEBS Lett.,* 94, 218, 1978.

110. **Sigrist-Nelson, K. and Azzi, A.,** The proteolipid subunit of the chloroplast adenosine triphosphatase complex. Reconstitution and demonstration or proton-conductive properties, *J. Biol. Chem.,* 255, 10638, 1980.

111. **Sierra, M. F. and Tzagoloff, A.,** Assembly of the mitochondrial membrane system. Purification of a mitochondrial product of the ATP-ase, *Proc. Natl. Acad. Sci. U.S.A.,* 70, 3155, 1973.

112. **Dianoux, A. Ch., Bof, M., and Vignais, P. V.,** The dicyclohexylcarbodiimide-binding protein of rat liver mitochondria as product of the mitochondrial protein synthesis, *Eur. J. Biochem.,* 88, 69, 1978.

113. **Blodin, G. A.,** Resolution of the mitochondrial N, N′ - dicyclohexylcarbodiimide binding proteolipid fraction into three similar sized proteins, *Biochem. Biophys. Res. Commun.,* 87, 1087, 1979.

114. **Sebald, W., Hoppe, J., and Wachter, E.,** Amino acid sequence of the ATPase proteolipid from mitochondria, chloroplasts and bacteria (wild type and mutants), in *Function and Molecular Aspects of Biomembrane Transport,* Quagliariello, E., Palmieri, F., Papa, S., and Klingenberg, M., Eds., Elsevier/North-Holland, Amsterdam, 1979, 63.

115. **Mao, D., Wachter, E., and Wallace, B. A.,** Folding of the mitochondrial proton adensoine triphosphatase proteolipid channel in phospholipid vesicles, *Biochemistry,* 21, 4960, 1982.

116. **Schmid, R. and Altendorf, K.,** The DCCD-reactive protein of the ATP-synthetase from *Escherichia coli* — its orientation in the membrane as determined by chemical modification, *EBEC Rep.,* 2, 101, 1982.

117. **Deckers, G., Schmid, R., Kiltz, H. H., and Altendorf, K.,** The DCCD-reactive protein of the ATP-synthetase from *Escherichia coli* — chemical modification of the tyrosine residues, *EBEC Rep.,* 2, 77, 1982.

118. **Wachter, E., Schmid, R., Deckers, G., and Altendorf, K.,** Amino acid replacement in dicyclohexyl-carbodiimide-reactive proteins from mutant strains of *Escherichia coli* defective in the energy transducing ATPase complex, *FEBS Lett.,* 113, 265, 1980.

119. **Hoppe, J., Schairer, H. U., and Sebald, W.,** Identification of amino-acid substitutions in the proteolipid subunit of the ATP synthase from dicyclohexylcarbodiimide-resistant mutants of *Escherichia coli, Eur. J. Biochem.,* 112, 17, 1980.

120. **Downie, J. A., Gibson, F., and Cox, G. B.,** Membrane adenosine triphosphatases of procaryotic cells, *Ann. Rev. Biochem.,* 48, 103, 1979.

121. **Loo, T. W. and Bragg, P. D.,** The DCCD-binding polypeptide alone is insufficient for proton translocation through F$_o$ in membrane of *Escherichia coli, Biochem. Biophys. Res. Commum.,* 103, 52, 1981.

122. **Guerrieri, F. and Papa, S.,** Effects of chemical modifiers of amino acid residues on H$^+$-conduction by the H$^+$-ATPase of mitochondria, *J. Bioenerg. Biomembr.,* 13, 393, 1981.

123. **Sone, N., Ikeba, K., and Kagawa, Y.,** Inhibition of proton conduction by chemical modification of the membrane moiety of proton translocating ATPase, *FEBS Lett.,* 97, 61, 1979.

124. **Enns, R. K. and Criddle, R. S.,** Affinity labeling of yeast mitochondrial adenosine triphosphatase by reduction with [^{3}H] borohydride, *Arch. Biochem. Biophys.,* 182, 587, 1977.

125. **Mahler, H. R. and Bastos, R. N.**, Coupling between mitochondrial mutation and energy transduction, *Proc. Natl. Acad. Sci. U.S.A.*, 71, 2241, 1974.

126. **Turner, G., Imam, G., and Küntzel, H.**, Mitochondrial ATPase complex of *Aspergillus nidulans* and the dicyclohexylcarbodiimide-binding protein, *Eur. J. Biochem.*, 97, 565, 1979.

127. **Yong, Y. de, Holtrop, M., and Kroon, A. M.**, The biogenesis of rat-liver mitochondrial ATPase. Evidence that the N, N′-dicyclohexylcarbodiimide-binding protein is synthesized outside the mitochondria, *Biochem. Biophys. Acta*, 606, 331, 1980.

128. **Berden, J. A. and Henneke, M. A.**, The uncoupler-binding protein in the proton-pumping ATPase from beef-heart mitochondria, *FEBS Lett.*, 126, 211, 1981.

129. **Kagawa, Y. and Racker, E.**, Partial resolution of the enzymes catalyzing oxidative phosphorylation i correlation of morphology and function in submitochondrial particles, *J. Biol. Chem.*, 241, 2475, 1966.

130. **Tragoloff, A. and Meagher, P.**, Assembly of the mitochondrial membrane system, *J. Biol. Chem.*, 246, 7328, 1971.

131. **Drozdovskaya, N. R., Kozlov, I. A., Milgrom, Ya. M., and Tsybovski, I. S.**, The membrane in submitochondrial particles protects F_1-ATPase from trinitrobenzosulfonate and dinitrofluorobenzole, *FEBS Lett.*, 150, 385, 1982.

132. **MacLennan, D. H. and Asai, J.**, Studies on the mitochondrial adenosine triphosphatase system. V. Localization on the oligomycin-sensitivity conferring protein, *Biochem. Biophys. Res. Commun.*, 33, 441, 1968.

133. **Dupuis, A., Satre, M., and Vignais, P. V.**, Titration of the binding sites for the oligomycin-sensitivity conferring protein in beef heart submitochondrial particles, *FEBS Lett.*, 156, 99, 1983.

134. **Gregory, R. and Hess, B.**, The sulphydryl content of yeast mitochondrial F_1-ATPase and the stoichiometry of subunits, *FEBS Lett.*, 129, 210, 1981.

135. **Bragg, P. D. and Hou, C.**, Cross-linking of minor subunits in adenosine triphosphatase of rat liver mitochondria, *Can. J. Biochem.*, 55, 1121, 1977.

136. **Wingfield, P. T. and Boxer, H.**, Cross-linking of isolated oligomycin-insensitive ATPase from ox heart mitochondria, *Biochem. Soc. Trans.*, 3, 763, 1975.

137. **Todd, R. D. and Douglas, M. G.**, A model for the structure of yeast mitochondrial adenosine triphosphatase complex, *J. Biol. Chem.*, 256, 6984, 1981.

138. **Stare, M., Klein, G., and Vignais, P. V.**, Structure of beef heart mitochondrial F_1-ATPase. Arrangement of subunits as disclosed by cross-linking reagents and selective labeling by radioactive ligands, *Biochim. Biophys. Acta*, 453, 111, 1976.

139. **Enns, R. and Criddle, R. S.**, Investigation of the structural arrangement of the protein subunits of mitochondrial ATPase, *Arch. Biochem. Biophys.*, 183, 742, 1977.

140. **Kozlov, I. A., Milgrom, Y. M., and Tsybovski, I. S.**, Isolation of α-subunits of factor F_1 from submitochondrial particles and the reconstitution of active ATPase from isolated α-subunits and β-subunits bound to the mitochondrial membrane, *Biochem. J.*, 192, 483, 1980.

141. **Bäumert, H. G., Mainke, L., and Zimmer, G.**, Crosslinking studies in oligomycin-sensitive ATPase from beef heart mitochondria, *FEBS Lett.*, 132, 308, 1981.

142. **Todd, R. D. and Douglas, M. G.**, Structure of yeast mitochondrial adenosine triphosphatase. Results of trypsin degradation, *J. Biol. Chem.*, 256, 6990, 1981.

143. **Klein, G., Satre, M., Dianoux, A. C., and Vignais, P. V.**, Photoaffinity labeling of mitochondrial adensine triphosphatase by an azido derivative of the natural adenosine triphosphatase inhibitor, *Biochemistry*, 20, 1339, 1981.

144. **Kagawa, Y., Sone, N., Yoshida, M., Hirata, H., and Okamoto, H.**, Proton translocating ATPase of a thermophilic bacterium. Morphology, subunits and chemical composition, *J. Biochem.*, 80, 141, 1976.

145. **Satre, M. and Zaccai, G.**, Small angle neutron scattering of *Escherichia coli* BF_1-ATPase, *FEBS Lett.*, 102, 244, 1979.

146. **Süss, K.-H., Damaschun, H., Damaschun, G., and Zirwer, D.**, Chloroplast coupling factor CF_1 in solution. Small-angle X-ray scattering and circular dichroism measurements, *FEBS Lett.*, 87, 265, 1978.

147. **Spitsberg, V. and Howorth, R.**, The crystallization of beef heart mitochondrial adenosine triphosphatase, *Biochim. Biophys. Acta*, 492, 237, 1977.

148. **Akey, C. W., Spitsberg, V., and Edelstein, S. J.**, Electron microscopy of beef heart F_1-ATPase crystals, *J. Biol. Chem.*, 258, 3222, 1983.

149. **Amzel, L. M., McKinney, M., Naragannan, P., and Pedersen, P.**, Structure of mitochondrial F_1-ATPase at 9 Å resolution, *Proc. Natl. Acad. Sci. U.S.A.*, 79, 5852, 1982.

150. **Esch, F. S. and Allison, W. S.**, On the subunit stoichiometry of the F_1-ATPase and the sites in it that react specifically with p-fluoro-sulfonylbenzoyl-5′-adenosine, *J. Biol. Chem.*, 254, 10740, 1979.

151. **Tiedge, H., Lücken, U., Weber, J., and Schäfer, G.**, High-affinity binding of ADP and of ADP analogues to mitochondrial F_1-ATPase, *Eur. J. Biochem.*, 127, 291, 1982.

152. **Yoshida, M., Sone, N., Hirata, H., Kagawa, Y., and Ui, N.,** Subunit structure of adenosine triphosphatase. Comparison of the structure in thermophilic bacterium PS 3 with those in mitochondria, chloroplasts, and *Escherichia coli*, *J. Biol. Chem.*, 254, 9525, 1979.

153. **Tiedge, H., Schäfer, G., and Mayer, F.,** An electron microscopic approach to the quarternary structure of mitochondrial F_1-ATPase, *Eur. J. Biochem.*, 132, 37, 1983.

154. **Sigrist-Nelson, K. and Azzi, A.,** The proteolipid subunit of chloroplast adenosine triphosphatase complex. Mobility, accessiblity, and interactions studied by a spin label technique, *J. Biol. Chem.*, 254, 4470, 1979.

155. **Kopecký, J., Guerrieri, F., and Papa, S.,** Interaction of dicyclohexlcarbodiimide with the proton-conducting pathway of mitochondrial H^+-ATPase, *Eur. J. Biochem.*, 131, 17, 1983.

156. **Hovmöller, S., Karlsson, B., Slanghter, M., Leonard, K., and Weiss, H.,** Crystalization of membrane proteins, *EBEC Rep.*, 2, 151, 1982.

157. **Sone, N., Takeuchi, Y., Yoshida, M., and Ohno, K.,** Formations of electrochemical proton gradient and adenosine triphosphate in proteolisones containing purified adenosine triphosphate and bacterio-rhodopsin, *J. Biochem.*, 82, 1751, 1977.

158. **Rogner, M., Ohno, K., Hamamoto, T., Sone, N., and Kagawa, Y.,** Net ATP synthesis in H^+-ATPase macroliposomes by an external electric field, *Biochem. Biophys. Res. Commun.*, 91, 362, 1979.

159. **Teissie, J., Knox, B. E., Tsong, T. Y., and Wehrle, J.,** Synthesis of adenosine triphosphate in respiration-inhibited submitochondrial particles induced by microsecond electric pulses, *Proc. Natl. Acad. Sci. U.S.A.*, 78, 7473, 1981.

160. **Hamamoto, T., Ohno, K., and Kagawa, Y.,** Net adenosine triphosphatase synthesis driven by an external electric field in rat liver mitochondria, *J. Biochem.*, 91, 1759, 1982.

161. **Skulachev, V. P.,** The localized $\Delta \mu H^+$ problem. The possible role of the local electric field in ATP synthesis, *FEBS Lett.*, 146, 1, 1982.

162. **Haines, T. H.,** Anionic lipid headgroups as a proton-conducting pathway along the surface of membranes: a hypothesis, *Proc. Natl. Acad. Sci. U.S.A.*, 80, 160, 1983.

163. **Brown, R. E. and Cunningham, C. C.,** Negatively charged phospholipid requirement of the oligomycin-sensitive mitochondrial ATPase, *Biochim. Biophys. Acta*, 684, 141, 1982.

164. **Houštěk, J. and Drahota, Z.,** Purification and properties of mitochondrial adenosine triphosphatase of hamster brown adipose tissue, *Biochim. Biophys. Acta*, 484, 127, 1977.

165. **Houštěk, J., Kopecký, J., and Drahota, Z.,** Specific properties of brown adipose tissue mitochondrial membrane, *Comp. Biochem. Physiol.*, 60B, 209, 1978.

166. **Svoboda, P., Houštěk, J., Kopecký, J., and Drahota, Z.,** Evaluation of the specific dicyclohexylcarbodiimide binding sites in brown adipose tissue mitochondria, *Biochim. Biophys. Acta*, 634, 321, 1981.

167. **Boyer, P. D., Kohlbrenner, W. E., McIntosh, D. B., Smith, L. T., and O'Neal, C. C.,** ATP and ADP modulation of catalysis by F_1 and Ca^{2+}, Mg^{2+}-ATPases, *Ann. N. Y. Acad. Sci.*, 402, 65, 1982.

168. **Mitchell, P. and Koppenol, W. H.,** Chemiosmotic ATPase mechanisms, *Ann. N. Y. Acad. Sci.*, 402, 584, 1982.

169. **Slater, E. C.,** From cytochrome to adenosine triphosphate and back, *Biochem. Soc. Trans.*, 2, 149, 1974.

170. **Alexandre, A., Reynafarje, B., and Lehninger, A. L.,** Stoichiometry of vectorial H^+ movements coupled to electron transport and to ATP synthesis in mitochondria, *Proc. Natl. Acad. Sci. U.S.A.*, 75, 5296, 1978.

171. **Cross, R. L. and Nalin, C. M.,** Adenine·nucleotide binding sites on beef heart F_1-ATPase evidence for three exchangeable sites that are distinct from three noncatalytic sites, *J. Biol. Chem.*, 257, 2874, 1982.

172. **Webb, M. R., Grubmayer, C., and Penefsky, H. S.,** The stereochemical course of phosphoric residue transfer catalysed by beef heart mitochondrial ATPase, *J. Biol. Chem.*, 255, 11637, 1980.

173. **Bossard, M. J. and Schuster, S. M.,** Catalysis of partial reactions of ATP synthesis by beef heart mitochondrial adenosine triphosphatase, *J. Biol. Chem.*, 256, 1518, 1981.

174. **Kohlbrenner, W. E. and Boyer, P. D.,** Catalytic properties of beef heart mitochondrial ATPase modified with 7-chloro-4-nitrobenzo-2-oxa-1,3-diazole, *J. Biol. Chem.*, 257, 3441, 1982.

175. **Andreoli, T. E., Lam, K. W., and Sanadi, D. R.,** Studies on oxidative phosphorylation. X. A coupling enzyme which activates reversed electron transfer, *J. Biol. Chem.*, 240, 2644, 1965.

176. **Schäfer, G.,** Differentiation of two states of F_1-ATPase by nucleotide analogs, *FEBS Lett.*, 139, 271, 1982.

177. **Kolarov, J., Kopecký, J., Kužela, Š., and Houštěk, J.,** The binding of dicyclohexylcarbodiimide to cytochrome oxidase in rat heart mitochondria, *Biochem. Int.*, 5, 609, 1982.

178. **Kolarov, J., Houštěk, J., Kopecký, J., and Kužela, Š.,** The binding of dicyclohexylcarbodiimide to uncoupling protein in brown adipose tissue mitochondria, *FEBS Lett.*, 144, 6, 1982.

179. **Beattie, D. S. and Clejan, L.,** The binding of dicyclohexylcarbodiimide to cytochrome b of complex III isolated from yeast mitochondria, *FEBS Lett.*, 149, 245, 1982.

180. **Kopecký, J., Dědina, J., Votruba, J., Svoboda, P., Houštěk, J., Babitch, S., and Drahota, Z.,** Stoichiometry of dicyclohexylcarbodiimide-ATPase interaction in mitochondria, *Biochim. Biophys. Acta*, 680, 80, 1982.

181. **Kopecký, J., Pavelka, S., Dědina, J., Siglerová, V., and Vereš, K.,** Stability of dicyclohexylcarbodiimide in aqueous medium. The effect of mitochondrial phospholipids, *Coll. Czech. Chem. Commun.,* 48, 662, 1983.

182. **Bertina, R. M., Steenstra, J. A., and Slater, E. C.,** The mechanism of inhibition by oligomycin of oxidative phosphorylation in mitochondria, *Biochim. Biophys. Acta,* 368, 279, 1974.

183. **Somlo, M.,** Effect of oligomycin on coupling in isolated mitochondria from oligomycin-resistant mutants of *Saccharomyces cerevisiae* carrying an oligomycin-resistant ATP phosphohydrolase, *Arch. Biochem. Biophys.,* 182, 518, 1977.

184. **Glaser, E., Norling, B., and Ernster, L.,** A study of the dicyclohexylcarbodiimide-binding component of the mitochondrial ATPase complex from beef heart, *Eur. J. Biochem.,* 115, 189, 1981.

185. **Drahota, Z., Kopecký, J., Svoboda, P., Pavelka, S., and Houštěk, J.,** The stoichiometry and differentiation of dicyclohexylcarbodiimide reactive sites on mitochondria H^+-ATPase, in *Vectoral Reactions in Electron and Ion Transport in Mitochondria and Bacteria,* Palmieri, F., Ed., Elsevier/North-Holland, Amsterdam, 1981, 223.

186. **Hoppe, J. and Sebald, W.,** An essential carboxyl group for H^+-conduction in the proteolipid subunit of the ATP synthase, in *Chemiosmotic Proton Circuits in Biological Membranes,* Skulchev, V. P. and Hinkle, P. C., Eds., Addison-Wesley, Reading, Mass., 1981, 449.

187. **Glaser, E., Norling, B., Kopecký, J., and Ernster, L.,** Comparison of the effects of oligomycin and dicyclohexylcarbodiimide on mitochondrial ATPase and related reactions, *Eur. J. Biochem.,* 121, 525, 1982.

188. **Palatini, P. and Bruni, A.,** Reversal by phospholipids of the oligomycin induced inhibition of membrane associated adenosintriphosphatases, *Biochim. Biophys. Res. Commun.,* 40, 186, 1970.

189. **Glaser, E. and Norling, B.,** Kinetics of interaction between the H^+-translocating component of the mitochondrial ATPase complex and oligomycin or dicyclohexylcarbodiimide, *Biochim. Biophys. Res. Commun.,* 111, 333, 1983.

190. **Vinogradov, A. D., Gyurova, Z. S., and Fitin, A. F.,** Effect of SH-reagents on the mitochondrial ATPase and induction of respiratory control in EDTA particles, *FEBS Lett.,* 54, 230, 1975.

191. **Zimmer, G., Mainka, L., and Ohlenschläger, G.,** Oligomycin-sensitive ATPase from beef heart: reaction with 2-mercaptopropionylglycine, *FEBS Lett.,* 94, 233, 1978.

192. **Mairouch, H. and Godinot, C.,** Location of protein(s) involved in oligomycin-induced inhibition of mitochondrial adenosintriphosphatase near the outer surface of the inner membrane, *Proc. Natl. Acad. Sci. U.S.A.,* 74, 4185, 1977.

193. **Kopecký, J., Glaser, E., Norling, B., and Ernster, L.,** Relationship between the binding of dicyclohexylcarbodiimide and the inhibition of H^+-translocation in mitochondrial particles, *FEBS Lett.,* 131, 208, 1981.

194. **Linnett, P. E., Mitchell, A. D., and Beechey, R. B.,** Changes in inhibitor sensitivity of mitochondrial ATPase activity after detergent solubilization, *FEBS Lett.,* 53, 180, 1975.

195. **Houštěk, J., Pavelka, S., Kopecký, J., Drahota, Z., and Palmieri, F.,** Is the mitochondrial dicyclohexylcarbodiimide-reactive protein of Mr 33,000 identical with phosphate transport protein?, *FEBS Lett.,* 130, 137, 1981.

196. **Mitchell, P. and Moyle, J.,** Acid-base titration across the membrane system of rat-liver mitochondria, *Biochem. J.,* 104, 588, 1967.

197. **Nicholls, D. G.,** The influence of respiration and ATP hydrolysis on the proton-electrochemical gradient across the inner membrane of rat-liver mitochondria as determined by ion distribution, *Eur. J. Biochem.,* 50, 305, 1974.

198. **Schönfeld, M. and Neumann, J.,** Proton conductance of the thylakoid membrane: modulation by light, *FEBS Lett.,* 73, 51, 1977.

199. **Sorgato, M. C., Branca, D., and Ferguson, S. J.,** The rate of ATP synthesis by submitochondrial particles can be independent of the magnitude of the proton motive force, *Biochem. J.,* 188, 945, 1980.

200. **Okamoto, H., Sone, N., Hirata, H., Yoshida, M., and Kagawa, Y.,** Purified proton conductor in proton translocating ATPase of a thermophilic bacterium, *J. Biol. Chem.,* 252, 6125, 1977.

201. **Negrin, R. S., Foster, D. L., and Fillingame, R. H. C.,** Energy-transducing H^+-ATPase of *Escherichia coli.* Reconstitution of proton translocation activity of the intrinsic membrane sector, *J. Biol. Chem.,* 255, 5643, 1980.

202. **Schneider, E. and Altendorf, K.,** ATP synthetase (F_1F_0) of *Escherichia coli* K-12. High-yield preparation of functional F_0 by hydrophobic affinity chromatography, *Eur. J. Biochem.,* 126, 149, 1982.

203. **Pick, U. and Racker, E.,** Purification and reconstitution of the N,N'-DCCD-sensitive ATPase complex from spinach chloroplast, *J. Biol. Chem.,* 254, 2793, 1979.

204. **Shchipakin, V., Chuchlova, E., and Evtodienko, Y.,** Reconstruction of mitochondrial H^+-transporting system in proteoliposomes, *Biochem. Biophys. Res. Commun.,* 69, 123, 1976.

205. **Papa, S., Guerrieri, F., and Rossi-Bernardi, L.,** Electrometric measurement of rapid proton transfer reactions in respiratory chains, *Methods Enzymol.,* 55, 614, 1979.

206. **Chance, B. and Scarpa, A.,** Acid-base indicator for the measurement of rapid changes in hydrogen ion concentration, *Methods Enzymol.,* 24, 336, 1972.

207. **Rottenberg, H.,** The measurement of membrane potential and pH in cells, organelles, and vesicles, *Methods Enzymol.,* 55, 547, 1979.

208. **Criddle, R. S., Pacler, L., and Snieh, P.,** Oligomycin-dependent ionophoric protein subunit of mitochondrial adenosinetriphosphatase, *Proc. Natl. Acad. Sci. U.S.A.,* 74, 4305, 1977.

209. **Nelson, N., Eytan, E., Notsani, B., Sigrist, H., and Sigrist-Nelson, K.,** Isolation of a chloroplast N,N'-dicyclohexylcarbodiimide-binding proteolipid, active in proton translocation, *Proc. Natl. Acad. Sci. U.S.A.,* 74, 2375, 1977.

210. **Konishi, T., Packer, L., and Criddle, R.,** Purification and assay of a proteolipid ionophore from yeast mitochondrial ATP synthetase, *Methods Enzymol.,* 55, 414, 1979.

211. **Cellis, H.,** 1-Butanol extracted proteolipid. Proton conducting properties, *Biochem. Biophys. Res. Commun.,* 92, 26, 1980.

212. **Schindler, H. and Nelson, N.,** Proteolipid of adenosintriphosphatase from yeast mitochondria forms proton-selective channels in planar lipid bilayers, *Biochemistry,* 21, 5781, 1982.

213. **Pansini, A., Guerrieri, F., and Papa, S.,** Control of proton conduction by the H^+-ATPase in the inner mitochondrial membrane, *Eur. J. Biochem.,* 92, 545, 1978.

214. **Sone, N., Yoshida, M., Hirata, H., and Kagawa, Y.,** Carbodiimide-binding protein of H^+-translocating ATPase and inhibition of H^+-conduction by dicyclohexylcarbodiimide, *J. Biochem.,* 85, 503, 1979.

215. **Ho, Y. K., Lin, C. J., Saunders, D. R., and Wang, J. H.,** Light dependence of the decay of the proton gradient in broken chloroplasts, *Biochim. Biophys. Acta,* 547, 149, 1979.

216. **Bulychev, A. A., Andrianov, V. K., and Kurella, G. A.,** Effect of dicyclohexylcarbodiimide on the proton conductance of thylakoid membranes in intact chloroplast, *Biochim. Biophys. Acta,* 590, 300, 1980.

217. **Ellenson, J. L., Pheasant, D. J., and Levine, R. P.,** Light/dark labeling differences in chloroplast membrane polypeptides associated with chloroplast coupling factor O, *Biochim. Biophys. Acta,* 504, 123, 1978.

218. **Prochaska, L. J. and Dilley, R. A.,** Chloroplast membrane conformational changes measured by chemical modification, *Arch. Biochem. Biophys.,* 187, 61, 1978.

219. **Prochaska, L. J. and Dilley, R. A.,** Site specific interaction of protons liberated from photosystem II. Interaction with a hydrophobic membrane component of the chloroplast membrane, *Biochem. Biophys. Res. Commun.,* 83, 644, 1978.

220. **Gräber, P., Burmeister, M., and Hortsch, M.,** Regulation of the membrane permeability of spinal chloroplasts by binding of adenine nucleotides, *FEBS Lett.,* 136, 25, 1981.

221. **Pedersen, P. L., Hullihen, J., and Wehrle, J. P.,** Proton adenosine triphosphatase complex of rat liver, *J. Biol. Chem.,* 256, 1362, 1981.

222. **Dewey, T. G. and Hammes, G. G.,** Steady state kinetics of ATP synthesis and hydrolysis catalyzed by reconstituted chloroplast coupling factor, *J. Biol. Chem.,* 256, 8941, 1981.

223. **Graf, T. and Sebald, W.,** The dicyclohexylcarbodiimide-binding protein of the mitochondrial ATPase complex from beef heart, *FEBS Lett.,* 94, 218, 1978.

224. **Macreadie, I. G., Novitski, C. E., Maxwell, R. J., John, U., Ooi, B.-G., Mc Mullen, G. L., Lukins, H. B., Linnane, A. W., and Nagley, P.,** Biogenesis of mitochondria: the mitochondrial gene (aap1) coding for mitochondrial ATPase subunit 8 in *Saccharomyces cerevisiae, Nucl. Acid Res.,* 11, 4435, 1983.

225. **Mariottini, P., Chomyn, A., Attardi, G., Tronato, D., Strong, D. D., and Doolittle, R. F.,** Antibodies against synthetic polypeptides reveal that the unidentified reading frame A6L, overlapping the ATPase 6 gene, is expressed in human mitochondria, *Cell,* 32, 1269, 1983.

Chapter 6

MOLECULAR ASPECTS OF THE STRUCTURE-FUNCTION RELATIONSHIP IN CYTOCHROME *c* OXIDASE

Angelo Azzi, Kurt Bill, Reinhard Bolli, Robert P. Casey, Katarzyma A. Nałęcz, and Paul O'Shea

TABLE OF CONTENTS

I. INTRODUCTION

Cytochrome c oxidase is the terminal oxygen-reducing enzyme of the respiratory chains of mitochondria and bacteria, conducing electrons, which may be derived from various metabolic substrates, between cytochrome c and molecular oxygen to produce water. While a family of enzymes exist in different organisms which can carry out this function (see, e.g., Poole[1] for a review) we shall consider here only the class of cytochrome c oxidases which have been studied most, i.e., the heme A-type cytochrome c oxidases and, in particular, the cytochrome c oxidase from mammalian mitochondria. This is a membrane-bound, multisubunit enzyme, having a molecular weight of approximately 200,000[2] and containing two heme A groups and two copper ions per catalytic unit. Heme A has an isoprenoid chain at position 2 and a carbonyl group at position 8 of the porphyrin ring, and the heme iron can make two further coordination bonds with axial ligands. Owing to the two heme groups having different environments they also have differing spectral properties and reactivity with ligands, enabling them to be resolved as hemes a and a_3. The heme groups absorb visible light to give absorbance spectra which are characteristic of the reduced and oxidized enzymes (see, e.g., Lemberg[3]). The main peaks for the reduced oxidase occur at 443 and 603 nm, whereas those for the oxidized enzyme are at 421 and 598 nm. While both heme types contribute equally to the Soret peak of the reduced enzyme, approximately 80% of the 603-nm peak is due to heme a absorbance.[4] Heme a has a lower midpoint potential (230 mV) than heme a_3 (380 mV), the measured midpoint potentials of both heme groups being pH-sensitive under certain conditions. Heme a_3, unlike heme a, can bind ligands which, apart from oxygen, may be, e.g., CO, HCN, and HN_3 (see Lemberg[3]). Owing to the paramagnetic properties of the heme irons, hemes a and a_3 may also be studied using ESR spectroscopy,[5] both groups giving rise to complex spectra. The two copper ions of cytochrome c oxidase also experience different environments within the enzyme and thus are also resolvable according to their spectral properties as Cu_A and Cu_B. Cu_A (midpoint potential = 240 mV) is in rapid electronic equilibrium with heme a and is predominantly responsible for the peak at 820 to 840 nm in the absorbance spectrum of the oxidized enzyme. Cu_A also gives a characteristic signal in the ESR spectrum of the enzyme.[5] Cu_B, on the other hand, is closely associated with heme a_3[6] and is intimately involved in O_2 reduction. Unlike Cu_A it gives no ESR signal.

While the exact electron transfer pathway in cytochrome c oxidase is as yet not totally clarified, there is fairly strong evidence derived from rapid kinetic measurements that electrons donated from cytochrome c pass first to heme a. Cu_A, while in rapid redox equilibrium with heme a, does not appear to be on the main electron transfer route. Electrons then pass to the heme a_3-Cu_B complex and then to oxygen (see Wikström[7] for a more detailed treatment).

As remarked above, cytochrome c oxidase is a membrane-bound enzyme and, according to the classical statements of the chemi-osmotic hypothesis,[8,9] catalyzes exclusively the transfer of electrons from cytochrome c to oxygen. From recent finding of Wikström and others, however, it seems that in addition to this activity, the oxidase couples the energy derived from electron flow to the outward translocation of protons across the mitochondrial membrane, leading to the formation of a transmembrane electrochemical gradient of H^+ which is used as an energetic intermediate in the synthesis of ATP.

The quaternary structure of cytochrome c oxidase is highly complex (see below) and, clearly, the assembly of the various components into the functional enzyme requires a highly integrated and controlled system of biogenesis. Mitochondria possess their own biosynthetic apparatus (see Tzagaloff[10]) having, for example, their own genome, ribosomes, and transfer RNA. In spite of its completeness, however, the mitochondrial biogenetic system produces relatively few mitochondrial proteins. Among these are subunits 1, 2, and 3 of cytochrome c oxidase (coded for by the Oxi 3, 1, and 2 genes, respectively, of the mitochondrial genome).

The remainder of the oxidase subunits are products of nuclear genes and there is evidence that at least some of these are synthesized as precursors[11,12] which then undergo an energy-dependent processing[13] involving specific mitochondrial proteases, during their integration into the membrane-bound form of the enzyme.

II. PURIFICATION

Starting material for the isolation of cytochrome *c* oxidase can be either microorganisms (bacteria, fungi) or mammalian tissues (heart, liver, muscle). Since cytochrome *c* oxidase is an integral part of the inner mitochondrial membrane, all conventional purification procedures are based on the solubilization of this membrane-bound enzyme with bile acids or nonionic detergents followed by its salt precipitation in order to remove other solubilized membrane components. Finally, the isolated enzyme is further purified by several salting-out and dissolution steps or by adsorption on hydrophobic[13] or ion-exchange columns.[14] The purity of the isolated enzyme can then be checked by a spectrum in the visible region (400 to 630 nm) and by electrophoretic analysis followed by an assay to determine its enzymatic activity.

For the initial solubilization of mitochondria or submitochondrial particles (Keilin-Hartree preparation[15], different detergents including Triton® X-100 and 114,[16—18] cholate,[19,20] de-oxycholate,[21—24] and Tween® 20 and 80[25] have been used. Often ionic detergents (cholate, deoxycholate) are most adequate for the first stages of isolation. The requirement of cytochrome *c* oxidase solubilization of higher detergent concentrations than most other components of the inner mitochondrial membrane is utilized to perform the purification in a sequential way. First, the more soluble reductases (complex I, II, and III) are extracted with a lower detergent concentration leaving the cytochrome *c* oxidase still bound to the membrane. This is called the "red-green split" and is the basis of all conventional preparative methods. After the solubilization of the cytochrome *c* oxidase with a higher detergent concentration, residual cytochromes (*b* and c_1) are then removed by repeated fractionation with ammonium sulfate in the presence of a detergent. The different methods for the purification of the cytochrome *c* oxidase are described by Hoechli and Hackenbrock.[26]

All these isolation procedures of such a delicate enzyme may easily disturb the native state of the complex. It is well known that long-term exposure of the cytochrome *c* oxidase to high concentrations of detergents can change its structural and functional integrity. Also, it is difficult to compare enzymes prepared by different methods since, e.g., different detergents used for solubilization can affect the subunit composition (stoichiometry and number of subunits) and repeated ammonium sulfate precipitations reduce the phospholipid content.[27] Moreover, one cannot exclude that proteins or fragments of the inner mitochondrial membrane are copurified with the oxidase. Indeed, heterogeneity or polydispersity is quite common in cytochrome *c* oxidase preparations.

Among the more recently developed isolation procedures affinity chromatography turned out to be the most useful and promising. The idea was to use immobilized horse heart cytochrome *c* as the affinity ligand.[28—30] Such an affinity chromatography was suggested by Weiss and Kolb[31] as effective only with cytochrome *c* oxidase from *Neurospora*[32] but not with yeast or mammalian cytochrome *c* oxidase. More recently Thompson et al.[33] have described conditions in which the cytochrome *c* oxidase from rat liver can be prepared by using an affinity chromatography procedure similar to that of Weiss and Kolb.[31]

This problem was recently solved by using as an affinity ligand cytochrome *c* from *Saccharomyces cerevisiae* covalently attached to an activated thiol-Sepharose-4B column via its cysteinyl-103 residue.[34] This cysteine residue, located close to the C terminus and being functionally not important, leaves the lysine residues of the binding site free for the interaction with cytochrome *c* oxidase and other cytochrome *c* binding enzymes.[35]

When a Triton® X-100 extract of beef heart mitochondria is loaded on such an affinity column specific binding of cytochrome c oxidase and b-c_1-complex occurs. The bound cytochrome c oxidase and bc_1-complex can be eluted separately by increasing salt concentrations.[36,37] This technique being very mild, specific, and rapid yields a highly pure and active enzyme and can be used also for microscale preparations. It has been applied for the isolation of a number of different bacterial oxidases[38—40] and for cytochrome c_1.[41]

III. STRUCTURE

The primary structure of the beef heart cytochrome c oxidase has been elucidated to a great extent. Together with the information available from electron diffraction analysis on two-dimensional crystals (cross-linking,[42—48] chemical labeling,[49—55] immunochemical methods[50,56—59]) it has been possible to construct a structural model of cytochrome c oxidase. However, for the full understanding of its different functions and its molecular reaction mechanism much more information is needed which can only derive from the elucidation of its three-dimensional structure. For this purpose X-ray crystallography is still the method of choice although not applicable at the moment to cytochrome c oxidase, which has not yet been crystallized in three dimensions.

Reports of crystalline cytochrome c oxidase have not been confirmed and no X-ray crystallographic data have been obtained from these crystals.[60—62] Cytochrome c oxidase is embedded in the mitochondrial inner membrane. Several highly hydrophobic segments of the protein are in contact with the alkane chains of the lipids and are thus to be considered an intrinsic part of the membrane, whereas the hydrophilic segments are exposed to the aqueous phases on each side of the membrane. This amphiphilic nature of the cytochrome c oxidase (common to all membrane proteins) is responsible for the difficulties in its purification, characterization, and crystallization. Since it is not soluble in aqueous buffers (in contrast to globular proteins) detergents have to be added for its isolation and purification. By removing these detergents, aggregation due to nonspecific hydrophobic interactions occurs. Above a certain concentration, detergents form micelles.[63] If cytochrome c oxidase is solubilized in an excess of a detergent, it is incorporated together with parts of the lipids. One can say that these micelles replace the lipid bilayer. The alkane chains of the detergents interact with the hydrophobic parts of the protein, whereas the polar regions of the detergents interact with the hydrophilic (polar) regions of the protein. Using charged detergents like triethylammonium or dodecylsulfate the protein complex can be changed or even denaturation of the protein may occur. Using "milder" detergents such as Triton® X-100 or β-D-octylglucopyranoside, it is more likely that the native conformation of the membrane protein is retained. The native state of the protein should be proven in any case by an enzymatic assay and a spectroscopic test before starting crystallization (which does not exclude that during crystallization the native conformation may change).

The more successful method up to now to get structural information was the three-dimensional reconstruction from two-dimensional crystals. These two-dimensional crystals have crystalline order but the third dimension contains only one layer of molecules. Henderson and Unwin,[64] combining low-dose electron microscopy with electron diffraction and tilting the specimen in the electron microscope, obtained a low resolution structure. A high resolution analysis was not possible, since the electron diffraction data at high tilt angles (greater than 60°) are difficult to interpret and, moreover, the specimen cannot be tilted at an angle of 90°.

Two-dimensional crystals of cytochrome c oxidase from bovine heart suitable for electron microscopy and image reconstruction[65—68] have been obtained. Using several different types of crystalline arrays for the three-dimensional reconstruction, Fuller et al.[67] determined the general size and shape of the cytochrome c oxidase molecule. The maximal resolution reached up to now is in the range of 20Å. This does not allow any information about the tertiary

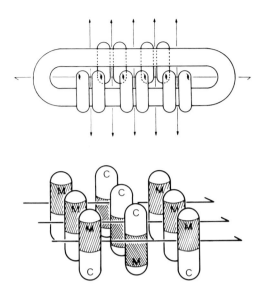

FIGURE 1. The orientation of cytochrome *c* oxidase molecules in Triton® X-100 and deoxycholate crystals. (From Henderson, R., Capaldi, R. A., and Leigh, J. S., *J. Mol. Biol.*, 112, 631, 1977. With permission.)

structure or the topology of the subunits to be obtained since for the identification of the amino acid residues a resolution higher that 3 Å is needed.

Two different types of crystals have been obtained and studied (Figure 1). Membranes of cytochrome *c* oxidase isolated in the presence of Triton® X-100 occasionally exhibit small crystalline areas.[56,66,69,70] In this form the enzyme is a dimer spanning the lipid bilayer; the preparation is vesicular and the vesicles are flattened, being stacked one upon another. Most of the protein protrudes towards the inside of these vesicles. The low resolution (25 Å) of the electron micrographs did not allow the identification of those parts of the enzyme in this crystalline form inserted into the membrane.

A more suitable crystalline form of cytochrome *c* oxidase is obtained by deoxycholate extraction of mitochondria. Negative staining with phosphotungstate or uranyl acetate shows that the enzyme is monomeric (two heme-complex) arranged in detergent-rich sheets with no continuous bilayer.[67,68,71] The monomer is packed in rows with opposite orientation.

For both crystal forms a low resolution structure for the cytochrome *c* oxidase has been determined.[67,69] There is a general agreement on the two structures of the overall size and shape of the protein. The cytochrome *c* oxidase monomer is seen as an asymmetric "Y" with a length of about 110 Å giving rise to three domains (Figure 2). The two arms of the Y (M_1- and M_2-domains) are approximately 50 Å in length with a center-to-center separation of 40 Å, each being surrounded by phospholipids. These two M-domains separated on the cytosolic side of the membrane are spanning the bilayer and protrude out of the inner membrane by about 15 to 20 Å on the matrix side. The stalk of the Y is a single domain (C-domain) and extends about 50 to 55 Å from the cytoplasmic side of the mitochondrial inner membrane. Such a conclusion on the orientation of the enzyme has been reached in the light of the finding by Frey et al.[57] that the enzyme domain that is exposed in the Triton® crystals reacts with antisubunit IV but not with antisubunit II. Since in mitochondria subunit II is on the cytosolic side (binding of cytochrome *c*), the enzyme in the Triton® vesicles is apparently "inside out".

Low-angle X-ray studies show reflections characteristic of helical structures perpendicular

FIGURE 2.   A model of cytochrome *c* oxidase obtained
from electron microscopic analysis of a crystal form
obtained with deoxycholate extraction of mitochondria.
(From Fuller, S. D., Capaldi, R. A., and Henderson,
R., *J. Mol. Biol.*, 134, 305, 1979. With permission.)

to the plane of the membrane. The M_1-domain with a surface area in projection of about
900 Å^2 is large enough to contain 8 to 12 helices of the same average dimensions as those
in bacteriorhodopsin. The smaller M_2-domain with about 650 Å^2 could contain from five to
eight helices. Labeling and cross-linking experiments have allowed the location of the
different subunits of the cytochrome *c* oxidase within the Y-shaped profile discussed
above.[50—54,73—75] The major part of the C-domain should be formed by the subunits I, II,
III, and V. Subunit I, II, and III having 10,[76] 2,[77] and 6[76] transmembrane stretches are likely
to contribute as well to the M_1- as to the M_2-domain which are both embedded in the lipid
bilayer. These conclusions are inferred from labeling experiments using
arylazidophospholipids[50,53] and iodonaphthylazide.[54] Subunit IV, VII$_{Ser}$, and VII$_{Ile}$, each
having a single stretch of hydrophobic amino acids,[78,79] are also labeled by arylazidophospholipids.[50,52] All these 21 transmembrane segments contribute to the bilayer-intercalated
part of the M_1- and M_2-domains. Since neither the M_1- nor the M_2-domain is large enough
to contain the 16 helices indicated in subunit I and III, Capaldi et al.[79] suggested placing
them in separate domains (subunit I in M_1 and subunit III in M_2).

Cytochrome *c* oxidase is a dimer in the vesicle crystal.[66,69] From low resolution structure
determination, it was seen that contact between monomers exclusively occurs through the
C-domain with a closest approach of 36 Å between the M_2-domains (center to center) of
each monomer. The 20-Å resolution so far achieved does not allow the location of individual
subunits. However, with a theoretically possible resolution of 7 Å with an unstained wet
specimen it might become possible in the future to establish the relative positions of subunits
and to some extent their tertiary structure.

A. Molecular Weight

Cytochrome *c* oxidases isolated from different sources contain different numbers of subunits (for review see Azzi[81]). In cytochrome *c* oxidase preparations from higher animals,

Table 1
ANALYTICAL ULTRACENTRIFUGATION ANALYSIS
OF CYTOCHROME *c* OXIDASE

	S-values		
Source of oxidase	**Monomeric**	**Dimeric**	**Polymeric**
Beef	—	11 (90%)	>17 (~10%)
Hammerhead shark	7 (>90%)	—	—
Dogfish	6 (40%)	11 (60%)	—
Rat	6 (85%)	—	>20 (15%)
Chicken	5 (40%)	10 (60%)	—
Camel	5.6 (>90%)	—	—

Note: Samples were centrifuged at a protein concentration of approximately 10 mg/mℓ in 10 mM phosphate buffer pH 7.4 containing 1% Tween® 80.

different numbers of subunits were observed, varying from 7 to 13 subunits, depending on the system of sodium dodecyl sulfate (SDS) electrophoresis applied. The enzymes, composed of at least seven subunits, were isolated from beef heart as well as from rat, chicken, camel, and hearts of several Elasmobranch fish.[53, 82—84] Recently a higher resolving procedure of SDS electrophoresis was developed[85,86] which allows the separation of cytochrome *c* oxidase from different mammalian sources (beef and pig heart, beef and rat liver) into 13 different polypeptides. The controversy about how many of the polypeptides in cytochrome *c* oxidase preparations are true subunits of the enzyme is still unresolved.

Irradiation of rat liver mitochondria with high energy electrons results in inhibition of the cytochrome *c* oxidase activity. Application of the target theory permitted calculation of the molecular weight of the minimum active unit of cytochrome *c* oxidase to be 50,000 ± 16,000 *in situ*, and 67,000 ± 8000 after purification.[87] Recently, by using the same method, a similar value of 70,000 was obtained by Thompson et al.[88]

An average molecular weight of complex IV of 230,000 was calculated from light scattering data,[89] after subtraction of the experimentally determined amount of protein-bound lipid. It was also concluded from ultracentrifugation data that the enzyme had a sedimentation coefficient of 12 S and that there are two heme prosthetic groups per unit of 230,000 M_r.

Ultracentrifuge studies on oxidases from a number of higher eukaryotic organisms have revealed that in nonionic detergents, such as Tween®-80 or Triton® X-100, the enzyme can be either monomeric or dimeric, or in both forms, depending on the organism (see Table 1[82—84]). From all those experiments, sedimentation coefficients of about 5 to 7 S were calculated for the monomeric and 10 to 13 S for the dimeric form, respectively. On the basis of these values the molecular weight of 100,000 to 120,000 was calculated for a monomer.[82,83,90] Capaldi and co-workers[84,91] correlated the sedimentation coefficient value of 10 to 13 S with the molecular weight of 320,000 to 345,000 for a dimer of beef heart oxidase after subtraction of the bound detergent. It should be taken into account that the sum of the molecular weights of the amino acids of the 12 to 13 polypeptides which are considered to be subunits of the oxidase[85,92,93] would give molecular weight of 210,000 for the whole oxidase. However, due to the fact that cytochrome *c* oxidase is a highly asymmetric protein and that the proteins used for calibration do not belong to the same conformational class, definite conclusions on the molecular weight of the enzyme are questionable.[94—96] In the case of asymmetric proteins a large discrepancy occurs between the Stokes radii measured by centrifugation and the apparent values derived from gel chromatography. The latter method seems to underestimate hydrodynamic radii[95] which limits the use of gel filtration as a quantitative measure of these values for asymmetric proteins, but still the method seems to be useful to establish the homogeneity of the preparation and to compare the relative molecular weights rather than to obtain absolute values.

The majority of the results on the molecular form of cytochrome c oxidase were obtained from experiments either on the enzyme from Elasmobranch fish or on that from bovine heart. In the first case the enzyme is mainly monomeric, while the aggregation state of beef heart oxidase depends on the experimental conditions, such as pH, salt, or detergent concentration and type.

Under conditions of alkaline pH (pH 9.5 and 10.5) the enzyme has been reported to dissociate into a monomeric species with a M_r of 100,000 and sedimentation coefficient 6 S.[90] A similar effect of high pH was reported by Georgevich et al.[84] after incubation of the oxidase at pH 8.5 in the presence of 5% Triton® X-100. From sedimentation equilibrium studies they obtained values in the range 6.5 to 7.5 S and molecular weights 129,000 and 160,000 after correction for the bound detergent, the heterogeneity being ascribed to some loss of subunits by the enzyme.

At neutral pH and in the presence of nonionic detergents the beef heart oxidase forms mainly dimers as was concluded from sedimentation equilibrium data. For the media containing Emasol® or Tween®-80, the sedimentation coefficient of 10 to 11 S was calculated.[83,90] Also, in the presence of Triton® X-100 the sedimentation coefficient was in the range of 10 to 13 S which corresponds to a molecular weight in the range of 326,000 to 385,000.[84,91,97]

Two or more independent species of oxidase were found during the analytical ultracentrifugation of the enzyme in deoxycholate. Estimations of the protein molecular weight gave values of 200,000 for a monomer and 360,000 for a dimer, respectively.[91]

This specific effect of detergents was confirmed by crystallization studies. In deoxycholate-derived crystals, the enzyme exists as a monomer in detergent-rich sheets, and in Triton®-derived crystals, the oxidase is a dimer inserted across the lipid bilayer (for references see Capaldi et al.[79]).

Gel filtration procedures were used as well to determine the hydrodynamic radii of the oxidase. The values of the calculated molecular weight in the presence of Tween®-80 or Triton® X-100 were in accordance with those obtained from centrifugation analyses.[237]

Using octylglucoside as detergent,[98] three molecular forms of beef heart cytochrome c oxidase were measured by gel filtration through Ultrogel® Ac A 34. The first, in the excluded volume, appeared to be aggregated enzyme, the second had a M_r of 280,000, and a broad shoulder of a lower molecular weight was also visible. Using dodecylmaltoside, the oxidase was eluted as a single peak of M_r 300,000 and the authors concluded that this form was a dimer. Recently, it was shown that in the presence of dodecylmaltoside, also, monomeric forms can be isolated under the conditions of low ionic strength and that the ionic strength can regulate the interconversion of the enzyme from monomeric to dimeric form.[99]

Some differences of the catalytical and physical properties between monomers and dimers were reported as well.

The pH optimum for the electron transfer reaction of the monomeric camel enzyme (pH 7.2) is significantly different from that of the dimeric form of the bovine enzyme (pH 6.2).[83] Some spectral differences, attributed to monomerization, were observed after treatment of the enzyme at high pH, namely, loss of the 444-nm shoulder and the 593-nm shoulder in the Fe^{2+}-CO spectrum, and changes in the ratios of peak heights ($\gamma[Fe^{2+}]/\gamma[Fe^{3+}]$ from 1.25 to 1.09; $\gamma[Fe^{2+}]/\alpha[Fe^{2+}]$ from 4.84 to 3.48, for the native and alkali-treated enzyme, respectively). The extinction coefficients of the Fe^{2+} and Fe^{2+}-CO forms in the Soret region are lower for the alkali-treated enzyme, which also reveals the increased reactivity with carbon monoxide.[100] The assignment of these spectral differences to the monomeric species is questionable.

Different data were reported as regards the activities of monomeric and dimeric forms of cytochrome *c* oxidase. It was found that monomers of the enzyme from different animal species are active in all the cases when obtained at neutral pH and in the presence of a nonionic detergent.[82—84] In the case of the dogfish oxidase, for instance, it was found that the monomeric form was more active than the other aggregation states.[82] When, however, the monomers of bovine heart oxidase were obtained by alkaline pH treatment, the situation was less clear. In the Emasol®-phosphate buffer system,[100] 83 to 97% of the activity remained after treatment of the enzyme at pH 11.0, but, since the activity was measured at neutral pH, the problem of reassociation of the enzyme to dimers could not be excluded.

In all the cases when monomers of beef heart oxidase were obtained by a treatment with different detergents (deoxycholate, octylglucoside, dodecylmaltoside), diminished activity of monomers was reported.[91,98,99] Removal of lipids by cholate or deoxycholate resulted in a preparation of oxidase which was considerably less active, even in the presence of Tween®, than the original sample. The reduced activity was also found in the deoxycholate-treated samples in which diphosphatidylglycerol was not removed.[91] When three species of oxidase were separated in octylglucoside, only dimers could be reactivated in the presence of dodecylmaltoside, which supports the concept that the dimer is the most active species of the beef heart cytochrome *c* oxidase.[98] Recently, the monomers of oxidase were prepared at low ionic strength in the presence of dodecylmaltoside.[99] The kinetics of the monomeric enzyme activity are significantly different from those of the dimeric form of the enzyme. The dimeric enzyme has a high and a low affinity site for cytochrome *c*, as it was reported for beef heart cytochrome *c* oxidase.[101] With monomeric enzyme, however, only the presence of the high affinity site was shown kinetically and the maximal molecular activity of the monomeric form was much lower than in the case of the dimers of oxidase. Thompson et al.[33] have isolated from rat liver a cytochrome *c* oxidase species which they claim to be monomeric. The kinetics of this preparation did not seem to be different from those of a dimeric enzyme.

The subunit pattern after monomerization is also slightly changed[84] and the preparations of oxidase depleted of the third largest subunit seem to be monomeric.[33,79,97] These results would imply that some of the subunits can control the conversion of the molecular form of cytochrome *c* oxidase.

Despite the intensive work on the isolated enzyme, it has not been established yet what is the state of aggregation of cytochrome *c* oxidase in membranes.

In the last few years, heme aa_3-type cytochrome *c* oxidases from a variety of prokaryotes have been isolated and purified (for a review see References 1,102). In general, they are much simpler complexes than the mammalian or the fungal oxidases and consist only of two (*Paracoccus denitrificans*,[103] *Thiobacillus novellus*,[104] *Pseudomonas spheroides*,[39] or three (*Bacillus subtilis*,[40] *Thermus thermophilus*,[105] PS3[106]) polypeptides (Table 2). For some of these subunits certain similarities with the three largest subunits of the eukaryotic enzymes have been shown.[102] Minimal molecular weights from SDS polyacrylamide gel electrophoresis were estimated to be between 55,000 and 136,000, depending on the organism. However, not many data on sedimentation analysis or gel filtration have been reported. A preliminary sedimentation study of the cytochrome oxidase from *Thiobacillus novellus* gave an M_r of around 115,000,[107] which corresponds to two sets of the two subunits analyzed by electrophoresis. The authors claimed, however, that this complex contained only two hemes (heme *a* and a_3) which could not be compatible with a homogenous dimer. Sedimentation analysis of the *Paracoccus denitrificans* enzyme gave an M_r of 86,000 ± 5000,[102] which would indicate a monomeric species.

Fungal cytochrome *c* oxidase isolated from *Neurospora crassa* or baker's yeast are composed of seven or eight subunits,[15,31,108—110] which correspond to a minimal M_r of 150,000. This is in good agreement with the molecular weight of 140,000 calculated from the heme

Table 2
SOME PROPERTIES OF PURIFIED CYTOCHROME c OXIDASES FROM BACTERIA

Properties	Paracoccus denitrificans	Thiobacillus novellus	Thermus thermophilus	PS3	Nitrobacter agilis	Rhodopseudomonas sphaeroides	Bacillus subtilis
Heme type(s)	(102) aa_3	(104) aa_3	(105) c_1/aa_3	(106) c/a_3	(107) aa_3	(39) aa_3	(40) aa_3
Heme a content (nmol/mg protein)	27	n.r.	10	11.3—13.5	20	14	17
Absorption peaks (reduced)	605,445	602,442	602,442	604,444	606,442	606,443	601,443
Copper content (natom/mg protein)	31	18	11.8	16—183	19.1	n.r.	n.r.
Number of subunits	2	3	3	3	2	2	3
M_r of subunits	45, 28	32, 23	52, 37, 29	56, 58, 22	51, 31	45, 37	57, 37, 21
Stoichiometry of subunits	1:1	1:1	1:1:1	1:1:1	1:1	n.r.	n.r.
Minimal M_r (from PAGE-SDS)	73	55	118	136	82	82	115
Sedimentation analysis	86 ± 5	115	n.r.	250	n.r.	n.r.	n.r.
Turnover number	90	40	60	*PS3*	*N. agilis*	300	66

carboxylate groups which are involved in the binding of cytochrome c at the level of subunit II are glutamate 18, aspartate 112, glutamate 114 and 198, as judged from the labeling experiments described above.[145] It is interesting to notice that a cluster of aromatic residues, -Trp-Tyr-Trp-Ser-Tyr-Glu-Tyr-, included in the beef heart sequence positions 104 to 110, is completely invariant. This special structure has been suggested to be involved in electron conduction.[79,136,145]

E. The Site of Interaction with Lipid

Photoactivatable arylazido derivatives have been employed to study which subunits of the yeast and of the bovine heart enzymes are in contact with phospholipids.[48,52,54,55] It has been found, especially using arylazidophospholipids,[52] iodonaphtylazide,[54] and adamantane diazirines,[54] that subunits I and III are largely in contact with the lipid, that subunits II, IV, and VII are also in contact with the membrane although to a lesser extent, while subunits V and VI have apprently no interaction with the lipid.

Critical information for the understanding of the assembly of the protein, as well as our understanding of the stability of such a protein, has been made with the localization by Bisson et al.[151] of the lipid binding domains of subunit II of the bovine oxidase. It was shown that the sequence 20 to 98 which contains two long stretches of uncharged amino acid residues is in contact with the phospholipids. On this basis of a scheme of the partial folding of the subunit II of cytochrome c oxidase has been proposed (Figure 4). Also, the arrangement of subunit IV has been explored by chemical labeling and protease digestion studies. It was concluded that the N-terminal segment[1—71] is in the water phase at the surface of the complex and located on the matrix side of the mitochondrial inner membrane. The residues 79 to 98 are buried in the membrane. The most likely structure would be that having the N and C termini on opposite sides of the membrane and with the sequence 79 to 98 spanning the membrane in a helical structure (alpha-helix or 3/10-helix).[59]

IV. THE PROTON PUMP

It is accepted almost universally by now that H^+-translocation by the electron transport chain is an important intermediate step in oxidative phosphorylation as proposed by Mitchell.[8,9] According to the original statements of the chemi-osmotic hypothesis, H^+-translocation by the respiratory chain occurs exclusively through a ligand-conduction mechanism where protons and electrons share the same pathway down "loops" of alternating hydrogen and electron carriers, cytochrome c oxidase being one of a series of electron carriers in the third proton-motive loop. Though this model was later modified to combine the second and third loops (the so-called Q-cycle, see Mitchell[152]), the function of cytochrome c oxidase remained simply to transport electrons from cytochrome c to oxygen. Mitchell and Moyle[153] also demonstrated an interesting feature of this electron-conducting activity; although the cytochrome c-binding site on cytochrome c oxidase is located on its external surface, the protons which are required to combine with the reduced oxygen atoms to produce water are taken from the intramitochondrial space. This entails that enzyme turnover will lead to the formation of an electrochemical proton gradient (inside negative and alkaline) across the mitochondrial membrane which will exert a "control" on the enzyme's activity which may be released by a protonophore. These latter two predictions were, indeed, confirmed[154,155] and this provided an important experimental support for the chemiosmotic model. It was proposed by Wikström,[156] however, that the chemiosmotic description depicts only one aspect of the energy-transducing activity of cytochrome c oxidase and that electron transfer through the enzyme is coupled to outwardly directed H^+-translocation. Such an H^+-pumping activity was first demonstrated by experiments with intact mitochondria or submitochondrial particles[156,157] where cytochrome c oxidase was isolated from the remainder of the oxidative

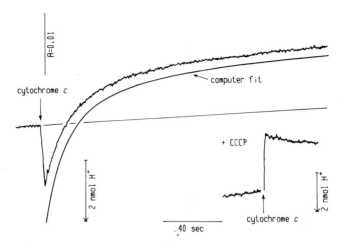

FIGURE 5.   The proton pump of cytochrome *c* oxidase reconstituted in phospholipid vesicles. The changes in pH of a suspension of cytochrome *c* oxidase vesicles (0.6 nmol of heme *a* in 75 m*M* choline chloride, 25 m*M* KCl, 50 m*M* phenol red, pH 7.4 (total volume 1.4 m*ℓ*) in the presence of 4.5 nmol of valinomycin were measured spectrophotometrically at 24°C as the change in phenol red absorbance. At the point indicated, 2.3 nmol of ferrocytochrome *c* were added. The curve below the trace is a computer-generated fit to the decay of the pulse of extruded H$^+$. The square-ended line at the end of the trace shows the change in absorbance given by 2 μ*ℓ* of 0.5 m*M* oxalic acid in the presence of 3.5 μ*M* carbonyl-cyanide-*p*-hydroxyphenylhydrazone.

phosphorylation apparatus by using inhibitors. It was shown that when electrons flowed through cytochrome *c* oxidase, there was an acidification of the extramitochondrial phase, consistent with one H$^+$ being extruded per electron (vice versa for the experiments with submitochondrial particles). These results were complemented by the finding that concomitantly with this H$^+$ appearance, two positive charges were taken up by the mitochondria per electron passing to oxygen,[158,159] cf. the predicted ratio of one charge per electron if cytochrome *c* oxidase operated according to the chemi-osmotic model. While the experiments with mitochondrial systems have been criticized,[160,161] these points have been answered[162,163] and a large amount of supportive evidence has come from experiments with purified cytochrome *c* oxidase reconstituted into phospholipid vesicles (see Casey and Azzi[164] for a review). Thus, despite some remaining objections,[165—167] at present it is widely held that cytochrome *c* oxidase is a redox-linked proton pump (Figure 5). There is some disagreement concerning the number of protons extruded per cytochrome *c* molecule oxidized. While most laboratories have found this value to be close to one, both for the enzyme in mitochondria[156,159] and in reconstituted vesicles,[168—170] Lehninger and co-workers have consistently reported values higher than this (see, e.g., Alexandre et al.[171,172]). From a thermodynamic viewpoint, the redox gap between cytochrome *c* and oxygen is approximately 550 mV and the proton-motive force formed by coupled mitochondria has been reported as being approximately 200 mV.[173] Thus, the transfer of one electron from cytochrome *c* to O$_2$ could account for the movement of two to three H$^+$s against the electrochemical proton gradient. As one of these is needed for water formation it follows that between one and two protons could be pumped per electron. In any consideration of H$^+$/e$^-$ ratios, however, it must be remembered that values for the redox potentials of hemes *a* and a_3 are highly dependent on the conditions of the measurements (see Dutton and Wilson[174] and that determinations of the proton-motive force in mitochondria are open to a number of experimental errors, particularly with regard to the electrical component. In addition, the H^+ pump may not have a fixed mechanistic H^+/e^- ratio, but this may vary and, as such, might depend on the conditions of measurement, for example, different values would be expected for determinations carried out under steady-state and pulse conditions. The mechanism of proton pumping is, as yet, relatively obscure. One requisite feature, however, of any H^+-translocating apparatus is the presence of one or more H^+ channels along which protons may pass through the hydrophobic core of the protein and thus from one side of the membrane to the other (see Nagle and Tristram-Nagle[175] for a review). A number of lines of evidence have indicated that in the case of the cytochrome *c* oxidase proton pump, subunit III may be involved in this function. The first indications came from studies with dicyclohexylcarbodiimide (DCCD).[176,177] These showed that when cytochrome *c* oxidase vesicles or mitochondria were incubated with DCCD, the H^+-pumping activity was strongly inhibited though there were only minor effects on the electron transfer rate. In addition, the DCCD-induced decrease in size of the cytochrome *c*-induced extra-vesicular acidification in oxidase vesicles was not caused by a decreased rate of H^+ extrusion but by a lowering of the number of H^+ translocated per electron (see Casey et al.[178]), i.e., the reaction with DCCD induced a molecular decoupling of the H^+ pump. It was also found that following incubation of cytochrome *c* oxidase in reconstituted vesicles or mitochondria with DCCD, there was a specific labeling of subunit III[177—181] at a single glutamic acid residue;[180] furthermore, the sequence of amino acids flanking this residue is remarkably similar to those of the DCCD-binding peptides from H^+-conducting proteolipids of a number of H^+-translocating ATPases (see Prochaska[180]). These results were interpreted as indicating that subunit III of cytochrome *c* oxidase is involved in some aspect of the coupling of electron flow to H^+-translocation, e.g., in H^+-channeling.

These proposals were strengthened by the observation[97] that removal of subunit III from cytochrome *c* oxidase had negligible effects on the electron transfer rate but led to loss of H^+-pumping activity following reconstitution. A further indication that the function of subunit III is not in electron transfer came from the finding[116] that the hemes and coppers of cytochrome *c* oxidase are not located in subunit III.

The cytochrome *c* oxidase from *Paracoccus denitrificans* contains only two subunits following isolation[103] and when reconstituted displays a proton-pumping activity, though with a considerably lower ratio of H^+ translocated per electron than that obtained with the reconstituted bovine enzyme.[38] This observation is consistent with the findings presented above assuming that this bacterial enzyme possesses a less efficient H^+ transfer than that in the bovine enzyme.

The drop in redox potential between the cytochrome *c* and oxygen couples is sufficiently large (approximately 500 mV) that electron transfer between these may be considered practically irreversible. This would appear to contradict the concept of the above-described control of the rate of electron transport exerted as a result of proton gradient formation. This apparent paradox is resolved by the fact that this electron transfer is divided into a number of steps (see Wikström et al.[7]), one or more of which occurs across sufficiently small redox gaps that they are subject to feedback control from the proton gradient. The question arises as to which of these partial electron transfer reactions are associated with proton translocation; clearly, the most likely candidates are those steps which may be under thermodynamic control by the proton gradient. It is almost certain that H^+-translocation is not associated with the terminal oxygen reducing reaction; the difference between the operating potential of heme a_3 in phosphorylating mitochondria and that of the oxygen/water couple is approximately 200 mV[182] and thus this process is effectively irreversible. It has been suggested, instead,[7] that the proton translocation event is associated with the transfer of electrons to and from heme *a*. The midpoint potential of this heme is decreased as the pH of the intramitochondrial space is increased, and it has been found[183] that depletion of subunit III from cytochrome *c* oxidase which leads to loss of H^+-pumping activity also results in the

Table 4
DEPENDENCE OF THE REACTION OF REDUCED CYTOCHROME *c*
WITH CYTOCHROME *c* OXIDASE ON THE CONCENTRATION OF
CYTOCHROME *c*

	Cytochrome *c*		Cytochrome oxidase		
				No. of electrons accepted	
	k_{obs}	No. of electrons	k_{obs}		
$[Cytc^{2+}]:[aa_3]$	s^{-1}	transferred	s^{-1}	Cyt *a*	Cu_A
8	630 ± 10	1.9 ± 0.2	600 ± 100	0.98 ± 0.1	0.9 ± 0.3
4	490 ± 120	1.4 ± 0.1	480 ± 90	0.83 ± 0.05	0.6 ± 0.02
2	310 ± 90	1.0 ± 0.1	310 ± 30	0.6 ± 0.1	0.3 ± 0.2
1	280 ± 60	0.73 ± 0.04	250 ± 60	0.4 ± 0.05	0.3 ± 0.1

loss of this pH dependence. DCCD, on the other hand, has no effect on the pH dependence of the E_m of heme *a*.[184] Earlier observations of a shift in the spectrum of heme a_3 of cytochrome *c* oxidase in mitochondria resulting from ATP hydrolysis which were interpreted as an effect on heme a_3 of the proton gradient[185] have since been shown to be fully explicable by reversed electron transport.[186]

Thus, a tentative model for the cytochrome *c* oxidase proton pump would be that reduction of heme *a* leads to the uptake by this heme or by a group in close proximity to it, of a proton which is in electrochemical equilibrium with the protons in the internal phase via a proton channel. Subsequent oxidation of heme *a* would lead to release of the proton into another group in equilibrium with the external space. While these proton pathways may traverse several subunits, the findings described above suggest that at least one important group is associated with subunit III. A notable feature of such a model is that it would require an intersubunit communication between subunit III and heme *a* which is located in subunits I or II (see above).

V. KINETICS OF ELECTRON TRANSFER

As was pointed out in the introduction, cytochrome *c* oxidase catalyzes the transfer of electrons from ferrocytochrome *c* to molecular oxygen; ferricytochrome *c* and water are yielded as the final products of this reaction. Although the reduction of oxygen to water requires four electrons, a concerted four-electron transfer from reduced cytochrome *c* to cytochrome *c* oxidase has never been suggested, even though the oxidase is a potential four-electron acceptor. The mechanism of electron transfer catalyzed by cytochrome *c* oxidase remains obscure. The kinetics of the reaction were investigated in several laboratories either under steady-state conditions or by presteady-state measurements. On the basis of these studies different models and mechanisms have been proposed.

A. Presteady-State Measurements
1. The Oxidation of Cytochrome c

When the oxidation of ferrocytochrome *c* by oxidized cytochrome *c* oxidase was followed under anaerobic conditions, two phases of the reaction could be observed. A rapid initial oxidation of cytochrome *c*, complete within 10 msec, was followed by a much slower electron transfer[187] (see Table 4). The absorbance changes measured at 550 as well as 445 and 605 nm indicate that the oxidation of cytochrome *c* during the rapid phase is correlated with the reduction of cytochrome *a*. Since the rapid changes were twice as big in the presence of cyanide, it was suggested that in the absence of this inhibitor the slow phase represents the reduction of cytochrome a_3.[187] Recently, Antalis and Palmer[188] confirmed this under carefully controlled anaerobic conditions. When oxidase was reduced in the presence of CO and then illuminated, the slow appearance of the second phase also indicated that the transfer of electrons from cytochrome a to a_3 represents the slow process.[187] The rate of this process was found to be independent of cytochrome c concentration with a first-order rate constant of about 3 sec^{-1}.[187,188] This extremely slow rate may suggest the existence of a kinetic barrier to the reduction of cytochrome a_3 under anaerobic conditions. In the case of the initial fast reaction between reduced cytochrome c and the oxidase, the second-order rate constant was calculated to be between 10^6 and 10^7 $M;^{-1}$sec^{-1}.[187,189—191]

Gibson et al.[187] suggested as well that also one of the coppers could be dynamically linked to the reduction of cytochrome a. The bleaching of the 830-nm band lagged behind or was simultaneous with the increase of absorption measured at 605 nm and did not depend on the cytochrome c concentration in presteady-state measurements. Addition of cyanide to the cytochrome c oxidase had no effect on the kinetics of the reduction of cytochrome a and "visible" copper (Cu_A), but inhibited electron transfer to the two other sites, cytochrome a_3 and Cu_B.[191]

There were several values reported for the stoichiometry of the initial rapid transfer of electrons. Andreasson[190] showed that when the reduced cytochrome c reacted with the oxidized enzyme, the number of moles of cytochrome c oxidized per mole of oxidase reduced approached the value of 1 at relatively high concentrations of cytochrome c, in the rapid initial reaction. Several other investigators have reported, however, values approaching 2 mol of cytochrome c oxidized per mole of cytochrome oxidase reduced.[188,191,192] When the same reaction was studied in the reverse direction between the reduced oxidase and oxidized cytochrome c, the number of electrons transferred reached a saturation level of two electrons at relatively high concentrations of cytochrome c.[193]

It was recently shown again that, using cytochrome c in excess, the number of electrons transferred from the reduced cytochrome c to the reduced oxidase reached a maximum of two. Since one of them was transferred to cytochrome a, the total number of electrons transferred was distributed equally between cytochrome a and Cu_A[188] (see Table 4). When the ratio of cytochrome c to oxidase equalled 1, only one electron was transferred per mole of oxidase and the amount of electrons was equally distributed between cytochrome a and Cu_A as well. An equilibrium constant of 1.0 proposed for the electron transfer between heme a and copper A[188] would indicate that these redox centers have very similar midpoint potentials. The latter conclusion seems to be contradictory, however, to the potentiometric studies,[194,195] showing a different midpoint potential for heme a and Cu_A of 210 and 245 mV, respectively.

Van Gelder et al.[196] have shown that the presteady-state kinetics of the reaction between reduced cytochrome c and oxidase were very dependent on ionic strength.

With increased ionic strength the first burst phase of the reaction between ferrocytochrome c and cytochrome c oxidase became markedly biphasic. The slow phase observed under these conditions is unrelated to the slow transfer of electrons to cytochrome a_3. Both redox centers, cytochrome a and Cu_A, have accepted one electron each, but the number of electrons transferred in the fast phase decreased gradually from two to one with increase of ionic strength, and, in parallel, the number of electrons transferred in the slow phase increased, approaching one as the maximal value. A simple two-step mechanism has been proposed for this biphasic reaction: the first molecule of cytochrome c reacts with the oxidase but the second molecule cannot react until dissociation of the first occurs.[188]

Contrary to this scheme, where a two-step electron transfer at one binding site occurs, the group of Van Gelder has interpreted their results in terms of two or even four binding sites.[197] When a cytochrome c-aa_3 complex was studied in the media of low ionic strength, different rate constants were observed for the initial burst phase of the reaction, depending which component, ferrocytochrome or cytochrome c-aa_3 complex, were in excess. The amount of reduced heme a was double when ferrocytochrome c was in excess,[198] and the authors concluded that it could indicate the existence of two cytochrome c binding sites on the enzyme molecule. The same authors[198] presented the effect of polylysine on the addition of cytochrome c (reduced by ascorbate/tetramethyl-p-phenylenediamine). In the presence of this positively charged polypeptide which inhibits competitively the steady-state activity of oxidase towards cytochrome c,[199] only little reduction of heme a occurs. Since polylysine stimulates the release of the tightly bound cytochrome c from the cytochrome c-aa_3 complex, the existence of negative cooperativity between the two binding sites has been proposed.[198] The possibility of anticooperativity between monomers was discussed as well by Wikström et al.,[7] who suggested that the enzyme may function as a dimer with a "half-of-the-sites" reactivity.

2. Reduction of Dioxygen

Molecular oxygen is reduced by reduced cytochrome c oxidase at room temperature with a second-order velocity constant of $10^8 \, M^{-1} sec^{-1}$.[187,200] Intermediates of this fast reaction could be detected only by a special low temperature technique ("triple trapping"[201]). Fully or partially reduced cytochrome c oxidase was blocked with CO, cooled and frozen in different steps. The CO-cytochrome a_3 complex was then split with an intense flash of light. The reaction was followed spectrophotometrically and could be stopped at any time by freezing in liquid nitrogen. Under anaerobic conditions, the first event detected was the decrease of the band at 590 nm and, simultanously, appearance of a band at 612 nm, the classical CO dissociation spectrum. The rebinding of CO was abnormally slow, which could be recently explained by assuming that there is no real dissociation of CO but rather a displacement from the a_3-iron to Cu_B.[202]

In the presence of O_2, the first detectable compound (compound A[201]) had a similar but clearly distinguishable spectrum from that of the Fe^{II}-CO complex of cytochrome a_3. The authors suggested binding of O_2 to ferrous cytochrome a_3 or to both a_3 and Cu_B. Whereas the fully reduced as well as the partially reduced (i.e., heme a_3 and Cu_B reduced) enzyme formed compound A,[201,203,204] the products followed kinetically had different properties. Compound C was the second product obtained by the half-reduced cytochrome c oxidase and had a high extinction at 605 nm but a very weak Soret band compared to reduced a_3.[205,206] The EPR spectrum showed clearly oxidized heme a and Cu_A.[201,204] Several structures were proposed for compound C, where the absorption at 605 nm was not attributed to ferrous a_3 but rather to a peroxide complex with ferric heme a_3 and Cu_B (II) or even ferryl a_3 (Fe^{VI}) and Cu_B(I).[7] It is also not excluded that compound C is two species formed sequentially after compound A[204,206] and which differ only in their conformation or degree of protonation.

If compound C really has catalytic importance, it should also be formed when the fully reduced enzyme reacts with O_2. It was, therefore, surprising that, in this case, a different compound (compound B) was observed, although other intermediates between compound A and B have also been demonstrated.[203] One of them, called "II", was associated with an increase in absorbance at 605 to 610 nm and resembled, in this respect, the peroxy compound C, but had reduced heme a_3 and Cu_A. "II" may be a transient form of compound C and compound B the result of an electron transfer from heme a and Cu_A. Indeed, compound B contained partially oxidized heme a and Cu_A, what could be clearly shown by EPR and optical difference spectra.[201,203] This leads to the conclusion that the enzyme is heterogenous in this state, if the triple-trapping method is not giving low temperature artifacts. Indications that the latter is not the case come from fast kinetic measurements of the oxidation of reduced cytochrome c oxidase by oxygen at room temperature.[187] The large extent of the fast kinetic phase of the absorbance decrease at 605 nm includes oxidation of heme a_3 as well as 30 to 50% oxidation of heme a. Oxidation of Cu_A had also biphasic kinetics at room temperature.[207]

When the absorbtion at 610 nm instead of 605 nm was recorded, an intermediate like the "II" of Clore et al.[203] could have been already interpreted (cf. Figure 5[207]). It can be concluded that the low temperature measurements give reliable data for the elucidation of the mechanisms occurring normally at room temperature.

Recently, intermediates beyond compound B were described.[208,209] Similar low temperature measurements were applied to the reduced cytochrome c oxidase but in the absence of extraneous electron donors, as, e.g., ferrocytochrome c. A new EPR signal as well as absorbtion at 580 nm were detected and attributed to an EPR visible form of Cu_B (II). Although the valence states of these latter intermediates were not unequivocal, these data supported similar conclusions coming from partial reversal cytochrome c oxidase reaction experiments,[187] from which a mechanism for the electron transfer from cytochrome a to oxygen was proposed. Cytochrome a may transfer one electron to the cytochrome a_3/Cu_B-complex which may be in the "peroxy" compound C state, which transfers then in a concerted reaction two electrons to oxygen (for more detailed information see Wikström[186]).

3. The Pulsed Oxidase

Cytochrome c oxidase, after its isolation, contains all the redox centers in the oxidized form (resting enzyme). If this enzyme is first fully reduced and pulsed with O_2 in the presence of reduced cytochrome c, it is transformed into a catalytically more active state (the O_2 pulsed enzyme.[210] The rate of ferrocytochrome c oxidation is stimulated four- to fivefold in the case of pulsed oxidase in comparison with the resting enzyme.[210—213] Other reducing agents (other than ferrocytochrome c/ascorbate) like dithionite or NADH and ferricyanide as oxidant instead of oxygen may be used to enhance the resting enzyme to a similar catalytically more active state.[214,215] Because the rates for electrons transferred from cytochrome c to cytochrome a in the pulsed and the resting enzyme are similar, the transfer of electrons from cytochrome a to the cytochrome a/Cu_B-center must be strongly stimulated.[212] Brunori et al.[212] measured the spectral characteristics of the resting and the pulsed oxidase as a function of the reaction time. In the presteady state, the pulsed oxidase showed a higher extinction coefficient of the 605-nm band and a slight red shift of the spectrum compared to the resting oxidase. Reaching the steady state both enzymes have similar spectra, resembling strongly those of the fully oxidized species. Thus, the pulsed enzyme is not the prevailing form during the steady state; a half-time of 100 msec is needed to convert the pulsed enzyme into the steady-state configuration.[212] Moreover, the data of Brunori et al.[212] did not exclude the possiblity that the pulsed oxidase is heterogenous, that may increase the difficulties in assigning a correct structure for the pulsed form of the enzyme. All data available until now favor the idea that a conformational change within the enzyme might be the most striking event to convert the resting state into a highly active form. The resting state might be a thermodynamically favorable conformation to which the enzyme is slowly dropping back if it is not stimulated by a pulse of reductant in the presence of a suitable oxidant. This conformational change might be a part of the still unknown function of the small subunits in the eukaryotic oxidase. However, recently it was shown that the two subunits containing cytochrome c oxidase from *Paracoccus denitrificans* could be pulsed under the same conditions as it was described for the beef enzyme.[216] One difference between the two enzymes is, however, that the rate of the transfer from the reduced resting to the reduced pulsed state in *P. denitrificans* is about 50 times higher, so that the resting state may scarcely be populated during turnover. This might imply that the bacterial oxidase is much less sensitive towards regulation by reductants.

B. Steady-State Kinetics

The oxidation of cytochrome c by cytochrome c oxidase was demonstrated by Keilin,[217] who described the apparent Michaelis kinetics of the reaction. Smith and Conrad[218] reported

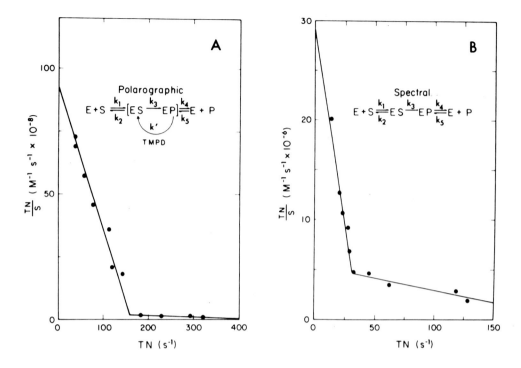

FIGURE 6.   The steady-state kinetics of cytochrome *c* oxidase as measured by two different assays. (From Ferguson-Miller, S., Brautigan, D. L., and Margoliash, E., *J. Biol. Chem.*, 253, 149, 1978. With permission.)

that the reaction between cytochrome *c* and the purified and solubilized oxidase showed unusual kinetics. It has been demonstrated[218—222] that the reaction of ferrocytochrome *c* with oxidase obeys the general rate law: velocity = k′ (oxidase) (ferrocytochrome *c*) where k′ (oxidase) is the pseudo-first-order rate constant k_{obs}. The constant k′ is a function of the total cytochrome *c* concentration,[218,222] and the nonlinear relationship was obtained when the data obtained over a wide range of cytochrome *c* concentrations were plotted as 1/k′ vs. cytochrome *c* concentration (analogous to v/[S] vs. [S] plot).[222] Furthermore, the apparent inhibition by cytochrome *c* is independent of the oxidation state of the cytochrome *c*. Smith and Conrad[218] postulated the formation of inhibitory complexes between cytochrome *c* and the enzyme as a possible explanation of this phenomenon.

Two different kinds of methodology have been recently utilized in studies of the steady-state kinetics of cytochrome *c* oxidase: either the oxidation of the ferrocytochrome *c* was followed spectrophotometrically, or ferricytochrome *c*-oxidase complex was reduced by ascorbate or ascorbate + *N,N,N′,N′*-tetramethylphenylenediamine and the O_2 uptake was measured polarographically (see the reaction schemes on Figure 6).

Ferguson-Miller et al.[101] when studying polarographically the activity of the purified cytochrome *c* oxidase, showed that the enzyme has biphasic kinetics with apparent K_m values of $5 \times 10^{-8} M$ and 0.35 to $1.0 \times 10^{-6} M$ for the high and low affinity phases, respectively (Figure 6). Those values were confirmed later by Nicholls et al.[224] and Smith et al.[225] Ferguson-Miller et al.[101] correlated these biphasic kinetics with two binding sites for cytochrome *c* on the oxidase molecule. Later, the same authors[223] showed, as well, that biphasic kinetics of cytochrome *c* oxidation were also obtained using the spectrophotometric assay. Although both the spectral and polarographic assays gave similar biphasic Eadie-Hofstee plots, different numerical values were obtained for the apparent kinetic constants. The apparent K_m values calculated from the spectral assay were $10^{-6} M$ and $4 \times 10^{-5} M$ for the low and high affinity phase, respectively. The explanation of the differences obtained using the two assays was that in the spectral assay, due to the very slow dissociation rate of cytochrome *c*, the high affinity reaction was not visible. This reaction could be observed in the polarographic system in which tetramethylphenylenediamine reduces bound cytochrome *c*. So the authors proposed that the first phase observed spectrally would correspond to the second phase observed polarographically and they considered the possibility of the binding of a third molecule of cytochrome *c*. Smith et al.[225] presented their kinetic data from polarographic measurements as smooth continuous curves in an Eadie-Hofstee plot. They also tried to compare both assay methods under the same experimental conditions with low concentrations of cytochrome *c*. Under all conditions tested and irrespective of the content of oxidase in the assays, they observed a sharp increase in the reaction rate with increasing concentrations of cytochrome *c* up to between 0.1 and 0.25 μM, and then a more gradual increase with higher concentrations of cytochrome *c*.

Ferguson-Miller et al.[101] conducted the binding experiments with ferricytochrome *c*, studying in this way the dissociation of the product of the reaction. The values of K_D obtained are similar to kinetic, "apparent K_m", constants, which could imply that either the binding of ferro- and ferricytochrome *c* are the same, or the dissociation of the product from the enzyme could be the controlling factor in the reaction.

Errede and Kamen[226] also assumed that the K_{eq} for the binding of the ferrocytochrome *c* to the oxidase equaled that for the dissociation of the ferricytochrome *c*-oxidase complex. They have reported[222,226] that the nonlinear dependence of the rate constant k' on cytochrome *c* concentration (c) closely fits the equation:

$$k' = \frac{\alpha_1 + \alpha_2(c)}{1 + \beta_1(c) + \beta_2(c)^2}$$

The denominator terms represent the distribution of the oxidase in its various forms at steady state; the quadratic term indicates that, under certain conditions two molecules of cytochrome *c* are bound to the oxidase. The four constants were approximated by satisfying two limiting conditions (very low and very high concentrations of cytochrome *c*). From this equation, any number of mechanisms can be devised. Errede et al.[222] modified the so-called "mechanism IV" of Minnaert[219] taking into account complexes involving two molecules of cytochrome *c* per oxidase. These are so-called "dependent site" and "dead-end" mechanisms, assuming either productive or nonproductive complexes, respectively. "Independent site" mechanism as well as a scheme in which free cytochrome *c* participates in electron transfer when the other cytochrome *c* molecule is bound ("exchange" mechanism) were also proposed.[226]

The steady-state reaction between cytochrome *c* and cytochrome *c* oxidase can be controlled by different factors. Ferguson-Miller et al.[101] reported a decrease of the high affinity phase of the reaction in the presence of different anions (with the effectiveness ATP > ADP > P_i). Some discrepancies exist concerning the effect of pH. Ferguson-Miller et al.[101] reported the increase of maximal velocities with pH, while Wilms et al.[227] showed the decrease of velocity and K_m with pH increase. Some interesting effects of cations were also reported by the group of Van Gelder.[198,227] The activity optimum was reported to occur at ionic strength of about 75 mM and the discontinuity of the rate constant k' against ionic strength was correlated with this optimum. Below this optimum, an increase of the amount of the complex between cytochrome *c* and the oxidase was observed.

Experiments performed under different conditions (various buffers and pH values) gave quite different apparent K_m values derived from spectrophotometric and polarographic assays[225] against the simple assumption that the conditions which increase the dissociation of the complex should increase the rate in the spectrophotometric assays and decrease the uptake of O_2 in polarographic ones.

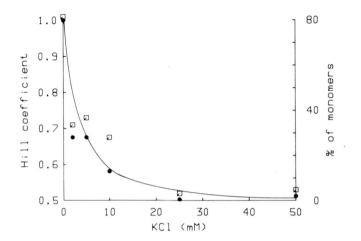

FIGURE 7. The relationship between negative cooperativity in cytochrome *c* oxidase and its dimeric structure. The Hill coefficient was measured from Hill plots (□) and the % of monomers (●) at the different salt concentrations was obtained after gel filtration. (From Nałęcz, K., Bolli, R., and Azzi, A., *Biochem. Biophys. Res. Commun.*, 114, 822, 1983. With permission.)

Smith et al.[225] suggest that complexes of different reactivity can be formed between cytochrome *c* and oxidase, depending on the experimental conditions, and that some especially reactive complexes can exist.

The concept of complexes with different molecular activity was developed further recently. Nałęcz et al.[99] reported that the kinetics of monomeric oxidase (prepared in dodecylmaltoside) differ from the kinetics of the dimeric form. The monophasic kinetics with low V_{max} and K_m values have been ascribed to monomers, while in the case of dimers the kinetic data fit with a model of homotropic, negative cooperativity. Under conditions where the amount of monomers decreased with the increase of ionic strength, more negative cooperativity was revealed (Figure 7). On the basis of this finding, a model has been proposed in which cytochrome *c* oxidase dimers contain two cytochrome *c* binding sites, namely, one per monomer, and the occupation by one molecule of cytochrome *c* of one binding site will hinder the occupation by the second molecule of the second site.

The model containing one binding site could have been questioned by the results of Ferguson-Miller et al.,[101] who proposed two cytochrome *c*-binding sites on the oxidase molecule, on the basis of their kinetic results. From the direct binding measurements the ratio of cytochrome *c* to oxidase equalled 1.4 at low ionic strength. Later on, however, the same authors, improving the gel filtration technique, obtained the value of 1 at the same ionic strength.[223,228] A complex with a stoichiometry 1:1 between cytochrome *c* and oxidase was also reported by several other authors.[229—232] Some secondary sites binding cytochrome *c* were ascribed by Nicholls[223] to the membrane parts, while the results of Vik et al.[234] indicated that diphosphatidylglycerol molecules bind the second molecule of cytochrome *c*, and this phenomenon did not occur in delipidated enzyme. Wilms et al.,[198] who obtained the same ratio of 1:1 for the cytochrome *c*-oxidase complex at low ionic strength, observed dissociation of this complex when salt concentration was increased. Recently, Margoliash and Bosshard[235] discussed the electrostatic influence of a single bound molecule of cytochrome *c* on the binding of the others by cytochrome *c* oxidase. In fact, they found the previous model (i.e., that assuming two catalytic sites on the enzyme) to be unlikely, since it can be shown that only binding of the first molecule of cytochrome *c* has an effect on the heme absorbance spectrum.

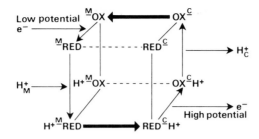

FIGURE 8. The "cubic model" of cytochrome *c* oxidase. (From Wikström, M. K. F., Krab, K., and Saraste, M., *Cytochrome Oxidase, A Synthesis*, Academic Press, London, 1981. With permission.)

The stoichiometry of this complex of one cytochrome *c* per oxidase monomer (as resulted from the binding studies of arylazidocytochrome *c* derivatives[44]) and the observed negative cooperativity in the dimers can lead to a "half-of-the-sites" reactivity. Moreover, it was recently reported[236] that the domain on the horse heart cytochrome *c* molecule which interacts with high affinity with beef heart cytochrome *c* oxidase was undistinguishable from one supposed to interact with a "low affinity site".

VI. CONCLUSIONS

The minimum scheme for a redox-driven proton pumping cycle has been described by the "cubic model" of Wikström et al.[7] (Figure 8). Although rather general, this model imposes some restrictions, such as the requirement of pH-dependent midpoint potentials of at least some of the redox centers associated with the enzyme and an asymmetry of their pH-dependent properties. It has been shown by Wikström et al.[7] that cytochrome oxidase adequately fulfills these criteria. Wikström et al.[7] have later suggested that an alternating site mechanism would better describe the proton pumping events associated with cytochrome oxidase. This latest model requires that the functional unit of the proton pumping cytochrome oxidase is a dimer.

The proton pumping cytochrome *c* oxidase, at present, remains a thermodynamic "black box". The details of the coupling between the redox reactions and the proton translocation are only at the very beginning of a molecular understanding. The study of Callahan and Babcock,[147] as already discussed in a previous section, suggests that the primary energy-conserving event may be associated with a change in hydrogen bond strength of a formyl group attached to cytochrome *a*. Structural and covalent modification studies,[177] together with subunit III depletion studies,[183] have shown that subunit III is an essential component of the proton pump. It is remarkable that removal of this subunit abolishes the pH dependency of the redox midpoint potential of cytochrome *a*.[183]

Although at an initial stage and very far from the detailed understanding of proteins such as hemoglobin, knowledge of the catalytic mechanisms of cytochrome *c* oxidase will rapidly increase. Cytochrome *c* oxidase research, in fact, represents not only a possiblity to clarify a complicated enzyme mechanism but it may also provide a basis for the understanding of the role played by different subunits in membrane-bound enzymes or enzyme-receptor complexes.

ACKNOWLEDGMENTS

This work has been rendered possible through the support of the Schweizerischer Nationalfonds (3.739-0.80).

REFERENCES

1. **Poole, R. K.,** Bacterial cytochrome oxidases. A structurally and functionally diverse group of electron-transfer proteins, *Biochim. Biophys. Acta,* 726, 205, 1983.
2. **Merle, P. and Kadenbach, B.,** On the function of multiple subunits of cytochrome *c* oxidase from higher eukaryotes, *FEBS Lett.,* 135, 1, 1981.
3. **Lemberg, M. R.,** Cytochrome oxidase, *Physiol. Rev.,* 49, 48, 1969.
4. **Wikström, M. K. F., Harmon, J. H., Ingledew, W. J., and Chance, B.,** A re-evaluation of the spectral, potentiometric and energy-linked properties of cytochrome *c* oxidase in mitochondria, *FEBS Lett.,* 65, 259, 1976.
5. **Malmström, B. G.,** Cytochrome *c* oxidase, structure and catalytic activity, *Biochim. Biophys. Acta,* 549, 281, 1979.
6. **Lindsay, J. G., Owen, C. S., and Wilson, D. F.,** The invisible copper of cytochrome *c* oxidase, *Arch. Biochim. Biophys.,* 169, 492, 1975.
7. **Wikström, M. K., F., Krab, K., and Saraste, M.,** *Cytochrome Oxidase. A Synthesis,* Academic Press, London, 1981.
8. **Mitchell, P.,** *Chemiosmotic Coupling in Oxidative and Photosynthetic Phosphorylation,* Glynn Research, 1966.
9. **Mitchell, P.,** *Chemiosmotic Coupling and Energy Transduction,* Glynn Research, 1968.
10. **Tzagaloff, A.,** *Mitochondria,* Plenum Press, New York, 1982.
11. **Lewin, A. S., Gregor, I., Mason, T. L., Nelson, N., and Schatz, G.,** Cytoplasmically made subunits of yeast mitochondrial F_1-ATPase and cytochrome *c* oxidase are synthesised as individual precursors, not as polyproteins, *Proc. Natl. Acad. Sci. U.S.A.,* 77, 3998, 1980.
12. **Mihara, K. and Blobel, G.,** The four cytoplasmically made subunits of yeast mitochondrial cytochrome *c* oxidase are synthesised individually and not as a polyprotein, *Proc. Natl. Acad. Sci. U.S.A.,* 77, 4160, 1980.
13. **Schatz, G. and Butow, R. A.,** How are proteins imported into mitochondria?, *Cell,* 32, 316, 1983.
14. **Rosen, S.,** Purification of beef-heart cytochrome c oxidase by hydrophobic interaction chromatography on octyl-sepharose CL-4B, *Biochim. Biophys. Acta,* 523, 314, 1978.
15. **Mason, T. L., Poyton, R. O., Wharton, D. C., and Schatz, G.,** Cytochrome c oxidase from baker's yeast. I. Isolation and properties, *J. Biol. Chem.,* 248, 1345, 1973.
16. **King, T. E.,** The Keilin-Hartree muscle preparation, in *Methods in Enzymology,* Estabrook, R. W. and Pullman, M. E., Eds., Academic Press, New York, 1967, 202.
17. **Jacobs, E. E., Andrews, E. C., Cunningham, W., and Crane, F. L.,** Membraneous cytochrome oxidase purification, properties and reaction chracteristics, *Biochem. Biophys. Res. Commun.,* 25, 87, 1966.
18. **Briggs, M., Kamp, P. F., Robinson, N. C., and Capaldi, N. C.,** The subunit structure of the cytochrome c oxidase complex, *Biochemistry,* 14, 5123, 1975.
19. **Sun, F. E., Prezbindowski, K. S., Crane, F. L., and Jacobs, E. E.,** Physical state of cytochrome oxidase. Relationship between membrane formation and ionic strength, *Biochim. Biophys. Acta,* 153, 804, 1968.
20. **Kuboyama, M., Yong, F. C., and King, T. E.,** Studies on cytochrome oxidase. VIII. Preparation and properties of cardiac cytochrome oxidase, *J. Biol. Chem.,* 247, 6375, 1972.
21. **Yu, C. A., Yu, L., and King, T. E.,** Studies on cytochrome c oxidase. Interaction of the cytochrome oxidase protein with phospholipids and cytochrome c, *J. Biol. Chem.,* 250, 1383, 1975.
22. **Fowler, L. R., Richardson, S. H., and Hatefi, Y.,** A rapid method for the preparation of highly purified cytochrome oxidase, *Biochim. Biophys. Acta,* 64, 170, 1962.
23. **Capaldi, R. A. and Hayashi, H.,** The polypeptides composition of cytochrome oxidase from beef heart mitochondria, *FEBS Lett.,* 26, 261, 1972.
24. **Downer, N. W. and Robinson, N. C.,** Characterization of a seventh different subunit of beef heart cytochrome c oxidase. Similarities between the beef heart enzyme and that from other species, *Biochemistry,* 15, 2930, 1976.
25. **Van Buuren, K. J. H., Nicholls, P., and van Gelder, B. F.,** Biochemical and biophysical studies on cytochrome aa_3. VI. Reaction of cyanide with oxidized and reduced enzyme, *Biochim. Biophys. Acta,* 256, 258, 1972.
26. **Hoechli, L. and Hackenbrock, C. R.,** Cytochrome c oxidase from rat-liver mitochondria: purification and characterization, *Biochemistry,* 17, 3712, 1978.
27. **Hartzell, Ch. R., Beinert, H., Van Gelder, B. F., and King, T. E.,** Preparation of cytochrome oxidase from beef heart, in *Methods in Enzymology,* Vol. 53, Fleischer, S. and Packer, L., Eds., Academic Press, New York, 1978, 54.
28. **Ozawa, T., Okumura, M., and Yagi, K.,** Purification of cytochrome oxidase by using Sepharose-bound cytochrome *c,* *Biochem. Biophys. Res. Commun.,* 65, 1102, 1975.

29. **Weiss, H. and Seebald, W.,** Purification of cytochrome oxidase from *Neurospora crassa* and other sources, in *Methods in Enzymology,* Vol. 53, Fleischer, S. and Packer, L., Eds., Academic Press, New York, 1978, 66.

30. **Rascati, R. J. and Parsons, P.,** Purification of cytochrome c oxidase from rat liver mitochondria, *J. Biol. Chem.,* 254, 1586, 1979.

31. **Weiss, H. and Kolb, H. J.,** Isolation of mitochondrial succinate: ubiquinone reductase, cytochrome c reductase and cytochrome *c* oxidase from *Neurospora crassa* using nonionic detergent, *Eur. J. Biochem.,* 99, 139, 1979.

32. **Weiss, H., Juchs, B., and Ziganke, B.,** Complex III from mitochondria of *Neurospora crassa:* purification, characterization and resolution, in *Methods in Enzymology,* Vol. 53, Fleischer, S. and Packer, L., Eds., Academic Press, New York, 1978, 98.

33. **Thompson, D. A. and Ferguson-Miller, S.,** Lipid and subunit III depleted cytochrome c oxidase purified by horse cytochrome c affinity chromatography in lauryl maltoside, *Biochemistry,* 22, 3178, 1983.

34. **Bill, K., Casey, R. P., Broger, C., and Azzi, A.,** Affinity chromatography purification of cytochrome c oxidase. Use of a yeast cytochrome c-thiol Sepharose 4B column, *FEBS Lett.,* 120, 248, 1980.

35. **Rieder, R. and Bosshard, H. R.,** The cytochrome c oxidase binding site on cytochrome c. Differential chemical modification of lysine residues in free and oxidase-bound cytochrome c, *J. Biol. Chem.,* 253, 6045, 1978.

36. **Bill, K., Broger, C., and Azzi, A.,** Affinity chromatography purification of cytochrome c oxidase and bc_1-complex from beef heart mitochondria. Use of a thiol-sepharose-bound *Saccharomyces cerevisiae* cytochrome *c,* *Biochim. Biophys. Acta,* 679, 28, 1982.

37. **Azzi, A., Bill, K., and Broger, C.,** Affinity chromatography purification of cytochrome c binding enzymes, *Proc. Natl. Acad. Sci. U.S.A.,* 79, 2447, 1982.

38. **Solioz, M., Carafoli, E., and Ludwig, B.,** The cytochrome c oxidase of *Paracoccus* pumps protons in a reconstituted system, *J. Biol. Chem.,* 257, 1579, 1982.

39. **Gennis, R. B., Casey, R. P., Azzi, A., and Ludwig, B.,** Purification and characterization of the cytochrome c oxidase from *Rhodopseudomonas sphaeroides, Eur. J. Biochem.,* 125, 189, 1982.

40. **De Vrij, W., Azzi, A., and Konings, W. W.,** Structural and functional properties of cytochrome c oxidase from *Bacillus subtilis* W23, *Eur. J. Biochem.,* 131, 97, 1983.

41. **Broger, C., Salardi, S., and Azzi, A.,** Interaction between isolated cytochrome c_1 and cytochrome c, *Eur. J. Biochem.,* 131, 349, 1983.

42. **Briggs, M. M. and Capaldi, R. A.,** Near-neighbour relationship of the subunits of cytochrome c oxidase, *Biochemistry,* 16, 73, 1977.

43. **Bisson, R., Azzi, A., Gutweniger, H., Colonna, R., Montecucco, C., and Zanotti, A.,** Interaction of cytochrome c with cytochrome c oxidase. Photoaffinity labeling of beef heart cytochrome c oxidase with arylazido-cytochrome *c, J. Biol. Chem.,* 253, 1874, 1978.

44. **Bisson, R., Jacobs, B., and Capaldi, R. A.,** Binding of arylazidocytochrome c derivatives to beef heart cytochrome c oxidase: cross-linking in the high- and low-affinity binding sites, *Biochemistry,* 19, 4173, 1980.

45. **Birchmeier, W., Kohler, C. E., and Schatz, G.,** Interaction of integral and peripheral membrane proteins: affinity labeling of yeast cytochrome oxidase by modified cytochrome *c, Proc. Natl. Acad. Sci. U.S.A.,* 73, 4334, 1976.

46. **Moreland, R. N. and Dockter, E.,** Interaction of the "back" of yeast iso-1-cytochrome c with yeast cytochrome c oxidase, *Biochem. Biophys. Res. Commun.,* 99, 339, 1981.

47. **Fuller, S. D., Darley-Usmar, V. M., and Capaldi, R. A.,** Covalent complex between yeast cytochrome c and beef heart cytochrome c oxidase which is active in electron transfer, *Biochemistry,* 20, 7046, 1981.

48. **Bisson, R. and Montecucco, C.,** Different polypeptides of bovine heart cytochrome c oxidase are in contact with phospholipids, *FEBS Lett.,* 150, 49, 1982.

49. **Eytan, G. D., Caroll, R. C., Schatz, G., and Racker, E.,** Arrangement of the subunits in solubilized and membrane-bound cytochrome c oxidase from beef heart, *J. Biol. Chem.,* 250, 8598, 1975.

50. **Chan, S. H. P. and Tracy, R. P.,** Immunological studies on cytochrome c oxidase: arrangements of protein subunits in the solubilized and membrane bound enzyme, *Eur. J. Biochem.,* 89, 595, 1978.

51. **Eytan, G. D. and Broza, R.,** Role of charge and fluidity in the incorporation of cytochrome oxidase into liposomes, *J. Biol. Chem.,* 253, 3196, 1978.

52. **Bisson, R., Montecucco, C., Gutweniger, H., and Azzi, A.,** Cytochrome c oxidase subunits in contact with phospholipids, *J. Biol. Chem.,* 254, 9962, 1979.

53. **Ludwig, B., Downer, N. W., and Capaldi, R. A.,** Labeling of cytochrome c oxidase with (^{35}S)-diazobenzenesulfonate. Orientation of this electron transfer complex in the inner mitochondrial membrane, *Biochemistry,* 18, 1401, 1979.

54. **Prochaska, L., Bisson, R., and Capaldi, R. A.,** Structure of the cytochrome c oxidase complex: labeling by hydrophilic and hydrophobic protein modifying reagents, *Biochemistry,* 19, 3174, 1980.

55. **Cerletti, N. and Schatz, G.,** Cytochrome c oxidase from baker's yeast. Photolabeling of subunits exposed to the lipid bilayer, *J. Biol. Chem.,* 254, 7746, 1979.

56. **Lake, J. A.,** Ribosomal subunit orientations determined in the monomeric ribosome by single and by double immune electron microscopy, *J. Mol. Biol.,* 161, 89, 1982.

57. **Frey, T. G., Chan, S. H. P., and Schatz, G.,** Structure and orientation of cytochrome c oxidase in crystalline membranes, *J. Biol. Chem.,* 253, 4389, 1978.

58. **Merle, P., Jarausch, J., Trapp, M., Schereka, R., and Kadenbach, B.,** Immunological and chemical characterization of rat liver cytochrome c oxidase, *Biochim. Biophys. Acta,* 669, 222, 1981.

59. **Malatesta, F., Darley-Usmar, V., de Jong, C., Prochaska, L. J., Bisson, R., Capaldi, R. A., Steffens, G. C. M., and Buse, G.,** Arrangement of subunit IV in beef heart cytochrome c oxidase proposed by chemical labeling and protease digestion experiments, *Biochemistry,* 19, 4405, 1983.

60. **Tanaka, M., Suzuki, H., and Ozawa, T.,** The crystallization of mitochondrial cytochrome oxidase-cytochrome c complex, *Biochim. Biophys. Acta,* 612, 295, 1980.

61. **Ozawa, T., Suzuki, H., and Tanaka, M.,** Crystallization of the mitochondrial electron transfer chain: cytochrome c oxidase-cytochrome c complex, *Proc. Natl. Acad. Sci. U.S.A.,* 77, 928, 1980.

62. **Ozawa, T., Tanaka, M., and Wakabayashi, T.,** Crystallization of mitochondrial cytochrome oxidase, *Proc. Natl. Acad. Sci. U.S.A.,* 79, 7175, 1982.

63. **Helenius, A., Mc Caslin, D. R., Fries, E., and Tanford, C.,** Properties of detergents, in *Methods in Enzymology,* Vol. 56, Fleischer, S. and Packer, L., Eds., Academic Press, New York, 1979, 734.

64. **Henderson, R. and Unwin, P. N. T.,** Three-dimensional model of purple membrane obtained by electron microscopy, *Nature (London),* 257, 28, 1975.

65. **Vanderkooi, G., Senior, A. E., Capaldi, R. A., and Hayashi, H.,** Biological membrane structure. III. The lattice structure of membraneous cytochrome oxidase, *Biochim. Biophys. Acta,* 274, 38, 1972.

66. **Henderson, R., Capaldi, R. A., and Leigh, J. S.,** Arrangement of cytochrome molecules in two-dimensional vesicle crystals, *J. Mol. Biol.,* 112, 631, 1977.

67. **Fuller, S. D., Capaldi, R. A., and Henderson, R.,** Structure of cytochrome c oxidase in deoxycholate-derived two-dimensional crystals, *J. Mol Biol.,* 134, 305, 1979.

68. **Fuller, S. D., Capaldi, R. A., and Henderson, R.,** Preparation of two-dimensional arrays from purified beef heart cytochrome c oxidase, *Biochemistry,* 21, 2525, 1982.

69. **Deatherage, J. F., Henderson, R., and Capaldi, R. A.,** Relationship between membrane and cytoplasmic domains in cytochrome c oxidase by electron microscopy in media of different density, *J. Mol. Biol.,* 158, 487, 1982.

70. **Deatherage, J. F., Henderson, R., and Capaldi, R. A.,** Three-dimensional structure of cytochrome c oxidase vesicle crystals in negative stain, *J. Mol. Biol.,* 158, 500, 1982.

71. **Seki, S., Hayashi, G., and Oda, T.,** Studies on cytochrome oxidase. I. Fine structure of cytochrome oxidase-rich submitochondrial membrane, *Biochim. Biophys. Acta,* 138, 110, 1970.

72. **Fry, M., Vande Zande, H., and Green, D. E.,** Resolution of cytochrome oxidase into two component complexes, *Proc. Natl. Acad. Sci. U.S.A.,* 75, 5908, 1978.

73. **Miller, K. R.,** A chloroplast membrane lacking photosystem I changes in unstacked membrane regions, *Biochim. Biophys. Acta,* 529, 143, 1980.

74. **Barber, J.,** An explanation of the relationship between salt-induced thylakoid stacking and the chlorophyll fluorescence changes associated with changes in spillover of the energy from photosystem II to photosystem I, *FEBS Lett.,* 118, 1, 1980.

75. **Andersson, S., Bankier, A. T., Barrel, B. F., de Bruijn, M. H. L., Coulson, A. R., Drouin, J., Eperon, I. C., Nierlich, D. P., Roe, B. A., Sanger, F., Schreier, P. H., Smith, A. J. H., Staden, R., and Yong, I. G.,** Sequence and organization of the human mitochondrial genome, *Nature (London),* 290, 457, 1981.

76. **Marres, C. M. and Slater, E. C.,** Polypeptide composition of purified QH$_2$: cytochrome c oxidoreductase from beef-heart mitochondria, *Biochim. Biophys. Acta,* 462, 531, 1977.

77. **Sacher, R., Steffens, G. J., and Buse, G. C.,** Studies on cytochrome c oxidase. VI. Polypeptide IV. The complete primary structure, *Hoppe Seyler's Z. Physiol. Chem.,* 360, 1385, 1979.

78. **Buse, G. and Steffens, G. J.,** Studies on cytochrome c oxidase. II. The chemical constitution of a short polypeptide from the beef heart enzyme, *Hoppe Seyler's Z. Physiol. Chem.,* 359, 1005, 1978.

79. **Capaldi, R. A., Malatesta, F., and Darley-Usmar, V. M.,** Structure of cytochrome c oxidase, *Biochim. Biophys. Acta,* 726, 135, 1983.

80. **Azzi, A. and Casey, R. P.,** Molecular aspects of cytochrome c oxidase: structure and dynamics, *Mol. Cell. Biochem.,* 28, 169, 1979.

81. **Azzi, A.,** Cytochrome c oxidase. Towards a clarification of its structure, interactions and mechanism, *Biochem. Biophys. Acta,* 594, 231, 1980.

82. **Wilson, M. T., Lalla-Maharajh, W., Darley-Usmar, V., Bonaventura, J., Bonaventura, C., and Brunori, M.,** Structural and functional properties of cytochrome c oxidases isolated from sharks, *J. Biol. Chem.,* 255, 2722, 1980.

83. **Darley-Usmar, V. M., Alizai, N., Al-Ayash, A. I., Jones, G. D., Sharpe, A., and Wilson, M. T.,** A comparison of the structural and functional properties of cytochrome c oxidase isolated from beef *(Bos tauros)*, camel *(Camelus dromedarius)*, chicken *(Gallus domesticus)*, and rat *(Rattus norvegicus)*, *Comp. Biochem. Physiol.*, 68B, 445, 1981.

84. **Georgevich, G., Darley-Usmar, V. M., Malatesta, F., and Capaldi, R. A.,** Electron transfer in monomeric forms of beef and shark heart cytochrome c oxidase, *Biochemistry*, 22, 1317, 1983.

85. **Merle, P. and Kadenback, B.,** The subunit composition of mammalian cytochrome c oxidase, *Eur. J. Biochem.*, 105, 499, 1980.

86. **Kadenbach, B., Jarausch, J., Harmann, R., and Merle, P.,** Separation of mammalian cytochrome c oxidase into 13 polypeptides by a sodium dodecyl sulfate-gel electrophoretic procedure, *Anal. Biochem.*, 129, 517, 1983.

87. **Kagawa, Y.,** Target size of components in oxidative phosphorylation. Studies with a linear accelerator, *Biochim. Biophys. Acta*, 131, 586, 1967.

88. **Thompson, D. A., Suarez-Villafañe, M., and Ferguson-Miller, S.,** The active form of cytochrome c oxidase. Effects of detergents, the intact membrane, and radiation in activation, *Biophys. J.*, 37, 285, 1982.

89. **Tzagoloff, A., Yang, P. C., Wharton, D. C., and Rieske, J., S.,** Studies on the electron-transfer system. LX. Molecular weights of some components of the electron-transfer chain in beef-heart mitochondria, *Biochim. Biophys. Acta*, 96, 1, 1965.

90. **Love, B., Chan, S. H. P., and Stotz, E.,** Molecular weight of two states of cytochrome c oxidase, *J. Biol. Chem.*, 245, 6664, 1970.

91. **Robinson, N. C. and Capaldi, R. A.,** Interaction of detergents with cytochrome c oxidase, *Biochemistry*, 16, 375, 1977.

92. **Steffens, G. and Buse, G.,** Studien an Cytochrom-c-Oxidase. I. Reinigung und Charakterisierung des Enzyms aus Rinderherzen und Identifizierung der im Komplex enthalten Peptidketten, *Hoppe-Seyler's Z. Physiol. Chem.*, 357, 1125, 1976.

93. **Verheul, F. E. A. M., Draijer, J. W., Dentener, I. K., and Muijsers, A. O.,** Subunit stoichiometry of cytochrome c oxidase of bovine heart, *Eur. J. Biochem.*, 119, 401, 1981.

94. **Tanford, Ch., Nozaki, Y., Reynolds, J. A., and Makino, S.,** Molecular characterization of proteins in detergent solutions, *Biochemistry*, 13, 2369, 1974.

95. **Nozaki, Y., Schechter, N. M., Reynolds, J. A., and Tanford, Ch.,** Use of gel chromatography for the determination of the Stokes radii of proteins in the presence and absence of detergents. A reexamination, *Biochemistry*, 15, 3884, 1976.

96. **Le Maire, M., Rivas, E., and Moller, J. V.,** Use of gel chromatography for determination of size and molecular weight of proteins: further caution, *Anal. Biochem.*, 106, 12, 1980.

97. **Saraste, M., Penttilä, T., and Wikström, M.,** Quaternary structure of bovine cytochrome oxidase, *Eur. J. Biochem.*, 115, 261, 1981.

98. **Rosevear, P., VanAken, T., Baxter, J., and Ferguson-Miller, S.,** Alkyl glycoside detergents: a simpler synthesis and their effects of kinetic and physical properties of cytochrome c oxidase, *Biochemistry*, 19, 4108, 1980.

99. **Nałecz, K., Bolli, R., and Azzi, A.,** Preparation of monomeric cytochrome c oxidase: its kinetics differ from those of the dimeric enzyme, *Biochem. Biophys. Res. Commun.*, 114, 822, 1983.

100. **Chan, S. H. P., Love, B., and Stotz, E.,** Absorption spectra and enzymatic properties of monomer and dimer states of cytochrome c oxidase, *J. Biol. Chem.*, 243, 6669, 1970.

101. **Ferguson-Miller, S., Brautigan, D. L., and Margoliash, E.,** Correlation of the kinetics of electron transfer activity of various eukaryotic cytochromes c with binding to mitochondrial cytochrome c oxidase, *J. Biol. Chem.*, 251, 1104, 1976.

102. **Ludwig, B.,** Heme aa_3-type cytochrome c oxidases from bacteria, *Biochim. Biophys. Acta*, 594, 177, 1980.

103. **Ludwig, B. and Schatz, G.,** A two subunit cytochrome oxidase (cytochrome aa_3) from *Paracoccus denitrificans*, *Proc. Natl. Acad. Sci. U.S.A.*, 77, 196, 1980.

104. **Yamaka, T. and Fijii, K.,** Cytochrome a-type terminal oxidase derived from *Thiobacillus novellus*. Molecular and enzymatic properties, *Biochim. Biophys. Acta*, 591, 53, 1980.

105. **Hon-nami, K. and Oshima, T.,** Cytochrome oxidase from an extreme thermophile, *Thermus thermophilus* HB8, *Biochem. Biophys. Res. Commun.*, 92, 1023, 1980.

106. **Sone, N. and Yanagita, Y.,** A cytochrome aa_3-type terminal oxidase of a thermophilic bacterium. Purification, properties and proton pumping, *Biochim. Biophys. Acta*, 682, 216, 1982.

107. **Yamaka, T., Fujii, K., and Kamita, Y.,** Subunits of cytochrome a-type terminal oxidases derived from *Thiobacillus novellus* and *Nitrobacter agilis*, *J. Biochem.*, 86, 821, 1979.

108. **Sebald, W., Machleidt, W., and Otto, J.,** Products of mitochondrial protein synthesis in *Neurospora crassa*. Determination of equimolar amounts of three products in cytochrome oxidase on the basis of amino acid analysis, *Eur. J. Biochem.*, 38, 311, 1973.

109. **Poyton, R. O. and Schatz, G.,** Cytochrome c oxidase from bakers' yeast. III. Physical characterization of isolated subunits and chemical evidence for two different classes of polypeptides, *J. Biol. Chem.*, 250, 752, 1975.

110. **Werner, S.,** Preparation of polypeptide subunits of *Neurospora crassa*, *Eur. J. Biochem.*, 79, 103, 1977.

111. **Rubin, M. S. and Tzagoloff, A.,** Assembly of the mitochondrial membrane system. Purification, characterization, and subunit structure of yeast and beef cytochrome oxidase, *J. Biol. Chem.*, 248, 4269, 1973.

112. **Matsuoka, M., Maeshima, M., and Asahi, T.,** The subunit composition of pea cytochrome c oxidase, *J. Biochem.*, 90, 649, 1981.

113. **Maeshima, M. and Asahi, T.,** Mechanism of increase in cytochrome c oxidase activity in potato root tissue during aging of slices, *J. Biochem.*, 90, 391, 1981.

114. **Buse, G., Steffens, G. C. M., Steffens, G. J., Meineke, L., Biewald, R., and Erdweg, M.,** Cytochrome c oxidase: present status of the sequence analysis, in *2nd Bioenerg. Conf.*, Lyon, LBTM-CNRS Edition, 1982, 163.

115. **Winter, D. B., Bruyninckx, W. J., Foulke, F. G., Grinisch, N. P., and Mason, H. S.,** Location of heme a on subunits I and II and copper on subunit II of cytochrome c oxidase, *J. Biol. Chem.*, 255, 11408, 1980.

116. **Kadenbach, B.,** Struktur und Evolution Atmungsferments Cytochrom-c-Oxidase, *Angew. Chem.*, 95, 273, 1983.

117. **Biewald, R. and Buse, G.,** Studies on cytochrome c oxidase. IX. The primary structure of polypeptide VIa, *Hoppe-Seyler's Z. Physiol. Chem.*, 363, 1141, 1982.

118. **Freedman, J. A. and Chan, S. H. P.,** Redox-dependent accessibility of subunit V of cytochrome oxidase: a novel use of ELISA as a probe of intact membranes, *J. Biol. Chem.*, 258, 5885, 1983.

119. **Kyte, J. and Doolittle, R. F.,** A simple method for displaying the hydropathic character of a protein, *J. Mol. Biol.*, 157, 105, 1982.

120. **Senior, A. E.,** Secondary and tertiary structure of membrane proteins involved in proton translocation, *Biochim. Biophys. Acta*, 726, 81, 1983.

121. **Argos, P., Rao, J. K. M., and Hargrave, P. A.,** Structural prediction of membrane bound proteins, *Eur. J. Biochem.*, 128, 565, 1982.

122. **Kuhn, L. A. and Leigh, J. S.,** A new technique for predicting membrane protein structure, *J. Mol. Biol.*, in press.

123. **Horie, S. and Morrison, M.,** Cytochrome c oxidase components. V. A cytochrome a preparation free cytochrome a_3, *J. Biol. Chem.*, 230, 1483, 1964.

124. **Phan, S. H. and Mahler, H. R.,** Studies on cytochrome oxidase: preliminary characterisation of an enzyme containing only four subunits, *J. Biol. Chem.*, 251, 270, 1976.

125. **Yu, C. and Yu, L.,** Identification of subunit of bovine heart cytochrome oxidase, *Biochim. Biophys. Acta*, 495, 248, 1977.

126. **Yu, C. A., Yu, L., and King, T. E.,** Isolation of a heme binding subunit from bovine heart cytochrome c oxidase, *Biochem. Biophys. Res. Commun.*, 74, 670, 1977.

127. **Fry, M. and Green, D. E.,** Ion channel component of cytochrome oxidase, *Proc. Natl. Acad. Sci. U.S.A.*, 76, 2664, 1979.

128. **Fry, M.,** Resolution and reconstitution of cytochrome oxidase, *Biochem. Biophys. Res. Commun.*, 90, 1119, 1979.

129. **Ozawa, T., Tada, M., and Suzuki, H.,** *Cytochrome Oxidase*, King, T. E., Orii, Y., Chance, B., and Okunuki, K., Eds., Elsevier/North-Holland, Amsterdam, 1979, 39.

130. **Tada, M., Suzuki, H., Masashi, T., and Ozawa, T.,** Resolution of cytochrome oxidase into catalytically active monomers, *Biochem. Int.*, 2, 495, 1981.

131. **Schatz, G., Groot, G. S. P., Mason, T., Rouslin, W., Warton, D. C., and Saltzgaber, J.,** Biogenesis of mitochondrial inner membrane in bakers' yeast, *Fed. Proc., Fed. Am. Soc. Exp. Biol.*, 31, 21, 1972.

132. **Yamamoto, T. and Orii, Y.,** The polypeptide compositions of bovine cytochrome oxidase and its "proteinase-treated derivative", *J. Biochem.*, 75, 1081, 1974.

133. **Boonman, J. C. P., van Beek, G. G. M., Muijsers, A. O., and van Gelder, B. F.,** Properties of protease-treated cytochrome c oxidase from beef heart, *Mol. Cell. Biochem.*, 26, 183, 1979.

134. **Steffens, G. C. M., Steffens, G. J., and Buse, G.,** Studies on cytochrome c oxidase. VIII. The amino acid sequence of polypeptide VII, *Hoppe-Seyler's Z. Physiol. Chem.*, 360, 1641, 1979.

135. **Darley-Usmar, V. M., Capaldi, R. A., and Wilson, M. T.,** Identification of cysteines in subunit II as ligands to the redox centers of bovine cytochrome c oxidase, *Biochem. Biophys. Res. Commun.*, 103, 1223, 1981.

136. **Gutteridge, S., Winter, D. B., Bruyninckx, W. J., and Mason, H. S.,** Location of Cu and heme a on cytochrome c oxidase polypeptides, *Biochem. Biophys. Res. Commun.*, 78, 945, 1977.

137. **Freedman, J. A., Tracy, R. P., and Chan, S. H. P.,** Heme associated subunit complex of cytochrome c oxidase identified by a new two-dimensional gel electrophoresis, *J. Biol. Chem.*, 254, 4305, 1979.

138. **Penttilä, T.,** Properties and reconstitution of a cytochrome oxidase deficient in subunit III, *Eur. J. Biochem.*, 133, 355, 1983.

139. **Bill, K. and Azzi, A.,** An active cytochrome c oxidase depleted of subunit III prepared by covalent chromatography on yeast cytochrome c, *Biochem. Biophys. Res. Commun.*, 106, 1203, 1982.

140. **Corbley, M. J. and Azzi, A.,** Resolution of bovine heart cytochrome c oxidase into smaller complexes by controlled subunit denaturation, *Eur. J. Biochem.,* 139, 535, 1984.

141. **Buse, G., Steffens, G. J., and Steffens, G. C. M.,** Studies on cytochrome c oxidase. III. Relationship of cytochrome oxidase subunits to electron carriers of photophosphorylation, *Hoppe-Seyler's Z. Physiol. Chem.,* 359, 1011, 1978.

142. **Stevens, T. H., Martin, C. T., Wang, H., Brudwig, G. W., Scholes, C. P., and Chan, S. I.,** The nature of Cu_A in cytochrome c oxidase, *J. Biol. Chem.,* 257, 12106, 1982.

143. **Powers, L., Chance, B., Ching, Y., and Angiolillo, P.,** Structural features and the reaction mechanism of cytochrome oxidase: iron and copper X-ray absorption fine structure, *Biophys. J.,* 34, 465, 1981.

144. **Fox, D. T. and Leaver, C. J.,** The *Zea mays* mitochondrial gene coding cytochrome oxidase subunit II has an intervening sequence and does not contain TGA codons, *Cell,* 26, 315, 1981.

145. **Millet, F., deJong, C., Paulson, L., and Capaldi, R. A.,** Identification of specific carboxylate groups on cytochrome c oxidase that are involved in binding cytochrome c, *Biochemistry,* 22, 546, 1983.

146. **Blasie, J. K., Erecinska, M., Samuels, S., and Leigh, J. S.,** The structure of a cytochrome oxidase-lipid model, *Biochim. Biophys. Acta,* 501, 33, 1978.

147. **Callahan, P. and Babcock, G. T.,** Origin of the cytochrome a absorption red shift: a pH-dependent interaction between its heme a formyl and protein in cytochrome oxidase, *Biochemistry,* 22, 452, 1983.

148. **Chakrabarti, P. and Khorana, H. G.,** A new approach to the study of phospholipid-protein interactions in biological membranes. Synthesis of fatty acids and phospholipids containing photosensitive groups, *Biochemistry,* 14, 5021, 1975.

149. **Bisson, R., Gutweniger, H., Montecucco, C., Colonna, R., Zanotti, A., and Azzi, A.,** Covalent binding of arylazido derivatives of cytochrome c to cytochrome oxidase, *FEBS Lett.,* 81, 147, 1977.

150. **Bisson, R., Steffens, G. C. M., Capaldi, R. A., and Buse, G.,** Mapping of cytochrome c binding site on cytochrome c oxidase, *FEBS Lett.,* 144, 359, 1982.

151. **Bisson, R., Steffens, G. C. M., and Buse, G.,** Localisation of lipid binding domain(s) of beef heart cytochrome c oxidase, *J. Biol. Chem.,* 257, 6716, 1982.

152. **Mitchell, P.,** Protonmotive redox mechanism of the cytochrome bc_1 complex in the respiratory chain: protonmotive ubiquinone cycle, *FEBS Lett.,* 56, 1, 1975.

153. **Mitchell, P. and Moyle, M.,** Proton-transport phosphorylation: some experimental tests, in *Biochemistry of Mitochondria,* Slater, E. C., Kaniuga, Z., and Wojtczak, L., Eds., Academic Press, London, 1967, 53.

154. **Hinkle, P. C., Kim, J. J., and Racker, E.,** Ion transport and respiratory control in vesicles formed from cytochrome c oxidase and phospholipids, *J. Biol. Chem.,* 247, 1338, 1972.

155. **Hinkle, P. C.,** Electron transfer across membranes and energy coupling, *Fed. Proc., Fed. Am. Soc. Exp. Biol.,* 32, 1988, 1973.

156. **Wikström, M. K. F.,** Proton pump coupled to cytochrome c oxidase in mitochondria, *Nature (London),* 266, 271, 1977.

157. **Wikström, M. K. F. and Saari, H. T.,** The mechanism of energy conservation and transduction by mitochondrial cytochrome c oxidase, *Biochim. Biophys. Acta,* 462, 347, 1977.

158. **Wikström, M. K. F.,** Cytochrome c oxidase, the mechanism of a redox-coupled proton pump, in *The Proton and Calcium Pumps,* Azzone, G. F., Ed., Elsevier/North-Holland, Amsterdam, 1978, 215.

159. **Sigel, E. and Carafoli, E.,** The proton pump of cytochrome c oxidase and its stoichiometry, *Eur. J. Biochem.,* 89, 119, 1978.

160. **Moyle, J. and Mitchell, P.,** Cytochrome c oxidase is not a proton pump, *FEBS Lett.,* 88, 268, 1978.

161. **Moyle, J. and Mitchell, P.,** Measurements of mitochondrial H^+/O quotients: effects of phosphate and N-ethylmaleimide, *FEBS Lett.,* 90, 361, 1978.

162. **Wikström, M. K. F. and Krab, K.,** Cytochrome c oxidase is a proton pump, *FEBS Lett.,* 91, 8, 1978.

163. **Wikström, M. K. F. and Krab, K.,** Generation of electrochemical proton gradient by mitochondrial cytochrome c oxidase, in *Frontiers of Biological Energetics: From Electrons to Tissues,* Dutton, P. L., Leigh, J. S., and Scarpa, A., Eds., Academic Press, New York, 1978, 551.

164. **Casey, R. P. and Azzi, A.,** An evaluation of the evidence of H^+ pumping by reconstituted cytochrome c oxidase in the light of recent criticism, *FEBS Lett.,* 154, 237, 1983.

165. **Mitchell, P. and Moyle, J.,** Alternative hypotheses of proton ejection in cytochrome oxidase vesicles, *FEBS Lett.,* 151, 167, 1983.

166. **Papa, S., Lorusso, M., Capitanio, N., and de Nitto, E.,** Characteristics of redox-linked proton ejection in cytochrome c oxidase reconstituted in phospholipid vesicles, *FEBS Lett.,* 157, 7, 1983.

167. **Papa, S., Guerrieri, G., Izzo, G., and Boffoli, D.,** Mechanism of proton translocation associated to oxidation of N,N,N',N'-tetramethyl-p-phenylenediamine in rat liver mitochondria, *FEBS Lett.,* 157, 15, 1983.

168. **Krab, K. and Wikström, M. K. F.,** Proton-translocating cytochrome c oxidase in artificial phospholipid vesicles, *Biochim. Biophys. Acta,* 504, 200, 1978.

169. **Casey, R. P., Chappell, J. B., and Azzi, A.,** Limited turnover studies on proton translocation in reconstituted cytochrome c oxidase-containing vesicles, *Biochem. J.,* 182, 149, 1979.

170. **Sigel, E. and Carafoli, E.,** The charge stoichiometry of cytochrome *c* oxidase in the reconstituted system, *J. Biol. Chem.,* 254, 10572, 1979.

171. **Alexandre, A., Reynafarje, B., and Lehninger, A. L.,** Stoichiometry of vectorial H^+ movements coupled to electron transport and to ATP synthesis in mitochondria, *Proc. Natl. Acad. Sci. U.S.A.,* 75, 5296, 1978.

172. **Reynafarje, B., Alexandre, A., Davies, P., and Lehninger, A. L.,** Proton translocation stoichiometry of cytochrome oxidase: use of a fast responding oxygen electrode, *Proc. Natl. Acad. Sci. U.S.A.,* 79, 7218, 1982.

173. **Nicholls, D. G.,** Hamster brown adipose tissue mitochondria: the control of respiration and the proton electrochemical potential gradient by possible physiological effectors and by the H^+ conductance of the inner membrane, *Eur. J. Biochem.,* 49, 573, 1974.

174. **Dutton, P. L. and Wilson, D. F.,** Measurements of redox potentials in the mitochondrial electron transport chain, *Biochim. Biophys. Acta,* 346, 165, 1974.

175. **Nagle, J. F. and Tristram-Nagle, T.,** Hydrogen-bonded chain mechanisms for proton conduction and proton pumping, *J. Membrane Biol.,* 74, 1, 1983.

176. **Casey, R. P., Thelen, M., and Azzi, A.,** Dicyclohexylcarbodiimide inhibits proton translocation by cytochrome *c* oxidase, *Biochem. Biophys. Res. Commun.,* 87, 1044, 1979.

177. **Casey, R. P., Thelen, M., and Azzi, A.,** Dicyclohexylcarbodiimide binds specifically and covalently to cytochrome *c* oxidase while inhibiting its H^+-translocating activity, *J. Biol. Chem.,* 255, 3994, 1980.

178. **Casey, R. P., Broger, C., Thelen, M., and Azzi, A.,** Studies on the molecular basis of H^+ translocation by cytochrome *c* oxidase, *J. Bioenerg. Biomembr.,* 13, 219, 1981.

179. **Casey, R. P., Broger, C., Ariano, B. H., and Azzi, A.,** Studies on vectorial aspects of cytochrome *c* oxidase structure and function, *Dev. Bioenerg. Biomembr.,* 5, 81, 1981.

180. **Prochaska, L. J., Bisson, R., Capaldi, R. A., Steffens, G. C. M., and Buse, G.,** Inhibition of cytochrome *c* oxidase function by dicyclohexylcarbodiimide, *Biochim. Biophys. Acta,* 637, 360, 1981.

181. **Püttner, I., Solioz, M., Carafoli, E., and Ludwig, B.,** Dicyclohexylcarbodiimide does not inhibit proton pumping by cytochrome *c* oxidase of *Paracoccus denitrificans, Eur. J. Biochem.,* 134, 33, 1983.

182. **Wilson, D. F., Dutton, P. L., and Wagner, M.,** Electrometric measurements on the mitochondrial respiratory chain, *Curr. Top. Bioenerg.,* 5, 233, 1973.

183. **Penttilä, T.,** Properties and reconstitution of a cytochrome oxidase deficient in subunit III, *Eur. J. Biochem.,* 133, 355, 1983.

184. **Casey, R. P., Bill, K., and Azzi, A.,** Structural and functional studies on subunit III of cytochrome c oxidase, in 2nd Eur. Bioenerg. Conf., LBTM-CNRS Edition, Lyon, 1982, 149.

185. **Erecinska, M., Wilson, D. F., Sato, N., and Nicholls, P.,** The energy dependence of the chemical properties of cytochrome *c* oxidase, *Arch. Biochem. Biophys.,* 151, 188, 1972.

186. **Wikström, M. K. F.,** Energy-dependent reversal of the cytochrome *c* oxidase reaction, *Proc. Natl. Acad. Sci. U.S.A.,* 78, 4051, 1981.

187. **Gibson, Q. H., Greenwood, C., Wharton, D. C., and Palmer, G.,** The reaction of cytochrome oxidase with cytochrome c, *J. Biol. Chem.,* 240, 888, 1965.

188. **Antalis, T. M. and Palmer, G.,** Kinetic characterization of the interaction between cytochrome oxidase and cytochrome c, *J. Biol. Chem.,* 257, 6194, 1982.

189. **Andreasson, L.-E., Malström, B. G., Strömberg, C., and Vanngard, T.,** The reaction of ferrocytochrome c with cytochrome oxidase: a new look, *FEBS Lett.,* 28, 297, 1972.

190. **Andreasson, L.-E.,** Characterization of the reaction between ferrocytochrome c and cytochrome oxidase, *Eur. J. Biochem.,* 53, 591, 1975.

191. **Wilson, M. T., Greenwood, C., Brunori, M., and Antonini, E.,** Kinetic studies on the reaction between cytochrome c oxidase and ferrocytochrome c, *Biochem. J.,* 147, 145, 1975.

192. **Van Buuren, K. J. H., Van Gelder, B. F., Wilting, J., and Braams, R.,** Biochemical and biophysical studies on cytochrome c oxidase. XIV. The reaction with cytochrome c as studied by pulse radiolysis, *Biochim. Biophys. Acta,* 333, 421, 1974.

193. **Petersen, L. Ch. and Andreasson, L.-E.,** The reaction between oxidised cytochrome c and reduced cytochrome *c* oxidase, *FEBS Lett.,* 66, 52, 1976.

194. **Schroedl, N. A. and Hartzell, C. R.,** Oxidative titrations of reduced cytochrome aa₃: influence of cytochrome c and carbon monoxide on the midpoint potential values, *Biochemistry,* 16, 4966, 1977.

195. **Wilson, D. F., Erecinska, M., and Owen, Ch. S.,** Some properties of the redox components of cytochrome c oxidase and their interactions, *Arch. Biochem. Biophys.,* 175, 160, 1976.

196. **Van Gelder, B. F., Van Buuren, K. J. H., Wilms, J., and Verboom, C. N.,** Effect of ionic strength on the oxidation of cytochrome c by cytochrome c oxidase, in *Electron Transfer Chains and Oxidative Phosphorylation,* Quagliariello, E., Papa, S., Palmieri, F., Slater, E. C., and Siliprandi, N., Eds., North-Holland, Amsterdam, 1975, 63.

197. **Veerman, E. C. I., Wilms, J., Casteleijn, G., and Van Gelder, B. F.,** The pre-steady state reaction of ferrocytochrome c with the cytochrome c-cytochrome aa₃ complex, *Biochim. Biophys. Acta,* 590, 117, 1980.

198. **Wilms, J., Veerman, E. C. I., König, B. W., Dekker, H. L., and Van Gelder, B. F.,** Ionic strength effects on cytochrome aa₃ kinetics, *Biochim. Biophys. Acta*, 635, 13, 1981.

199. **Smith, L. and Minnaert, K.,** Interaction of macroions with the respiratory chain system of mitochondria and heart-muscle particles, *Biochim. Biophys. Acta,* 105, 1, 1965.

200. **Chance, B. and Erecinska, M.,** Flow flash kinetics of the cytochrome a₃-oxygen reaction in coupled and uncoupled mitochondria using the liquid dye laser, *Arch. Biochem. Biophys.*, 143, 675, 1971.

201. **Chance, B., Saronio, C., and Leigh, J. S.,** Functional intermediates in the reaction of membrane-bound cytochrome oxidase with oxygen, *J. Biol. Chem.*, 250, 9226, 1975.

202. **Alben, R., Altschuld, F., Fiamingo, F., and Moh, P.,** *Electron Transport and Oxygen Utilization,* Ho, C. and Eaton, W. C., Eds., Elsevier/North-Holland, Amsterdam, 1982.

203. **Clore, G. M., Andreasson, L.-E., Karlsson, B., Aasa, R., and Malström, B. G.,** Characterization of the low-temperature intermediates of the reaction of fully reduced soluble cytochrome oxidase with oxygen by electron-paramagnetic-resonance and optical spectroscopy, *Biochem. J.*, 185, 139, 1980.

204. **Clore, G. M., Andreasson, L.-E., Karlsson, B., Aasa, R., and Malström, B. G.,** Characterization of the intermediates in the reaction of mixed-valence-state soluble cytochrome oxidase with oxygen at low temperatures by optical and electron-paramagnetic-resonance spectroscopy, *Biochem. J.*, 185, 155, 1980.

205. **Chance, B., Saronio, C., and Leigh, J. S.,** Compound C₂, a product of the reaction of oxygen and the mixed-valance state of cytochrome oxidase. Optical evidence for a type-I copper, *Biochem. J.*, 177, 931, 1979.

206. **Denis, M.,** Resolution of two compound C-type intermediates in the reaction with oxygen of mixed-valence state membrane-bound cytochrome oxidase, *Biochim. Biophys. Acta*, 634, 30, 1981.

207. **Gibson, Q. H. and Greenwood, C.,** Kinetic observation on the near infrared band of cytochrome c oxidase, *J. Biol. Chem.*, 240, 2694, 1965.

208. **Shaw, R. W., Hansen, R. E., and Beinert, H.,** The oxygen reactions of reduced cytochrome c oxidase. Position of a form with an unusual EPR signal in the sequence of early intermediates, *Biochim. Biophys. Acta*, 548, 386, 1979.

209. **Karlsson, B. and Andreasson, L.-E.,** The identity of a new copper (II) electron paramagnetic resonance signal in cytochrome c oxidase, *Biochim. Biophys. Acta*, 635, 73, 1981.

210. **Antonini, E., Brunori, M., Colosimo, A., Greenwood, C., and Wilson, M. T.,** Oxygen 'pulsed' cytochrome c oxidase: functional properties and catalytic relevance, *Proc. Natl. Acad. Sci. U.S.A.*, 74, 3128, 1977.

211. **Wilson, M. T., Colosimo, A., Brunori, M., and Antonini, E.,** Pre-steady state and steady-state studies of resting and pulsed cytochrome c-oxidase, in *Frontiers of Biological Energetics. Electrons to Tissues,* Dutton, P. L., Leigh, J. S., and Scarpa, A., Eds., Academic Press, New York, 1978, 843.

212. **Brunori, M., Colosimo, A., Rainoni, G., Wilson, M. T., and Antonini, E.,** Functional intermediates of cytochrome oxidase. Role of 'pulsed' oxidase in the pre-steady state and steady state reactions of the beef enzyme, *J. Biol. Chem.*, 254, 10769, 1979.

213. **Sarti, P., Colosimo, A., Brunori, M., Wilson, M. T., and Antonini, E.,** Kinetic studies on cytochrome c oxidase inserted into liposomal vesicles. Effect of ionophores, *Biochem. J.*, 209, 81, 1983.

214. **Brunori, M., Colosimo, A., Sarti, P., Antonini, E., and Wilson, M. T.,** 'Pulsed' cytochrome oxidase may be produced without the advent of dioxygen, *FEBS Lett.*, 126, 195, 1981.

215. **Rosen, S., Branden, R., Vanngard, T., and Malström, B. G.,** EPR evidence for an active form of cytochrome c oxidase different from the resting enzyme, *FEBS Lett.*, 74, 25, 1977.

216. **Reichardt, J. K. V. and Gibson, Q. H.,** Turnover of cytochrome c oxidase from *Paracoccus denitrificans*, *J. Biol. Chem.*, 258, 1504, 1983.

217. **Keilin, D.,** Cytochrome and intracellular oxidase, *Proc. R. Soc. London, Ser. B*, 106, 418, 1930.

218. **Smith, L. and Conrad, H.,** A study of the kinetics of the oxidation of cytochrome c by cytochrome c oxidase, *Arch. Biochem. Biophys.*, 63, 403, 1956.

219. **Minnaert, K.,** The kinetics of cytochrome c oxidase. I. The system: cytochrome c-cytochrome oxidase-oxygen, *Biochim. Biophys. Acta*, 50, 23, 1961.

220. **McGuinness, E. T. and Wainio, W. W.,** Cytochrome c oxidase. I. Nature of the inhibition by cytochrome c, *J. Biol. Chem.*, 237, 3273, 1962.

221. **Yonetani, T. and Ray, G. S.,** Studies on cytochrome oxidase. VI. Kinetics of the aerobic oxidation of ferrocytocrome *c* by cytochrome oxidase, *J. Biol. Chem.*, 240, 3392, 1965.

222. **Errede, B., Haight, G. P., and Kamen, M. D.,** Oxidation of ferrocytochrome c by mitochondrial cytochrome c oxidase, *Proc. Natl. Acad. Sci. U.S.A.*, 73, 113, 1976.

223. **Ferguson-Miller, S., Brautigan, D. L., and Margoliash, E.,** Definition of cytochrome c binding domains by chemical modification. III. Kinetics of reaction of carboxydinitrophenyl cytochromes c with cytochrome c oxidase, *J. Biol. Chem.*, 253, 149, 1978.

224. **Nicholls, P., Hildebrandt, V., Hill, B. C., Nicholls, F., and Wrigglesworth, J. M.,** Pathways of cytochrome c oxidation by soluble and membrane-bound cytochrome aa₃, *Can. J. Biochem.*, 58, 969, 1980.

225. **Smith, L., Davies, H. C., and Nava, M. E.,** Studies of the kinetics of oxidation of cytochrome c by cytochrome c oxidase: comparison of spectrophotometric and polarographic assays, *Biochemistry,* 18, 3140, 1979.

226. **Errede, B. and Kamen, M. D.,** Comparative kinetic studies of cytochromes c in reactions with mitochondrial cytochrome c oxidase and reductase, *Biochemistry,* 17, 1015, 1978.

227. **Wilms, J., van Rijn, J. L. M. L., and van Gelder, B. F.,** The effect of pH and ionic strength on the steady-state activity of isolated cytochrome c oxidase, *Biochim. Biophys. Acta,* 593, 17, 1980.

228. **Dethmers, J. K., Ferguson-Miller, S., and Margoliash, E.,** Comparison of yeast and beef cytochrome c oxidases. Kinetics and binding of horse, fungal, and *Euglena* cytochromes c, *J. Biol. Chem.,* 254, 11973, 1979.

229. **Orii, Y., Sekuzu, I., and Okonuki, K.,** Studies on cytochrome c_1. II. Oxidation mechanism of cytochrome c_1 in the presence of cytochromes a and c_1, *J. Biochem.,* 51, 204, 1962.

230. **Kuboyama, M., Takemori, S., and King, T. E.,** Reconstitution of respiratory chain enzyme systems. IX. Cytochrome c-cytochrome oxidase complex of heart muscle, *Biochem. Biophys. Res. Commun.,* 9, 534, 1962.

231. **Nicholls, P.,** Observations on the oxidation of cytochrome c, *Arch. Biochem. Biophys.,* 106, 25, 1964.

232. **Petersen, L. Ch.,** Cytochrome c-cytochrome aa_3 complex formation at low ionic strength studied by aqueous two-phase partition, *FEBS Lett.,* 94, 105, 1978.

233. **Nicholls, P.,** Cytochrome c binding to enzymes and membranes, *Biochim. Biophys. Acta,* 346, 261, 1974.

234. **Vik, S. B., Georgevich, G., and Capaldi, R. A.,** Diphosphatidylglycerol is required for optimal activity of beef heart cytochrome c oxidase, *Proc. Natl. Acad. Sci. U.S.A.,* 78, 1456, 1981.

235. **Margoliash, E. and Bosshard, H. R.,** Guided by electrostatic, a textbook protein comes of age, *Trends Biochem. Sci.,* 8, 316, 1983.

236. **Veerman, E. C. J., Wilms, J., Dekker, H. L., Muijsers, A. O., Van Buuren, K. J. H., Van Gelder, B. F., Osheroff, N., Speck, S. H., and Margoliash, E.,** The presteady state reaction of chemically modified cytochromes c with cytochrome oxidase, *J. Biol. Chem.,* 258, 5739, 1983.

237. **Bolli, R., Nałecz, K. A., and Azzi, A.,** The interconversion between monomeric and dimeric bovine heart cytochrome *c* oxidase, *Biochimie,* 67, in press.

Chapter 7

MOLECULAR ASPECTS OF STRUCTURE-FUNCTION RELATIONSHIPS IN MITOCHONDRIAL ADENINE NUCLEOTIDE CARRIER

Pierre V. Vignais, Marc R. Block, François Boulay, Gérard Brandolin, and Gùy J.-M. Lauquin

TABLE OF CONTENTS

I. INTRODUCTION

The adenine nucleotide carrier is an intrinsic protein located in the inner mitochondrial membrane: its function in living cells is to mediate the transmembrane electrogenic exchange between the cytosolic ADP and the mitochondrial ATP generated by oxidative phosphorylation. The outer membrane is permeable to molecules with a molecular weight up to 5000,

Table 1
GENERAL PROPERTIES OF THE ADP/ATP CARRIER[a]

Localization	Inner membrane of mitochondria
Specificity	Among natural nucleotides, only ADP, ATP, dADP, and dATP are transported
Kinetics	Saturation kinetics of Michaelis-Menten type
	Exchange diffusion process with a 1:1 stoichiometry
	K_m ADP 1 to 5 μM in coupled and uncoupled mitochondria
	K_m ATP 1 to 5 μM in uncoupled mitochondria, up to 200 μM in coupled mitochondria
	Competition of ATP against ADP for transport
	Rate of exchange influenced by the size of the internal nucleotide pool and the internal ADP/ATP ratio
Temperature dependence	Break in the Arrhenius plot at 10—14°C
	Activation energy 11—12 kcal above the transition, 30—50 kcal below the transition
pH-Dependence	No effect on the rate of transport between pH 6 and 8
Specific ADP/ATP binding sites	1—2 mol/mol cytochrome *a* in liver and heart mitochondria
Turnover at 18—20°C	600—2000/min in liver and heart mitochondria
Specific inhibitors	Atractyloside: nonpermeant, competitive inhibitor
	Carboxyatractyloside: nonpermeant, quasiirreversible inhibitor
	Bongkrekic acid and isobongkrekic acid: permeant inhibitors at pH < 7, apparently uncompetitive

Note: For all above inhibitors K_d is 5—20 nM, number of inhibitor sites, 1—2 mol/mol of cytochrome *a* in liver and heart mitochondria.

[a]   For review see Klingenberg[3] and Vignais.[4]

due to the presence of pores created by assembly of subunits of a protein, called porin.[1,2] It follows that for small molecules, like nucleotides, the inner mitochondrial membrane is the true barrier between the matrix space and the cytosol of the cell. The adenine nucleotide carrier is, therefore, a link between the nucleotide pool of the matrix space of mitochondria and that of the cytosol. The adenine nucleotide carrier protein has been isolated and purified. Its amino acid sequence has been determined. The carrier protein is capable of assuming conformational changes that can be probed by two specific categories of inhibitors, atractyloside (ATR) and carboxyatractyloside (CATR) on one hand, and bongkrekic acid (BA) and isobongkrekic acid (isoBA) on the other.

Much of the early literature on adenine nucleotide transport was reviewed in 1976 by Klingenberg[3] and Vignais.[4] The data acquired at that time are summarized in Table 1. Since then, a number of shorter reviews have dealt with specific aspects of this transport system, namely, its asymmetry,[5] its energy dependency,[6] its role in the cell economy,[7] the topography of the carrier protein in the mitochondrial membrane,[8] and the mechanism of action of inhibitors.[9,10] The present review will focus on molecular aspects of the ADP/ATP carrier. However, to place the discussion in perspective, we have chosen to present first a brief survey of physiological features of the ADP/ATP carrier and recent data on its biogenesis.

II. GENETICS AND BIOGENESIS OF THE ADP/ATP CARRIER

Demonstration that the ADP/ATP carrier is coded by a nuclear gene stemmed from early studies on oxidative phosphorylation in mutants of the yeast *Saccharomyces cerevisiae*. One of the most studied *S. cerevisiae* mutants was op$_1$,[11-13] called also p$_9$.[14] It is a leaky mutant which carries out ADP/ATP transport at low rate, so that ATP export out of mitochondria is no longer able to cope with the energy requirement of the cell when growth is performed in a medium supplemented with oxidizable substrates. A defect in ADP/ATP transport in op$_1$ was first suggested on the basis that the low rate of ATP synthesis by mitochondria isolated from op$_1$ could be markedly increased at high concentrations of added ADP.[12,14] This was corroborated by the following findings: (1) ADP/ATP transport in op$_1$ mitochondria had a K$_M$ for ADP 100 times higher than that found with mitochondria of the wild type;[13] (2) op$_1$ mitochondria are characterized by a two- to four-fold decrease in the number of strong CATR sites,[15,16] by a four-fold decrease in the affinity for CATR,[15] and by a 20-fold decrease in the affinity for BA.[16,17] Other lines of evidence that the ADP/ATP carrier is coded by a nuclear gene are as follows: (1) *S. cerevisiae* ρ$^-$ mutants which lack mitochondrial protein synthesis possess, nonetheless, a mitochondrial ADP/ATP carrier;[17,18] this also holds for promitochondria from *S. cerevisiae* grown in anaerobiosis;[17] (2) resistance to BA in a *S. cerevisiae* mutant has been shown to be determined by a single nuclear gene;[19] (3) ethidium bromide, which at low concentration binds to the mitochondrial DNA and inhibits selectively the expression of the mitochondrial genome, does not inhibit the synthesis of the ADP/ATP carrier.[17]

The yeast nuclear OP$_1$ gene which is believed to code for the ADP/ATP carrier protein was selected by genetic complementation.[20,21] This test is based on the cure of the op$_1$ mutation; the transformants have the ability to grow on a glycerol medium in contrast to the op$_1$ mutant cells which require a glucose-supplemented medium. These transformants contain a higher amount of a protein of 30,000 to 32,000 daltons which cross-reacts with specific antiserum to the mitochondrial ADP/ATP carrier.[20]

Experiments with the op$_1$ mutant have led to puzzling observations which point to a primordial function of the ADP/ATP carrier in the yeast, even in cells utilizing cytosolic ATP as the sole source of energy: (1) the cytoplasmic respiratory mutation ρ$^-$ is abortive in the op$_1$ mutant;[14] (2) growth of ρ$^-$ mutants on glucose is arrested by BA.[22] BA was apparently interacting with the product of the OP$_1$ gene. A possible explanation of the above data is that mitochondria must contain ATP to ensure metabolic functions essential to life, even in cells that are able to live essentially on cytosolic ATP (ρ$^-$ mutants). Since the mitochondria of ρ$^-$ mutants do not synthesize ATP, their ATP must be imported from the cytosol (glycolytic ATP), and limitation of entry of cytosolic ATP into mitochondria either by BA or op$_1$ mutation results in the arrest of growth.

That the ADP/ATP carrier protein is synthesized outside the mitochondria on cytoplasmic ribosomes was confirmed by the demonstration that this synthesis is sensitive to cycloheximide, but not to chloramphenicol.[23] The carrier protein, therefore, has to be imported by mitochondria. The import is of the posttranslation type.[24] Interaction of the carrier protein with so-called receptor sites could be an early step in the import pathway.[25] In contrast to most other mitochondrial proteins, the precursor of the ADP/ATP carrier does not appear to differ in molecular weight from the mature membrane-integrated protein. This was shown for *Neurospora crassa*[24] and hepatoma cells.[26] Import of the ADP/ATP carrier protein into the mitochondrial membrane requires a transmembrane potential.[27] However, the mechanism by which this potential acts is still unclear.

III. THE ROLE OF ADP/ATP TRANSPORT IN THE CELL ECONOMY

A. Transmembrane ADP/ATP Exchange and Other Movements of Nucleotides in Mitochondria

Because the adenine nucleotide transport mediated by the ADP/ATP carrier is a transmembrane exchange strictly specific for ADP and ATP,[28,29] the adenine nucleotide pool of the matrix space is maintained at constant concentration during oxidative phosphorylation. A criterion often used to characterize the ADP/ATP carrier is its high sensitivity to the specific inhibitors, the atractylosides and the bongkrekic acids (for review see Klingenberg[3] and Vignais[4]). Besides the adenine nucleotide transmembrane exchange denoted as ADP/ATP transport, there are unidirectional processes of net release of adenine nucleotides from mitochondria or net uptake. For example, intramitochondrial adenine nucleotides can be released by exchange with extramitochondrial pyrophosphate;[30] since this process is sensitive to ATR, it could be mediated by the ADP/ATP carrier. The physiological impact of the pyrophosphate/adenine nucleotide exchange is, however, probably negligible; in fact, the pyrophosphate concentration required for efficient release of intramitochondrial adenine nucleotides is in the millimolar range, i.e., two orders of magnitude higher than the concentration of pyrophosphate in living cells. The pyrophosphate-adenine nucleotide exchange results in depletion of the intramitochondrial nucleotide pool. Intramitochondrial adenine nucleotides can also be released upon incubation with an oxidizable substrate, P_i and Mg^{2+};[31,32] this release is, however, insensitive to ATR. Addition of P_i and Ca^{2+} results in release of adenine nucleotides; in this case the release is inhibited by BA.[33-35] Nucleotide release from mitochondria may occur in vivo, under pathological conditions, for example, during cardiac ischemia.[36] An uncoupler-induced efflux of nucleotides from tumor mitochondria has been reported.[36a]

Besides release of intramitochondrial adenine nucleotides, a net uptake of ADP or ATP insensitive to ATR has been demonstrated in animal and plant mitochondria. For example, rat liver mitochondria that have been depleted of intramitochondrial adenine nucleotides in vitro can accumulate ADP or ATP from the extramitochondrial space, provided P_i be present in the medium;[37] absence of inhibition by ATR indicated that accumulation did not occur via the ADP/ATP carrier. Corn mitochondria also accumulate ADP in the presence of P_i via a mechanism insensitive to ATR.[38] In both cases, requirement of P_i suggested that the P_i carrier could play a significant role in the accumulation mechanism. Although large fluctuations in the adenine nucleotide content of mitochondria during the cell life are not likely to occur, because of the homeostasis of the cell medium, there are situations where a net uptake of adenine nucleotides plays a physiological role; this is the case of liver mitochondria in newborn rats[39,40] or rabbits.[41] The increase in intramitochondrial nucleotides just after birth is accompanied by an increase in the rate of ADP/ATP transport; possibly the concentration in internal nucleotides determines a gradient which provides the driving force favorable for the transmembrane exchange with external adenine nucleotides.

A specific transport system was proposed by which external ATP taken up by mitochondria would be exchanged against internal ADP and P_i.[42] This type of coupling has not been confirmed.[43,44]

B. Participation of ADP/ATP Transport to the Cellular ADP/ATP Cycle

1. Capacity of Transport of the ADP/ATP Carrier

When the ADP/ATP carrier operates under conditions of oxidative phosphorylation, for any molecule of ADP which enters the matrix space of the mitochondria, one molecule of ATP is exported to the cytosol. The transmembrane exchange between external ADP and internal ATP is part of a cycle, the ADP/ATP cycle, which consists of vectorial processes located in the inner mitochondrial membrane and scalar reactions occurring on each site of the membrane. The vectorial processes are the proton ejection that results from mitochondrial respiration and generates a membrane potential, the reentry of protons by the F_0 sector of the F_1-F_0 complex and the P_i carrier, and the ADP/ATP transmembrane exchange. The scalar reactions consist of ATP synthesis from ADP and P_i on the F_1 sector of the mitochondrial ATPase complex and ATP hydrolysis into ADP and P_i by energy-consuming reactions in cytosol. Based on a rough correlation between the caloric expenditure and oxygen consumption, the turnover of ATP can be approximated to 40 to 60 kg/day in a human adult body, which means that 40 to 60 kg of ADP and ATP are exchanged every day between mitochondria and the rest of the cell through the ADP/ATP carrier.[7]

The turnover of ADP/ATP transport (rate of transport per carrier unit) can be calculated from the velocity and the amount of functional carrier units per milligram of protein. For better accuracy, kinetic assays with mitochondria and inside-out submitochondrial particles are usually carried out a low temperature (0 to 10°C). At higher temperature, the kinetic resolution is no longer satisfactory because of the high velocity which depends on both the turnover and the density of carrier units in the inner mitochondrial membrane. However, values between 600 and 2000/min per carrier dimer at 18 to 20°C have been reported[4,6] (the carrier dimer is supposed to be the functional transport unit [see Section V.B and VI.A]). The reconstituted transport system offers the possibility to determine with good accuracy the turnover value of transport; due to the fact that liposomes in the reconstituted system contain a few carrier molecules, the velocity of transport is low and, therefore, easy to measure. On the basis of binding data with radioactive ATR, and kinetic data in the presence of ATR or CATR, the percentage of competent carrier protein in reconstituted vesicles was determined and, thereby, the turnover number for ADP/ATP transport was calculated.[45] The turnover number per carrier dimer at 20 to 25°C is between 1000 and 2000/min. By extrapolation, values around 4000 to 5000 at 37°C can be derived assuming a Q_{10} of 2 above 20°C.

In spite of its relatively modest turnover, the ADP/ATP carrier has, nonetheless, a capacity high enough to cope with the energy requirement of the cell. This is made possible by the high concentration of carrier protein in the inner mitochondrial membrane, which may amount to 10% of the protein content of this membrane.

2. Microcompartmentalization of the ADP/ATP Carrier

At low temperature (0 to 5°C), the ADP/ATP carrier and the mitochondrial ATPase appear to be functionally linked in such a way that the extramitochondrial ADP which enters the matrix space reacts more rapidly with the ATPase complex to produce ATP than it mixes with the intramitochondrial ADP.[46,47] This does not mean that there is a physical compartmentation of the intramitochondrial adenine nucleotides; the entire pool of adenine nucleotides is, in fact, exchangeable.[48] Slow diffusion in the viscous matrix gel[49] explains the slow equilibration between intra- and extramitochondrial nucleotides. Furthermore, the lateral diffusion of the intrinsic proteins of the inner mitochondrial membrane may allow random contacts between the ADP/ATP carrier and the ATPase complex giving rise to direct phosphorylation of the transported ADP. These results, which were obtained in experiments performed at 0 to 5°C, have been recently confirmed;[50] however, raising the temperature to 25°C speeds up equilibration of intra- and extramitochondrial ADP and P_i, which is consistent with a higher fluidity of the matrix medium at higher temperature.

It is likely that part of the ATP which is exported by the ADP/ATP carrier reacts directly with the phosphokinases attached to the inner or the outer mitochondrial membranes,[7] for example, creatine phosphokinase attached to the outer face of the inner mitochondrial membrane;[51,52] conversely, the ADP generated by the creatine phosphokinase reaction in the intermembrane space enters mitochondria in preference to cytosolic ADP.[53]

3. Is ADP/ATP Transport a Rate-Limiting Reaction in the Cellular ADP/ATP Cycle?

The pioneering work of Lardy and Chance (for review see Chance[54]) emphasized the dependence of mitochondrial respiration on externally added ADP. Later, Wilson et al.[55-58] provided experimental evidence that the first two phosphorylation sites of the respiratory chain in mitochondria either isolated from pigeon heart,[56] rat liver [57] and rat heart,[58] or *in situ* in the cell[55] are in equilibrium with the extramitochondrial phosphate potential, i.e., the (ATP)/(ADP) (P_i) ratio. Since ADP, P_i and ATP are the substrates and product of the oxidative phosphorylation system within mitochondria, it follows that the ADP/ATP carrier which is the link between the intra- and extramitochondrial pool of adenine nucleotides cannot be rate limiting in the oxidative phosphorylation of extramitochondrial ADP. In contrast to Wilson et al.,[55-58] Davis et al.[59,60] and Küster et al.[61] found that the mitochondrial respiration is governed by the extramitochondrial (ATP)/(ADP) ratio; they observed no effect of P_i and, therefore, no relationship between the rate of respiration and the extramitochondrial phosphate potential (ATP)/(ADP)(P_i), suggesting that ADP/ATP transport functions out of equilibrium and is rate limiting. In the above experiments, the phosphorylation state of the mitochondrial matrix had not been taken into account. Later, using rat liver mitochondria which are able to perform a number of energy-consuming reactions involved in the cell metabolism, it was recognized by the groups of Kunz[62] and Tager[63] that respiration is primarily related to the intramitochondrial (ATP)/(ADP) ratio.

As judiciously recalled by Tager et al.,[64] a reaction which operates out of equilibrium in a multistep pathway is not necessarily rate limiting. To assess the degree of control exerted by a given reaction on pathway flux, a kinetic approach is required which consists of measuring the decrease in the flux by blocking this reaction with a strong inhibitor, for example, CATR in the case of the ADP/ATP carrier, or cyanide in the case of cytochrome oxidase. From the inhibitory effects of CATR and cyanide on the rates of glucose and urea synthesis, it was found a few years ago[65] that CATR and cyanide, although inhibiting two different reactions affected not so differently the overall flux along the ADP/ATP cycle. It was concluded that ADP/ATP transport is not more rate limiting than cytochrome oxidase and the other reactions of the ADP/ATP cycle, and that the overall flux is governed by the rates at which ADP, P_i, and oxidizable substrates are fed into the respective reactions of the cycle. Kacser and Burns[66] and Heinrich and Rapoport[67] have provided a method to quantify the control exerted by an enzyme in a reaction pathway (control strength). Based on this method, the group of Tager[68,68a] and that of Kunz[69] have found that liver mitochondria control is, indeed, distributed among the different reactions of the ADP/ATP cycle and depends on metabolic conditions. This also holds for heart mitochondria.[69,69a] Interestingly, the rapid increase in the number of ADP/ATP carrier units in guinea pig liver just after birth is probably responsible for the lower control strength exerted by the ADP/ATP carrier in the adult compared to the newborn.[69c]

C. Electrogenicity of ADP/ATP Transport and Its Role in the Maintenance of a High Cytosolic ATP/ADP Ratio

Under conditions of oxidative phosphorylation, the cytosolic ADP^{3-} is exchanged against the mitochondrial ATP^{4-} without concomitant charge compensation by proton movement.[70-72] For this reason, ADP/ATP transport is said to be electrogenic. The electrogenic nature ADP^{3-}/ATP^{4-} transport in mitochondria[70] and inside-out submitochondrial particles[72] was clearly demonstrated with a fluorescent probe of membrane potential, the cationic dye dipropylthiodicarbocyanine. In these experiments, ATPase and respiration were blocked by appropriate inhibitors. Upon addition of ATP to ADP-loaded particles, a transient decrease in the fluorescent intensity of the dye was observed which was sensitive to inhibitors of ADP/ATP transport. This indicated hyperpolarization (internal side more negative), and was explained by the entry of ATP^{4-} accompanied by the release of ADP^{3-} through the ADP/ATP carrier; no simultaneous movements of protons was observed with neutral red, a pH indicator.[72] In other experiments, no concomitant movement of protons was observed when ADP/ATP exchange was tested in the presence of valinomycin and K^+.[71] Although largely accepted, the electrogenic nature of ADP/ATP transport has been recently controverted on the basis that the value of the intramitochondrial (free Mg-ATP)/(free Mg-ADP) ratio exceeds the extramitochondrial value.[72a]

The accumulation of negative charges outside mitochondria which results from the electrogenic $ADP^{3-}_{ex}/ATP^{4-}_{in}$ exchange must be neutralized in order to maintain a constant and rapid flux of ATP from mitochondria to cytosol; this is accomplished by means of the membrane potential generated by respiration. The influence of the membrane potential on the kinetic parameters of ADP/ATP transport will be examined later in this review. At present, it is sufficient to recall that the tenfold difference in (ATP)/(ADP) ratio between the cytosol and the mitochondria[73] is satisfactorily explained by the asymmetric exchange of mitochondrial ATP for cytosolic ADP, and that this asymmetric exchange consumes energy: about one third of the energy of the respiratory chain is required to maintain the membrane potential at constant value when ADP/ATP transport is operating.[74]

ATP competes with ADP for entry into mitochondria. It is remarkable, therefore, that the asymmetric exchange of cytosolic ADP for mitochondrial ATP, which is driven, as mentioned above, by the mitochondrial membrane potential, is not perturbed by the high (ATP)/(ADP) ratio in cytosol. This is explained by the low affinity of respiring mitochondria for ATP (K_M 100 to 200 μM[75]) compared to ADP (K_M 1 to 5 μM[4]) and by the fact that the true substrates for the carrier are the free (noncomplexed) forms of ADP and ATP.[76] The (free ATP)/(free ADP) ratio in cytosol is much lower than the (Mg-ATP)/(Mg-ADP) ratio. In fact, the association constant of Mg^{2+} with ATP is nearly ten times higher than that of Mg^{2+} with ADP. In the case of rat liver cells, for example, the concentrations of the free forms of ADP and ATP are 115 and 170 μM as compared to the concentrations of total ADP and total ATP, 315 and 2730 μM, respectively.[77] It was recently reported that part of the free ADP, but not free ATP, is bound to protein either in mitochondria[78] or in cytosol.[79] Even taking into account protein binding, the actual concentration of available free ADP in cytosol, 115 μM, remains much higher than the K_M for ADP, 1 to 5 μM, whereas the concentration of free ATP, 170 μM, is in the same range as the K_M for ATP.

D. The ADP/ATP Carrier and Its Lipid Environment in the Inner Mitochondrial Membrane

1. Is ADP/ATP Transport Dependent on the Lipid Nature of the Membrane?

Preliminary indications that the fluidity of the membrane could affect the turnover of the ADP/ATP carrier stemmed from experiments on yeast.[80] The degree of saturation of yeast phospholipids is amenable to modifications by changing growth conditions. When the phospholipid hydrocarbon chains were more unsaturated, resulting in a more fluid membrane, the rate of ADP/ATP transport was markedly increased; there was no change, however, in the number of nucleotide sites, in their affinity for ADP, or in the size of the internal nucleotide pool. Fluidity of the membrane, therefore, affects only the rate of transport per site (turnover number). A higher unsaturation index of liver mitochondrial lipids caused by cold adaptation was also found to result in a substantial increase in the rate of ADP transport.[81]

Reconstitution experiments[45] have shown that the ADP/ATP carrier protein embedded in a phosphatidylcholine (PC) bilayer exhibits a very low rate of transport. A considerable increase in velocity was obtained with liposomes made of cardiolipin, phosphatidylethanolamine (PE), lyso-PE, and PC. Omission of PC further increased the transport rate; for example, ADP/ATP transport in proteoliposomes made of 92% PE and 8% cardiolipin was seven times faster than in proteoliposomes made of 92% PC and 8% cardiolipin. An enhancing effect of PE on the reconstituted glucose transport activity from LM cells has also been reported.[82]

The considerable enhancement of transport rate by PE[45] poses the interesting possibility of a specific interaction of PE with the ADP/ATP carrier protein; alternatively, PE might generate some perturbation in the organization of the lipid bilayer. Along this line, it has been found that PE induces isotropic phospholipid motion by formation of nonbilayer structures in membranes.[83] A direct contact of the carrier protein with the lipid core of the membrane is consistent with the results of experiments with spin-labeled palmityl CoA[84] (see Section VI.B). Finally, it must be borne in mind that transport efficiency depends on the appropriate insertion and orientation of the carrier protein in the lipid bilayer, and this probably depends, in turn, on the charged and uncharged phospholipid distribution in the liposome vesicles (and in the inner mitochondrial membrane in the case of import of newly synthesized ADP/ATP carrier protein). This is in agreement with the asymmetrical topography of the embedded carrier protein in the case of the inner mitochondrial membrane[4] and reconstituted proteoliposomes,[45] as shown by the difference in accessibility of ATR and BA and also by freeze fracture electromicroscopy (see Section VI.A). The increased rate of transport by incorporation of cholesterol into reconstituted vesicles[85] has no physiological significance since the inner mitochondrial membrane is devoid of cholesterol.[86]

2. Is ADP/ATP Transport Controlled by Long-Chain AcylCoAs?

Long-chain acylCoAs are known to inhibit competitively ADP/ATP transport[87-89] with high affinity, the K_i value being as low as 0.2 μM.[89] Maximal efficiency was for carbon chain lengths of 14 to 16 C, a result in accordance with theoretical considerations on the relation between the acyl chain length of acylCoAs and their partition coefficient in organic solvent/water systems.[4] Clearly, the inhibitory potency of acylCoAs depends on the solubility of the hydrocarbon chain in the mitochondrial membrane. Long-chain acylCoAs also inhibit ADP/ATP transport in inside-out submitochondrial particles from heart with high efficiency (K_i = 1 to 2 μM).[90] This, of course, is of interest from the point of view of physiological regulation, since most of the acylCoAs in the cell are located in mitochondria.[91]

Long-chain acylCoAs inhibit competitively not only ADP/ATP transport, but also malate, citrate, and phosphate transport, although their affinities for these transport systems is lower than for ADP/ATP transport.[4] Comparison of the respective K_i values for palmitylCoA to K_M values for ADP led to the surprising conclusion that the K_M/K_i ratios for the above carriers were in the same range of values, 25 to 50 (for review see Vignais[4]). For example, the K_M ADP and the K_i palmitylCoA for the ADP/ATP carrier are about 5 and 0.2 μM respectively, so that the K_M/K_i ratio is 25. A close value was found in the case of citrate transport, for which the K_M citrate is 90 μM and the K_i palmitylCoA 4 μM. In other words, the relative inhibitory efficiency of palmitylCoA and other long-chain acylCoAs is virtually the same for transport of ADP/ATP, malate, citrate, and phosphate. A specific effect of long-chain acylCoAs on ADP/ATP transport[92] is, therefore, unlikely.

That long-chain acylCoAs act as negative effectors of ADP/ATP transport in mitochondria is not entirely supported by experiments on isolated cells. In rat liver cells, for example, it has been demonstrated that under conditions leading to a doubling of the concentration of long-chain acylCoAs, there is only a slight inhibitory effect on the provision of mitochondrial ATP for ATP-utilizing reactions in the cytosol.[93] In contrast, a net decrease in cytosolic ATP/ADP ratio has been found to occur in hepatocytes on addition of oleate.[93a] In ischemic heart, it has been proposed that accumulation of long-chain acylCoAs[91] could be responsible for the decrease in the rate of ADP/ATP transport;[94,94a] however, mitochondria from ischemic heart lose their adenine nucleotides and the lower rate of ADP/ATP transport may be due to the lower concentration of intramitochondrial adenine nucleotides.[36] Finally, theoretical considerations illustrate clearly the paradox of long-chain acylCoAs acting as negative effectors of ADP/ATP transport; as pointed out by Stubbs,[10] such an inhibition is incompatible with the fact that in situations like starvation or long-term exercise where fats are important fuels, the ADP/ATP carrier has to work at maximal capacity to provide the mitochondrial respiratory chain with ADP and to export ATP to cytosol. How the ADP/ATP carrier escapes inhibition by long-chain acylCoAs is not clear. It is possible, however, that the mitochondrial long-chain acylCoAs are trapped by the phospholipid bilayer of the mitochondrial membrane and by the proteins of the matrix space; in this respect, one may recall the very efficient removal of palmitylCoA bound to mitochondrial membranes by added serum albumin.[89] For more information in this field, readers could refer to a recent review on the interaction of long-chain acylCoAs with membranes.[95] The lower V_{max} and the higher K_m found for ADP/ATP transport in liver mitochondria from hypothyroid rats is likely explained by indirect effects of the thyroid hormone,[95a,95b] possibly secondary to changes in the lipid matrix of the inner mitochondrial membrane.[95b]

IV. THE USE OF SPECIFIC LIGANDS AS TOOLS FOR STUDY OF THE TOPOGRAPHY OF THE ADP/ATP CARRIER PROTEIN

A. The Free Forms of ADP and ATP are the True Substrates for the ADP/ATP Carrier

Until recently, it was difficult to decide on the basis of experiments with intact mitochondria whether the true substrates of the ADP/ATP carrier were the free nucleotides of the Mg^{2+} complexed forms. A direct demonstration of the inhibitory effect of Mg^{2+} on ADP/ATP transport came from experiments with both the isolated carrier protein in detergent and the reconstituted carrier. The intrinsic fluorescence of the protein was modified upon addition of ADP or ATP,[76] and this modification was sensitive to ATR and BA, reflecting conformational events linked to nucleotide binding (see Section VI.E). The fact that the intrinsic fluorescence change was abolished by Mg^{2+} afforded direct evidence that free ADP and ATP, and not the Mg-nucleotide complexes, are the true substrates for the carrier. Furthermore, the carrier inserted in liposomes was found to function at maximal capacity without Mg^{2+}. Addition of Mg^{2+} above 1 mM led to a decrease in transport rate, due to the formation of Mg-nucleotide complexes and also to the decrease of the concentration of internal nucleotide by Mg^{2+}-induced release.[45]

B. Nucleotide Analogs. Stereochemical Requirement for Binding and Transport

It was early recognized by means of radiolabeled nucleotides that among natural nucleotides, only ADP, ATP, dADP, and dATP are transported across the mitochondrial membrane by a carrier specifically inhibitable by ATR or CATR.[48,96] However, once the radiolabeled ADP and ATP have been transported into the matrix space, they are transphosphorylated via specific mitochondrial kinases or dephosphorylated via the F_1-ATPase. For this reason, the stoichiometry of the exchange and the size of the exchangeable nucleotide pool in the matrix space were difficult to assess with ADP or ATP; this could be done, however, with transportable nucleotide analogs that are not metabolized.[97-99] Experiments carried out with the methylene analog of ADP, AOPCP, showed that the stoichiometry of exchange with intramitochondrial adenine nucleotides is 1 and that all nucleotides present within the matrix space are exchangeable, the internal AMP being slowly converted to ADP by transphosphorylation with ATP prior to exchange.[48,97]

With other synthetic analogs of nucleotides, it has become possible to recognize the two-step sequence of the transport process, namely, the nucleotide binding to a specific site of the carrier, followed by the vectorial process of transport, and to characterize each step in terms of stereospecificity. Binding requires a lower degree of specificity than transport. Indeed, a number of adenine nucleotide analogs can bind to the ADP/ATP carrier with high affinity, which leads to the recognition of a specific binding site; yet, they are not transported. These analogs include 8 Br-ADP and 8 Br-ATP,[100] and 3'-O esters like 3'-arylazido ADP,[101] and naphthoyl ADP.[102,103] As these nontransportable derivatives are not metabolized, their binding responds essentially to the mass action law. On the other hand, conformational changes of the carrier protein that have been found only with transportable nucleotides are probably inherent in the vectorial phase of transport (see Section V.D and VI.E).

The stereochemistry of a large number of nucleotide analogs in their free forms has been investigated by X-ray diffraction, nuclear magnetic resonance, optical spectroscopy, and circular dichroism. Three features in terms of stereochemistry play a decisive function in binding and transport of ADP and ATP by the ADP/ATP carrier.[104] They are (1) the orientation of the planar purine base with respect to the quasi planar ribose moiety, (2) the location of the C2' and C3' atoms in the ribose moiety, (3) the rotation of the C5'-O5' bond (for review see Ts'o[105] and Figure 1 for illustration). The orientation of the base with respect to the sugar is defined by the angle ϕ_{CN} between the plane of the base and the O1'-C1'

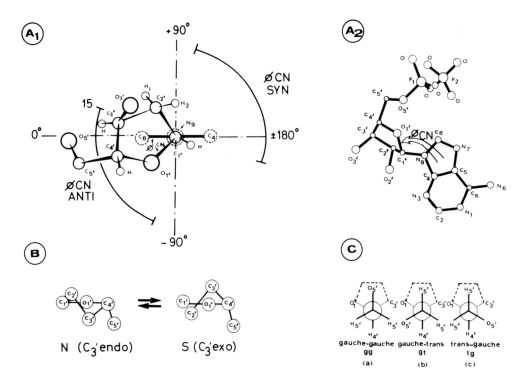

FIGURE 1.   Stereochemistry of the ADP/ATP molecule. (A1) Newman projection showing the torsion angle $\varnothing_{CN}$ O1′-C1′-N9-C8 which characterizes the position of the adenine ring relative to ribose; (A2) position of adenine relative to ribose in the ADP molecule (anticonformation) — the N9-C8 bond of adenine projects preferentially onto the sugar ring; (B) equilibrium between the S and N conformers by pseudorotation of the ribose ring; (C) Newman projection along the exocyclic C5′-C4′ bond showing the three staggered conformations gauche-gauche, gauche-trans, and trans-gauche. ([A1] Adapted from Haschemeyer, A. E. V. and Rich, A., *J. Mol. Biol.*, 27, 369, 1967. With permission. [A2] Viswamitra, M. A., Hosur, M. V., Shakked, Z., and Kennard, O., *Nature (London)* 262, 234, 1976. With permission.)

bond[106] (Figure 1A1). Nucleosides can adopt two conformations, *anti* and *syn*. The *anti* conformation corresponds to a $\varnothing_{CN}$ angle of $-30 \pm 45°$ and the *syn* conformation to an angle of $+150 \pm 45°$ (Figure 1A).[107] In the *anti* conformation, it is the N9-C8 bond of the adenine which projects onto the ribose ring (Figure 1 A2);[108] in the *syn* conformation, it is the N9-C4 bond. In solution, ADP and ATP are preferentially in the *anti* conformation, with a certain freedom of rotation around the C1′-N9 bond.[109] This nonrigid *anti* conformation is required for binding and transport.[104] The 8-Br-ADP analog[110] that is blocked in the *syn* conformation due to the bulky bromo group binds to the ADP/ATP carrier but is not transported.[100] Likewise, the 8-azido ADP, also in the *syn* conformation, binds to the carrier and is not transported.[111] On the other hand, ADP-1-*N*-oxide and ATP-1-*N*-oxide,[112] tubercidin di- or triphosphate,[113,114] and formycin di- or triphosphate,[113,115] that are preferentially in the *anti* conformation, bind to the carrier and are transported. None of the naturally occurring ribonucleotide 5′-di- and triphosphates with a pyrimidine base bind to the carrier.

By X-ray diffraction of nucleotide crystals it has been shown that a plane is defined by the C1′, O1′, and C4′ atoms of the sugar moiety. The C2′ and C3′ atoms may be located on the same side of this plane (conformation endo) or on the opposite side (conformation exo) with respect to the exocyclic atom C5′ (Figure 1B). The NMR technique applied to nucleotides in solution has allowed the demonstration of a rapid equilibrium by pseudo rotation between two conformers, the S type which corresponds to the C2′ endo form and the N type which corresponds to the C3′ endo form; the S type conformation is predominant (60%) in the case of ADP and ATP.[109] Transport but not binding is probably inhibited when the N⇌S transition is impeded either by cyclization of the vicinal 2′, 3′ alcohol groups or by opening of the cycle by oxidation.[104]

Due to the free rotation of the C4′-C5′ bond, there exists three possible rotamers (gauche-gauche, gauche-trans, and trans-gauche) with respect to the exocyclic group C5′-O5′ orientation (Figure 1C), ADP and ATP are characterized by a predominance of rotamers with gauche-gauche conformation, which results in the localization of the α phosphate above the sugar ring close to the base.[109] Only gauche-gauche rotamers are able to carry out transport; arabino nucleotides which are predominantly in the gauche-trans conformation[116] are not transported.[104]

The 3′-O derivatives of ADP and ATP[117] bind with high affinity, but are not transported.[101,102,104] In the series of the 3′-derivatives, the affinity increases with the degree of hydrophobicity; for example, the following derivatives 3′-O acetyl ADP, 3′-O arylazido ADP, and 3′-O naphthoyl ADP that are listed in the order of increasing hydrophobicity are characterized by Kd values of 500, 10, and 2.5 μM, respectively. The exceptionally high affinity of 3′-O naphthoyl ADP may be explained by a sandwich conformation in which a large part of the nucleotide molecule is free in spite of the large naphthoyl residue.[118]

The steric requirement concerning the base and the sugar moieties of the nucleotides can be summarized as follows:[104] (1) binding of nucleotide to the ADP/ATP carrier protein requires an *anti*- or *syn*-conformation of the purine base with respect to the sugar, and a S- or N-type of conformation for the sugar; (2) binding followed by transport requires a nonfixed *anti*-conformation, a free equilibrium between the N and S types of conformation, and a gauche-gauche orientation of the exocyclic C5′-O5′ group.

In contrast with the base and ribose moieties, modification of the phosphate chain has only moderate influence on specific binding and transport provided the number of negative charges along the phosphate moiety is unchanged.[104,119] The nucleotides used include the methylene phosphonate analogs of ADP and ATP,[97] the hypophosphate analog of ADP,[98] and the imidophosphate analog of ATP.[99] These modified nucleotides are transported, although they exhibit a lower affinity and velocity with respect to the natural nucleotides. In fact, replacement of the O bridge by a NH group or a CH_2 group in a phosphate chain does not alter significantly either the distance between the phosphorus atoms which increases from 2.92 to 3.00 Å, respectively, or the P-X-P angles which are 130° for POP bond angle, 127° for the PNP bond angle, and 117° for the PCP bond angle.[120] On the other hand, substitution at the Pγ of ATP, by an ethyl group[104] results in loss of transport, but not of binding. A free phosphate extremity appears, therefore, necessary for transport. This requirement for free phosphate groups together with the fact that Mg^{2+} binds to the terminal phosphate groups of ADP and ATP[121] may explain why Mg-ADP and Mg-ATP are not transported.[45,76]

C. Inhibitors

There are two families of specific inhibitors of ADP/ATP transport that differ by their ability to recognize two different conformations of the carrier protein, namely, ATR and CATR, on one hand, and BA and isoBA on the other. All of them are natural products. The atractyloside are extracted from a thistle, *Atractylis gummifera* and the bongkrekic acids are fermentation products from *Pseudomonas cocovenenans*. ATR was first recognized as an inhibitor of oxidative phosphorylation of ADP added to mitochondria[122—126] and more precisely as an inhibitor of the binding of external ADP to mitochondria.[126] These data together with the finding that ATR inhibit phosphorylation of intramitochondrial ADP[127—129] led to the recognition of a transmembrane exchange between intra- and extramitochondrial adenine nucleotides as a target reaction of ATR.[28,29] ADP/ATP transport is competitively inhibited by ATR,[130] while the carboxylated derivative of atractyloside, CATR, called initially gummiferin,[130,132] behaves as a virtually irreversible inhibitor in saline medium.[130] The atractyligenin moiety of ATR is a competitive inhibitor of ADP/ATP transport although its inhibitory efficiency is 100 times less than that of ATR; it is clear that the glucose disulfate moiety of ATR potentiates the activity of the genin.[133] The specificity of ATR and CATR for the ADP/ATP carrier is remarkable. Although CATR inhibits citrate transport, its affinity for this transport system is 500 times less than for the ADP/ATP carrier.[89] In contrast, agaric acid, a cetyl derivative of citric acid, inhibits both ADP/ATP transport[134] and citrate transport.[135]

The atractylosides are nonpermeant inhibitors; when added to mitochondria, they bind to a region of the ADP/ATP carrier protein accessible from the outside. In contrast, the bongkrekic acids are permeant inhibitors at slightly acid pH;[136] they are supposed to diffuse as protonated acids through the lipid core of the inner mitochondrial membrane and to attack the carrier protein from the inside of the membrane.[137,138] Although the atractylosides and bongkrekic acids have a quite different structure, they both contain anionic groups that may interact with cationic groups in a strategic region of the ADP/ATP carrier protein. The atractylosides and bongkrekic acids inhibit ADP/ATP transport in mitochondria from animals and plants. It is interesting to note, however, that the affinity of ATR for plant mitochondria is five to ten times lower than for mammalian mitochondria.[139]

Radiolabeled atractylosides and bongkrekic acids have been prepared for binding studies. The first preparations were radiolabeled biosynthetically;[140,141] a much higher specific radioactivity, however, was obtained by chemical radiolabeling.[137,138,142,143,143a] Radiolabeling of CATR by (³H) by oxidation and reduction of the primary alcohol was recently reported;[143a] (³H) CATR was also used,[144] but its preparation has never been described. We addressed ourselves to (¹⁴C)acetyl CATR, a CATR derivative prepared from (¹⁴C)acetic anhydride by attachment of the (¹⁴C)acetyl group to the primary alcohol of the glucose disulfate moiety.[143] (¹⁴C)acetyl CATR has the same inhibitory potency and specificity as CATR; it is, therefore, a suitable substitute for radiolabeled CATR. In general, the ATR and CATR derivatives obtained by addition of residues on the reactive primary alcohol group of the glucose disulfate moiety exhibit the same inhibitory efficiency and binding affinity as the original molecules. On this basis, photoactivable,[145] spin-labeled,[146] and fluorescent[147] derivatives of ATR have been synthetized (Figure 2). These derivatives have been of great benefit in structural studies of the carrier protein. For example, it is by means of radiolabeled derivatives of ATR that the ADP/ATP carrier protein has been characterized in yeast[101,148] and plants,[149] and the ATR site in the beef heart carrier protein has been mapped.[150,150a] Spin-labeled acyl-ATR has been used to probe the shape and the lipid environment of the ADP/ATP carrier.[146] Fluorescent derivatives of ATR are currently used for conformational studies.[150]

A new class of inhibitors, the anthraquinone dyes, have been recently reported;[151] their specificity remains to be determined.

V. TOPOGRAPHY OF THE ISOLATED ADP/ATP CARRIER

A. Purification Procedures

To keep its ability to bind substrates and specific inhibitors, and to transport ADP and ATP after insertion into liposomes, the ADP/ATP carrier protein has to be extracted from the mitochondrial membrane by appropriate detergents and purified in the presence of these detergents. For example, the ADP/ATP carrier protein extracted by the nonionic detergents, Triton® X-100 (Ter.octylphenylpolyoxyethylene)[152,153] and laurylamido-N-N-dimethyl propylaminoxide (LAPAO),[45] can catalyze an ATR and BA inhibitable nucleotide exchange after incorporation into liposomes. Extraction by Triton® X-100 is improved by high salt concentrations.[154] This effect could be due to the decrease of the CMC value by the high

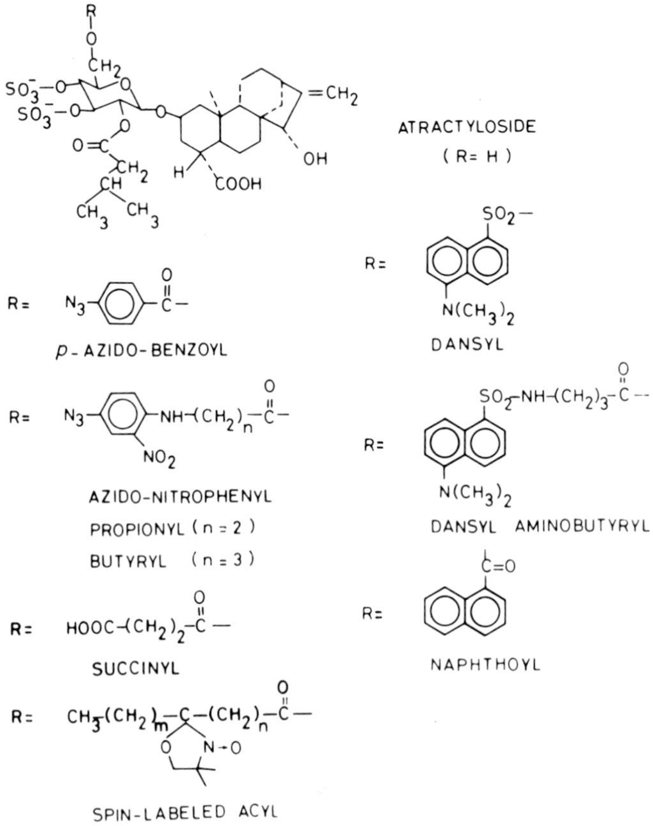

FIGURE 2. Structure of ATR and its derivatives. The ATR derivatives are obtained by substitution at the primary alcohol of the glucose disulfate moiety.

ionic strength of the medium. The anionic detergents cholate and deoxycholate irreversibly abolish the binding capacity of CATR, a reflection of serious structural alteration of the carrier protein.[154]

After extraction, the ADP/ATP carrier can be purified either by affinity chromatography using a column of agarose substituted with the succinyl derivative of ATR,[142] by chromatography on sulfopropyl Sephadex® and quaternary aminoethyl Sephadex®[148] or by hydroxyapatite chromatography.[153] In the latter method, the carrier protein is directly recovered in the pass-through fraction with good yield, and 70 to 80% purity; elimination of small molecular weight contaminants is readily achieved by chromatography on Ultrogel® AcA 202.[155,156] All purification steps are carried out in the presence of detergent. To investigate chemical aspects relevant to the proteinaceous nature of the carrier, such as the primary structure, it is convenient to remove the detergent attached to the protein by precipitation of the protein with acetone,[148,157] and the lipids by a mixture of formic acid, ethanol, and diethylether.[158]

B. Molecular Organization

1. Ultracentrifugation Data

By ultracentrifugation, the CATR-carrier complex extracted from beef heart mitochondria with Triton® X-100 was found to have total mass of 178,000 daltons, out of which the protein represented 60,000 daltons. The rest of the micelle consisted of 150 molecules of Triton® X-100 and 18 molecules of phospholipid.[159] A large portion of the surface of the carrier protein is enveloped, because of its hydrophobic nature, by Triton® and phospholipids. The rather large Stokes radius, 63Å, estimated by gel filtration was explained by the strong hydration of the detergent.

2. Neutron Scattering Data

The molecular weights of the beef heart CATR-carrier and BA-carrier complexes in LAPAO have been assessed by small angle neutron scattering, using a D_2O/H_2O ratio of 10% which annuls the contribution of LAPAO to the forward intensity of scattering.[160] The two carrier complexes had the same molecular weight, 61,000, assuming no exchangeable hydrogen in the protein, and 56,000 assuming 65% exchangeable hydrogen.

Since the minimum molecular weight of the beef heart carrier calculated from the amino acid sequence[161] is close to 32,000, it can be concluded from ultracentrifugation and neutron scattering data that the CATR-carrier and BA-carrier complexes in detergent are dimers.

C. Primary Structure and Theoretical Predictions on Secondary Structure

The beef heart ADP/ATP carrier protein contains 297 amino acid residues.[161] At the N terminus, the amino group of serine is blocked by an acetyl residue; the amino acid at the C terminus is valine. The large excess of positively charged amino acid residues, namely, 23 lysine and 17 arginine residues, with respect to the negatively charged aspartic acid (6 residues) and glutamic acid (15 residues), explains the alkaline isoelectric point of the carrier protein. The intrinsic fluorescence of the ADP/ATP carrier is due to the presence of tryptophanyl residues; binding of ADP or ATP results in changes in the environment of these residues that are reflected by modification of the intrinsic fluorescence (see Section VI.E). The seven methionine residues present in the carrier molecule are located near the C terminus; three of them, Met 237, 238 and 239, form an unusual cluster. This Met cluster is preceded by another unusual cluster of three arginine residues, Arg 234, 235, and 236. It may also be noted that the ADP/ATP carrier contains four cysteinyl residues Cys 56, 128, 159, and 256, that are spread all over the sequence. By chemical modification, it has been shown that a restricted number of arginine, cysteine, and tryptophanyl residues play a strategic role in the functioning of the carrier. Theoretical treatment of the sequence data has led to the following comments:[162] (1) the ADP/ATP carrier sequence shows a weak homology near the C terminus (residues 275 to 297) with that of other nucleotide-binding proteins, suggesting that this part of the sequence might be involved in forming a nucleotide binding pocket.[163] This, however, does not exclude other regions of the carrier as potential binding sites for nucleotides; a likely candidate is the segment 159-200 that is photolabeled by azido derivatives of ADP and ATR (see Section VI.B); (2) there is a threefold repeat in the sequence of the ADP/ATP carrier. Each repeat corresponds to about 100 amino acid residues and contains two hydrophobic segments linked by an extensive hydrophilic region; (3) six hydrophobic segments of the ADP/ATP carrier protein have a sufficient length to traverse the mitochondrial membrane. The three segments 105-137, 170-202, and 205-234 are hydrophobic enough to make α helices; further, they possess, close to their extremity, charged groups which are responsible for maintaining these α helices in the membrane and for two of the proline residues, Pro 202 and 204, which make tight bends in the peptide chain. The other three segments, which correspond to the sequences 9-39, 64-90, and 265-294, are only partially hydrophobic; for these segments to be buried in the lipid core of the membrane, it would be required that their hydrophilic residues, acid and basic, be associated by ion pairing resulting in charge neutralization. Predominantly hydrophilic segments correspond to the amino acid stretches, 40-63, 138-169, and 235-261. Each of these hydrophilic segments contains a cysteinyl residue at position 56, 159, and 256, respectively; (4) the two consecutive hydrophilic segments 138-169 and 235-261 are probably oriented to the same side of the membrane. The *Neurospora crassa* ADP/ATP carrier is longer than the beef heart ADP/ATP carrier by 15 amino acids; hemology between the two carrier proteins was found in 148 positions.[163a]

D. Nucleotide Binding Sites

With a purified ADP/ATP carrier protein in the detergent LAPAO, the binding parameters of ADP/ATP sites have been recently investigated by means of naphthoyl-ATP (N-ATP),[155] a fluorescent analog of ATP, capable of binding to the carrier most likely to the same sites of ATP or ADP, but not transportable.[102] The specific binding of N-ATP was assessed by displacement with CATR. By recording the fluorescence changes following additions of CATR to the specified carrier protein preincubated in the presence of increasing concentrations of N-ATP, it was possible to identify two classes of N-ATP binding sites on the basis of binding affinity. Each functional carrier unit contained an equal number of high affinity sites (Kd $<$ 10 nM) and low affinity sites (Kd $>$ 0.4 μM). A pair of N-ATP sites made of one high affinity site and one low affinity site interacted with one single CATR binding site, suggesting that the high and low affinity sites are interdependent.

Titration assays carried in parallel with N-ATP and formycin triphosphate (FTP),[156] a transportable nucleotide gave the same number of sites for FTP and N-ATP per functional carrier unit. However, instead of two classes of sites as found for N-ATP, four classes of sites were identified for FTP. One of them had a much higher affinity for FTP (Kd $<$ 10 nM) than the other three (Kd $=$ 0.5 to 2 μM). Upon high affinity binding of CATR to the carrier protein, FTP was removed from two of the four FTP sites, namely, the high affinity FTP site and one of the low affinity FTP sites. On the other hand, high affinity binding of BA to the carrier protein resulted in the chase of bound FTP from one of the three low affinity FTP sites. These data are summarized in Table 2. The striking difference in the binding parameters for the transportable FTP and the nontransportable N-ATP most likely reflects a different conformational state of the carrier at rest (binding without transport) compared to the functioning carrier (binding with transport). The transition from the resting to the active state would result in the emergence of a strong negative cooperativity between sites; one of the two strong sites of the resting carrier would lose its high affinity when it binds a transportable nucleotide. Considering the membrane-bound carrier and assuming that the functional protein is a dimer embedded in the membrane, one can imagine that each of the two subunits has two nucleotide sites exposed to the outside and two exposed to the inside, which corresponds to a total number of four potential sites per functional carrier unit. All sites are not functioning at the same time, but a sequential mechanism is plausible (see Sections VIII and IX).

VI. TOPOGRAPHY OF THE MEMBRANE-BOUND CARRIER IN MITOCHONDRIA AND LIPOSOMES

A. Molecular Organization

There are no available data concerning the oligomeric state of the carrier protein in the mitochondrial membrane, apart those obtained by freeze fracture of functional proteoliposomes made by insertion of the carrier protein into unilamellar liposomes.[45] By this technique the particles in the detergent LAPAO and those embedded in the lipid bilayer were found to have the same size spectrum compatible with an average molecular weight of 60,000. There remains, however, some uncertainty as to whether the inserted particles consist essentially of dimers. A higher degree of organization cannot be ruled out at present. It may be added that no more than 10% of the added carrier protein was functional in the reconstituted system in spite of the fact that about 50% of the carrier in detergent was reactive towards substrate and inhibitory ligands.

Table 2
BINDING PROPERTIES OF THE ISOLATED
ADP/ATP CARRIER. BINDING PARAMETERS
OF THE NONTRANSPORTABLE NUCLEOTIDE
ANALOG N-ATP, THE TRANSPORTABLE
ANALOG FTP, AND THE INHIBITORS CATR
AND BA

Substrate	Ligands investigated	No. of sites	Kd
N-ATP	N-ATP	2 (High affinity)	<10 nM
		2 (Low affinity)	0.5 μM
	CATR	2 (High affinity)	<10 nM
	BA	(α)	(α)
FTP	FTP	1 (High affinity)	<10 nM
		3 (Low affinity)	0.1—2 μM
	CATR	1 (High affinity)	<10 nM
		1 (Low affinity)	5 μM
	BA	1 (High affinity)	<10 nM
		n.d. (Low affinity)	—

Note: The total number of CATR sites is equal to half the total number of nucleotide (N-ATP and FTP) sites. The number of sites has been normalized on the basis of CATR effect on FTP and N-ATP binding.[155,156] (α) = no apparent effect of BA on N-ATP displacement.

The asymmetrical insertion of the ADP/ATP carrier in the inner membrane of intact mitochondria is well established on the basis of binding studies with the radiolabeled atractylosides and bongkrekic acids showing that the CATR(ATR) site is accessible from the outside and the BA(isoBA) site from the inside (see Section IV.C). Freeze cleavage of reconstituted proteoliposomes has also revealed an asymmetrical distribution of the incorporated carrier particles with a majority of the particles more exposed to the outside than to the inside.[45] This asymmetric distribution is possibly related to the asymmetric reactivity of the carrier to the inhibitory ligands.[45,164] In fact, the ADP/ATP carrier in reconstituted proteoliposomes is inhibited from the outside by the nonpermeant ATR and CATR, and from the inside by BA and isoBA. This situation is similar to that encountered in mitochondria. The asymmetric distribution of particles in proteoliposomes was not significantly modified after addition of CATR; in contrast, it changed drastically after addition of BA, leading to more particles exposed to the inside; this translational inward-directed movement was explained by a BA-induced unmasking of polar groups in the inner region of the carrier, and possibly also by a masking of polar groups at the opposite site.[45]

B. Mapping Studies

Different approaches have been used to investigate the arrangement of the peptide chain of the ADP/ATP carrier in the mitochondrial membrane. They include covalent photolabeling by photoactivable derivatives of ATR[145] and ADP (or ATP),[101] covalent binding of specific reagents to accessible amino acid residues in exposed hydrophilic segments of the chain,[165] and, finally, probing of the hydrophobic position of the carrier embedded in the lipid core by spin-labeled long-chain acylCoAs[84] or by spin-labeled long-chain acyl ATR.[146]

1. Photolabeling of the ATR Binding Site

Radioactively labeled long-chain and short-chain azido derivatives of ATR have been

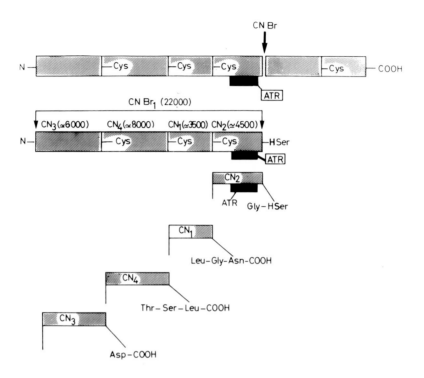

FIGURE 3.   Steps in the mapping study of the ATR site in the membrane-bound ADP/ATP carrier. After covalent photolabeling of the ADP/ATP carrier in beef heart mitochondria by a photoactivable derivative of ATR, the carrier protein is extracted, purified, and cleaved by CNBr. The large 22,000-M_r fragment resulting from CNBr cleavage is subsequently cleaved at the level of its three cysteinyl residues.(Reprinted with permission from Reference 158. Copyright 1983 American Chemical Society.)

used to photolabel covalently the ATR site in the ADP/ATP carrier from heart mitochondria. All of them had the same affinity and specificity as ATR when the binding was assayed under reversible conditions, i.e., in the absence of light.[145] Upon photoirradiation, covalent binding to the membrane-bound carrier occurred. The covalently radiolabeled carrier protein was extracted and purified. The amino acid sequence(s) which participate(s) to the ATR binding site and which, therefore, are expected to bind covalently azido ATR after photoirradiation, can be, in principle, determined after chemical cleavage of the photolabeled protein by appropriate reagents. The different steps of cleavage are summarized in Figure 3. The first step consisted in cleavage by cyanogen bromide at methionyl residues; it yielded one large peptide of M_r 22 kdaltons referred to as $CNBr_1$ and several small unlabeled peptides.[157] This was because all of the seven methionyl residues present in the beef heart carrier are located close to the C terminus. Since the $CNBr_1$ peptide contains three cysteinyl residues, distributed over its whole sequence, it was possible to investigate the photolabeled site by cleavage of the $CNBr_1$ peptide at cysteinyl residues. Cleavage was made by cyanide at alkaline pH, following reaction of the free sulfhydryl of the cysteinyl residues with dithiobisnitrobenzoate[158] (Figure 4). In a double labeling experiment, using a (^{3}H) azido derivative of ATR and (^{14}C)cyanide, only one peptide fragment was labeled both by (^{14}C) and (^{3}H). It corresponded to the cysteinyl peptide, Cyst 159-Met 200. Photolabeling restricted to a single fragment is somewhat unexpected in the case of a peptide chain that traverses the mitochondrial membrane several times. In the model of Saraste and Walker[162] that depicts the arrangement of the carrier peptide chain in the inner mitochondrial membrane (see above),

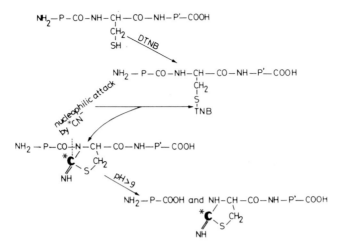

FIGURE 4. Principle of cleavage of a peptide at cysteinyl residues by cyanide at alkaline pH. (^{14}C)cyanide can be used to identify the fragments containing the cysteinyl residues. The N-terminal peptide is not labeled.

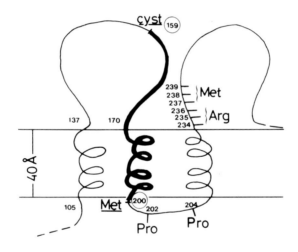

FIGURE 5. Position of the cysteinyl peptide covalently labeled by (^{3}H)azido-ATR (fragments 159 to 200) in beef heart ADP/ATP carrier. (The scheme illustrating the arrangement of the peptide chain is adapted from Saraste, M. and Walker, J. E., *FEBS Lett.*, 144, 250, 1982. With permission.)

the photolabeled peptide corresponds to an α helice that traverses the inner mitochondrial membrane, and a hydrophilic segment that belongs to one of the large hydrophilic loop (Figure 5).

2. Accessible Lysine Residues in Mitochondria and Inside-Out Submitochondrial Particles
The hydrophilic (^{3}H)pyridoxal-phosphate was used to localize the lysine residues of the ADP/ATP carrier exposed to the outside of the inner membrane in intact mitochondria, or to the inside using inverted submitochondrial particles.[165] The resulting Schiff base was reduced by Na borohydride. Only 8 out the 23 lysine residues present in the molecule of

beef heart carrier reacted with (^{3}H)pyridoxal-phosphate. Lysine residues 95, 198, 205, and 267 reacted only in intact mitochondria, which means that they are probably exposed to the outside. Lysine 146 that was labeled only in inside-out submitochondrial particles, is, therefore, expected to be exposed to the inside. Scrutiny of the labeling data, however, indicated that the yield of labeling by (^{3}H)PLP was surprisingly low, at the best one hundredth of that which would be expected for a 1:1 stoichiometry.

3. Access of SH Groups to Permeant and Nonpermeant SH Reagents

When mitochondria are supplemented with a small concentration of ADP or ATP, addition of *N*-ethylmaleimide (NEM) substantially depresses the ability of the mitochondria to carry out ADP/ATP transport or to bind ATR or CATR. Binding of BA is virtually unaltered.[166,167] This effect is specific for transportable nucleotides and is overcome by ATR. The NEM inhibition is enhanced upon generation of a proton motive force by respiration and is decreased by addition of uncouplers.[167] NEM is a penetrant−SH reagent;[168] it enters the mitochondrial membrane and, therefore, can have access to the hydrophobic region of the carrier embedded in the lipid core; fuscin, another penetrant SH reagent has the same effect as NEM; *N*-(*N*-acetyl-4-sulfamoylphenyl) maleimide (ASPM), which is penetrant at slightly acid pH, also behaves as NEM and fuscin do. On the contrary, nonpenetrant SH reagents including mersalyl, dithiobisnitrobenzoic acid, dithiobisnicotinic acid, and diamide were not inhibitory.

Twice as much (^{14}C)NEM as (^{3}H)ATR was found to bind rat liver mitochondria;[168] more recent data have confirmed that 2 mol of NEM must be incorporated into the carrier to fully inhibit the binding of 1 mol of CATR.[169] Since 1 mol of ATR or CATR binds per mole of carrier dimer, this suggests that each subunit contains one NEM-reactive SH group. Using the same approach as that described for the mapping of ATR site, the NEM-reactive cysteine was identified as cys 56.[150,150a]

4. Probing of the Shape of the Carrier by Spin-Labeled AcylCoAs and Spin-Labeled AcylATR

Spin-labeled palmityl CoA[84] and spin-labeled palmityl ATR[146] have been used to explore the interaction of the nitroxide radical placed at difference positions of the acyl chain with the hydrophobic portion of the carrier protein embedded in the membrane, and, thereby, the shape of the carrier protein. Interaction was reflected by a decreased motion of the probe. This interaction was specific since addition of any of the specific ligands ADP, ATP, ATR, CATR, and BA resulted in the mobilization of the probe in the lipid core. The spin-labeling data were consistent with a more globular shape of the carrier protein at the surface of the membrane than in the middle of the lipid core. Further, no significant annulus of lipids around the carrier protein could be demonstrated as shown by free access of the spin-labeled acyl chain to the portion of the carrier protein embedded in the membrane.

A short-chain spin-labeled acylATR with 5 C atoms in the acyl chain also exhibited an immobilized spectrum which became totally mobile upon addition of specific substrate or inhibiting ligands. In contrast to palmityl ATR, the short-chain acylATR was released to the medium.[146]

5. Motion of the Carrier

Not only the shape, but also the motion of the carrier in the mitochondrial membrane has been measured with spin-labeled ATR. By saturation transfer *esr* spectroscopy, the time constant of rotation was approximated to 10^{-4} to 10^{-5} sec.[170] A high degree of immobilization was also found recently for spin-labeled CATR bound to the carrier.[171] Another way to study the motion of the membrane-bound ADP/ATP carrier consisted in labeling the carrier with a triplet probe, the eosin-5-maleimide (EMA);[172] rotational diffusion of the carrier was measured by observing the flash-induced absorption anisotropy of EMA. It was concluded

Table 3
HIGH AFFINITY NUCLEOTIDE
BINDING SITES IN DIFFERENT
MITOCHONDRIAL MEMBRANE
PREPARATIONS

	High affinity Kd (μM)	
Mitochondrial preparations	**ADP**	**ATP**
Nucleotide-depleted mitochondria from rat liver[173]	0.1	n.d
Inner membrane from rat liver mitochondria[141]	0.04	n.d.
Nucleotide-depleted mitochondria from beef heart[173]	0.3	0.6
Beef heart mitochondria[174,188a]	4	7
Inside-out sonic particles from beef heart mitochondria[174,188a]	6	4

that the carrier rotates with a time constant of 2.10^{-4} sec at 5°C and 10^{-4} sec at 37°C; by this method, a significant fraction of the carrier units was found to be immobile, especially at low temperature.

C. Nucleotide Binding Sites

When a transportable nucleotide is incubated with mitochondria, it enters the matrix space through the ADP/ATP carrier; it also binds to the carrier protein (specific sites) and to nucleotide-binding proteins not related to ADP/ATP transport (unspecific sites). The amount of specific nucleotide sites is readily investigated by a differential method involving two types of incubation under equilibrium:[173] (1) incubation of mitochondria with (^{14}C)ADP; this results in (^{14}C)ADP uptake into the matrix space and in binding of (^{14}C)ADP to specific and unspecific sites; (2) incubation of mitochondria with (^{14}C)ADP followed by ATR; ATR removes (^{14}C)ADP bound to specific sites; since (^{14}C)ADP transport has already occurred, the incorporated radioactivity corresponds to the (^{14}C)ADP transported into the matrix space and the (^{14}C)ADP bound to unspecific sites. By difference, the number of specific sites is calculated. The specific sites in rat liver mitochondria and in rat and beef heart mitochondria amount to 1 mol/mol of cytochrome *a* and 2 mol/mol of cytochrome *a* respectively. They fall in two classes depending on affinity, the high affinity sites being in equal number or slightly less than the low affinity sites. The binding affinity is significantly higher in mitochondria which have been depleted from their internal adenine nucleotides (Table 3), and in the purified ADP/ATP carrier (see Section V.D).

D. Parameters Influencing the Topography of the Membrane-Bound Carrier

That the energy state of mitochondria influences the topography of the ADP/ATP is suggested by the following experiments. The first one concerns the effect of the proton motive force on the unmasking of SH group(s) elicited by addition of micromolar amounts of ADP or ATP to mitochondria. The proton motive force developed by respiring mitochondria markedly accelerates the −SH groups reactivity. The pH gradient is the component of the proton motive force responsible for this effect.[175] The pH gradient may influence an −SH unmasking quite indirectly, for example, through the protonation of charged groups in the vicinity of the reactive cysteine.

In another experiment, it was found that valinomycin significantly decreases the ATR binding capacity of heart mitochondria but not the ATR binding affinity. Because this effect was counteracted by nigericin and also by cationic surfactants, metal ions, and acidic pH, it was explained by a modification of the surface potential of mitochondria.[176] The manner by which a change in the surface potential influences the ATR binding capacity is only speculative at the present time; it may concern the conformation of the carrier protein, or its degree of insertion in the lipid core of the membrane. One may also imagine that, due to their ability to diffuse in the lateral plane of the membrane, the carrier units may be either dispersed or stacked, and that the ATR binding capacity per carrier unit depends on the oligomerization state of the carrier.

The conformational change of the ADP/ATP carrier is also under the control of temperature. For example, the transition between the CATR and BA conformations of the carrier as probed by the fluorescent analog N-ADP (see Section VI.E) is linearly dependent on temperature between 10 and 30°C with an activation energy of about 10 kcal.[103] The transition temperature of the rate of ADP/ATP transport in the Arrhenius plot was 10 to 13.5°C in rat liver mitochondria with an activation energy of 33 to 49 kcal below the transition and 13 to 14 kcal above;[48,177] it was 14°C in beef heart mitochondria.[177] It was thought, at first, that this transition temperature was correlated with the lipid composition of the membrane and, thereby, with its fluidity. However, since there is no lipid phase transition at 10 to 14°C in mitochondria, it has been later suggested[177] that the temperature break found for the ADP/ATP transport reflects a conformational change of the carrier protein rather than a modification in the ordering of the phospholipid molecules in the membrane.

E. CATR(ATR) and BA(isoBA) Binding Sites in the ADP/ATP Carrier
1. Evidence for CATR and BA Conformations

The terms of CATR and BA conformations that are used below are intended to designate two different native conformational states of the ADP/ATP carrier which are recognized by CATR and BA, respectively, but which are not obligatorily identical to the carrier conformations in the CATR- and BA-carrier complexes; it is possible, in fact, that upon binding of CATR or BA to the carrier in the CATR or BA conformation, a further change of conformation is induced.

Antagonistic effects of CATR and BA were perceived in early binding studies and were interpreted by an allosteric model in terms of two interchangeable conformations of the ADP/ATP carrier, one favored by CATR, the other by BA (and ADP).[178,179] These antagonistic effects were corroborated by double-labeling experiments showing that prebound (^{14}C)-acetyl CATR, a substitute for (^{14}C)CATR (Section IV.C), is released upon binding of (^{3}H)BA and vice versa.[143] It is, therefore, clear that CATR and BA cannot bind to the same transport unit at the same time.

Distinct conformations in the CATR- and BA-carrier complexes were revealed by the immunological approach;[180] less straightforward indications concerned the differential accessibility of the membrane-bound carrier to iodine, the labeling being higher when the mitochondria are incubated in the presence of CATR than in the presence of BA,[181] and the sensitivity of the isolated carrier to proteases, which is decreased in the presence of CATR and increased in the presence of BA. Although quite informative, the above data did not permit to decide whether the conformations of the carrier assumed in the presence of CATR and BA are inherent in the transport mechanism, or correspond to secondary events not related to transport and brought about by an induced-fit mechanism upon binding of CATR and BA.

2. Evidence for an ADP/ATP-Induced Transition between the CATR and BA Conformations

That the CATR and BA conformations occur in transport and that the transition between these two conformations is triggered by ADP or ATP (or transportable nucleotide analogs)

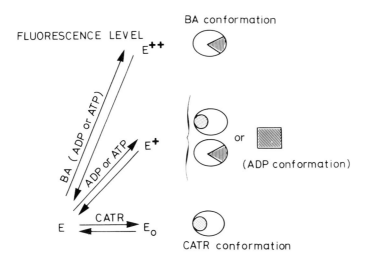

FIGURE 6.   Intrinsic fluorescence levels observed after addition of CATR, ADP or ATP, and BA plus ADP or ATP to a purified ADP/ATP carrier protein in detergent (LAPAO). The fluorescence data are interpreted in terms of CATR and BA conformational states.

that stem from chemical modification studies on the ADP/ATP induced-unmasking of $-SH$ groups[166,167,182] will be examined later (Section VI.E.4). Those that stem from fluorescence studies carried out with the isolated carrier protein in detergent[76] and the membrane-bound carrier[103] will be discussed now.

The ADP/ATP carrier protein in the detergent LAPAO exhibits an intrinsic fluorescence characteristic of tryptophanyl residues. Using a band pass centered at 355 nm to select the emitted light, it was observed that addition of ADP or ATP resulted in a fluorescence rise. The effect was saturable and reached a plateau for micromolar concentrations of ADP or ATP; it was not given by nontransportable nucleotides. Addition of CATR prior to ADP or ATP prevented the fluorescence rise; addition of CATR during the rise in fluorescence resulted in the return of the fluorescence to the basal level. Most interesting, when BA was added together with a saturating concentration of ADP or ATP, a further rise in fluorescence was observed. The data were interpreted as follows (Figure 6). The two extreme levels of fluorescence, i.e., the basal level stabilized by CATR and the highest level stabilized by BA, correspond to the CATR-carrier and BA-carrier complexes, respectively; the inter-mediate level may correspond to an ADP- or ATP-activated state, or to a mixture of the CATR and BA conformation following a preactivation of the carrier by ADP or ATP. A small concentration of ADP or ATP, lower than that of the carrier protein, was still susceptible to induce a rise of fluorescence.[183] Although this rise was slower than that observed with a saturating concentration of nucleotide, the plateau attained was virtually the same.

Direct evidence in favor of an ADP- or ATP-induced transition between two native conformational states recognizing CATR and BA comes from experiments with the carrier protein in its membrane-bound state.[102,103] In these experiments, heart mitochondria or inside-out heart submitochondrial particles were loaded with the fluorescent N-ADP, and the removal of N-ADP upon addition and binding of CATR and BA was followed by fluorescence changes. Addition of CATR to N-ADP-loaded mitochondria resulted in the rapid release of a fraction of bound N-ADP (between 40 and 70%), the remaining bound N-ADP being released much more slowly. The slow phase was markedly accelerated either by addition of BA or, more interestingly, by addition of a micromolar concentration of ADP or ATP

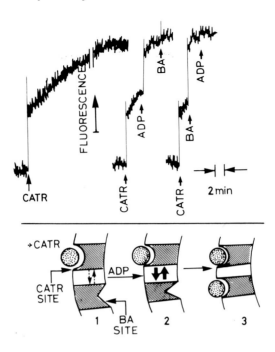

FIGURE 7.   Biphasic release of bound naphthoyl-ADP in beef heart mitochondria in the presence of CATR or BA, and acceleration of the slow phase in the presence of CATR upon addition of ADP. The experimental data are interpreted by an ADP-induced transition between CATR and BA conformations. (Adapted from Block, M. R., Lauquin, G. J. M., and Vignais, P. V., *Biochemistry*, 22, 2202, 1983. With permission.)

(Figure 7). Starting by the addition of BA led to a similar two-step release of the bound N-ADP consisting of a rapid phase of partial release of bound N-ADP followed by a slow one that was accelerated by CATR or ADP. All these data taken together are readily interpreted by the existence of two populations of membrane-bound carrier, one in the CATR conformation, the other in the BA conformation. The relative size of the two populations was found to depend on mitochondrial preparations; for example, in inside-out submitochondrial particles, most of the carrier units are in the BA conformation; at opposite, in intact mitochondria, a larger fraction of carrier units is in the CATR conformation; in frozen-thawed mitochondria, half of the carriers are in the CATR conformation and the other half in the BA conformation.

In the absence of external ADP or ATP, the CATR and BA conformations are not readily exchangeable. Addition of micromolar concentrations of ADP or ATP, however, facilitates the transition between the two conformations. For example, when CATR is added first, followed by ADP or ATP, the carriers already in the CATR conformation bind CATR immediately; for the other carriers in the BA conformation, the transition towards the CATR conformation becomes possible in the presence of traces of external ADP or ATP. Thereby, the carriers that were stabilized in the BA conformation can undergo the transition to the CATR conformation; because of the presence of CATR they are trapped in the CATR conformation. Nontransportable nucleotides are ineffective, which suggest that the CATR-BA conformational transition is linked to the vectorial reaction of transport.

The ADP-induced transition to the CATR conformation might well account for a number of results which had not received explanation at the time of their finding. This is the case of the ADP-induced enhancement of the CATR binding capacity of vesicles of inner mitochondrial membrane[130] and also of the increased sensitivity of plant mitochondria to inhibition by ATR and CATR upon preincubation with ADP.[184] At this point, it is appropriate to recall the original observation by Stoner and Sirak[185,186] who showed that micromolar concentrations of ADP added to beef heart mitochondria induce a pronounced contraction which is inhibited by ATR and that the inhibitory effect of ATR is antagonized by BA. The contraction elicited by ADP is even enhanced upon addition of BA.[187] The opposed effects of ATR and BA on the ADP-induced mitochondrial contraction are possibly related to the ADP-induced conformational changes at the level of the ADP/ATP carrier molecule itself; how such molecular events are amplified to generate gross morphological changes in the mitochondrial membrane is an open problem.

3. c State and m State, Two States of Orientation of a Single-Site Carrier

The theory of the single reorientable site (same site used for CATR, ATR, BA, isoBA, ADP, and ATP) was proposed to explain accumulation of externally added (^{14}C)ADP into nucleotide-depleted mitochondria upon addition of BA.[188] The data were interpreted as follows. In the absence of external ADP, the empty substrate site of the carrier is exposed to the outside. When (^{14}C)ADP is added, the site becomes filled with (^{14}C)ADP and moves to the inside (matrix space). When both (^{14}C)ADP and BA are present in the medium, the (^{14}C)ADP-loaded carrier is blocked with the substrate site exposed to the mitochondrial matrix, and the bound (^{14}C)ADP is released into the matrix and replaced by BA. On this basis, two different conformations were postulated; they were referred to as cytosolic or *c* state, the carrier site being exposed to the outside, and the matrix or *m* state, the carrier site being turned to the inside. Whereas the *c* and *m* states may be envisaged as conformations of the carrier protein equivalent to the CATR and BA conformations, respectively, the single reorientable site theory which underlies the concept of the *c* and *m* states is questionable. In fact, for the sake of clarity, it is worth mentioning that the amount of accumulated (^{14}C)ADP in the presence of BA is much lower (0.1 to 0.3 nmol/mg protein) than the theoretical amount expected to accumulate for a single turnover of the carrier ($\approx$1.5 nmol/mg protein). It may be added that the single-site theory cannot explain some of the results obtained with N-ADP.[103] For example, external ADP stimulates the slow release of N-ADP following the rapid release induced by CATR (see Figure 7); according to the reasoning developed in the preceding paragraphs, the carriers that do not respond immediately to CATR are in the BA conformation and upon addition of ADP, they undergo the transition towards the CATR conformation; it is clear that, for this transition to occur, external ADP has to be recognized by the carrier in the BA conformation. External ADP also stimulates the slow release of N-ADP following the rapid release induced by BA; it must, therefore, be recognized by the carrier in the CATR conformation. As a consequence the carrier protein in both the CATR and BA conformations can bind external ADP, a conclusion that is contradictory with the single site theory which postulates that the BA conformation of the carrier binds only internal nucleotides. Moreover, the single site theory cannot explain either the presence of several nucleotide binding sites per carrier unit[102,155,156] or the results concerning the chemical inactivation of ATR and BA binding. Finally, as mentioned above, externally added (^{14}C)ADP is accumulated into nucleotide-depleted mitochondria upon addition of BA.[188] However, it was recently found that binding of BA to nondepleted mitochondria displaces bound (^{14}C)ADP.[188a] These opposite effects of BA can be readily explained by assuming that BA abolishes negative cooperativity between external (cytosolic) ADP binding sites of the membrane-bound carrier.[188a]

4. Chemical Modification Data

Chemical modifications of amino acid residues in the ADP/ATP carrier protein have been carried out in an attempt to investigate whether modifications affect the binding parameters of ATR and BA. To avoid permeability barriers, only permeant modifiers were chosen. In assessing the effect of chemical modifiers on ATR and BA binding, it must be borne in mind that the modified site chain groups may be involved directly at the ATR or BA sites or indirectly at distance from the ATR or BA sites. For all chemical reagents used, the inhibitory effects were all-or-none, i.e., the binding capacity was decreased, but not the binding affinity; this is typical of an inactivation process. In brief, the results fall into two classes, depending on inactivation of both ATR and BA binding or preferential inactivation of ATR binding.

a. Inactivation of Both ATR and BA Binding

The arginyl modifiers, butanedione and phenylglyoxal, inactivated both ATR and BA binding.[189] Preincubation with ATR and BA prevented the inactivating effect of the reagents on the binding of ATR and BA, respectively. The order of the reaction of inactivation by butanedione was equal to one for both ATR and BA binding. When butanedione was replaced by phenylglyoxal, the order of the inactivation reaction was one in the case of ATR binding and two in that of BA binding. Inactivation by phenylglyoxal was well correlated with binding of (^{14}C)phenylglyoxal. Full inactivation of ATR binding required the binding of 1 mol of phenylglyoxal per mole of carrier dimer, which may be explained by a half-site reactivity mechanism, whereas full inactivation of BA binding required the incorporation of about 2 mol of phenylglyoxal per carrier dimer. Clearly, distinct arginyl residues of the ADP/ATP carrier are involved in ATR and BA binding.

b. Differential Inactivation of ATR and BA Binding

When mitochondria were subjected to UV light irradiation[190] or to incubation with 2-hydroxy-5-nitrobenzyl bromide (which preferentially reacts with tryptophan),[189] NEM (a well-known SH reagent),[182] or N-ethoxycarbonyl-2-ethoxy-1, 2-dihydroquinoline (a specific reagent of glutamic and aspartic acids),[182] their capacity to bind ATR or CATR was decreased, whereas binding of BA was not significantly altered. A direct explanation for the higher sensitivity of ATR binding is that binding of ATR to its specific site requires a higher degree of molecular organization of the carrier protein than BA binding to its own site. This implies that ATR (CATR) and BA (isoBA) bind to at least partially distinct sites.

An alternative explanation for the differential inactivation of ATR and BA binding by the above modifiers is that they readily react with the carrier in the BA conformation, but not with the carrier in the CATR conformation; upon modification, the carrier would be blocked in the BA conformation which still allows the binding of BA, but not that of ATR. The validity of the latter explanation can be tested with the following predictions: (1) the inactivation process of ATR binding should be accelerated by ligands like ADP or ATP, BA or isoBA which facilitate the transition between the CATR and BA conformation; (2) binding of intramitochondrial nucleotides should escape inactivation as BA binding does. The first prediction was fulfilled for NEM inactivation.[182] On the other hand, in the case of EEDQ or butanedione inactivation ADP or isoBA had no effect or even a protection effect against inactivation of ATR binding (isoBA was prefered to BA because of its lower affinity that makes its release by acetyl CATR easier).[182] The second prediction concerning the binding of intramitochondrial nucleotides was not either fulfilled. This prediction was tested with inside-out submitochondrial particles and the fluorescent and nontransportable ADP analog, N-ADP (see paragraph 2 in this section); upon UV light irradiation or treatment by butanedione, there was no inhibition of BA binding while the N-ADP specific binding was rapidly inactivated.[102]

c. Interpretation of Inactivation Data

The above results, except those concerning NEM inactivation, provide evidence for separate sites for ATR and BA binding. This does not preclude the possibility that the two sites may overlap. The term active center is commonly used in the case of an enzyme to designate amino acid residues involved not only in binding, but also in catalysis.[191] When applied to the ADP/ATP carrier, it would designate not only the substrate site, but also those regions of the carrier which control catalysis of transport, i.e., the ATR and BA binding sites. It is in this sense that we conclude that although the ATR and BA binding sites are at least partially distinct, they probably belong to the same active center.

VII. KINETICS OF ADP/ATP TRANSPORT

A. Is the ADP/ATP Transport of the Ping-Pong Type or Sequential Type?

Because ADP/ATP transport is a compulsory one-to-one exchange between the adenine nucleotides from the matrix space of mitochondria and those from the extramitochondrial space, it can be considered as a two-substrate reaction. In the single reorientable site carrier theory, it was proposed that the ADP/ATP carrier possesses a single accessible binding site for extramitochondrial and intramitochondrial adenine nucleotides. The carrier would then exist in two forms, one being loaded by an extramitochondrial substrate, the other by an intramitochondrial substrate; since the exchange between the two forms requires the substrate-loaded carrier, the mechanism involved is equivalent to ping-pong. As discussed by Duy-ckaerts et al.,[192] another possible model of transport is a sequential mechanism in which the external substrate and the internal substrate bind to the carrier before the exchange occurs; this implies the formation of the ternary complex, external substrate-carrier-internal substrate. Initial rates of transport at various concentrations of both internal and external substrates have been measured with rat heart mitochondria[192] and rat liver mitochondria.[193] The results disagreed with the ping-pong mechanism, but were consistent with the sequential mechanism, implying the formation of a ternary complex via a termolecular reaction and a strong positive cooperativity in the binding of the two substrates.

B. Effect of the Energy State of Mitochondria on Kinetics of ADP/ATP Transport

As previously discussed, free ADP^{3-} is exchanged against free ATP^{4-} without charge compensation, i.e., the ADP^{3-}/ATP^{4-} transport is electrogenic. Maintenance of a high rate of ADP/ATP transport requires neutralization of the excess of negative charges transported outside. This can be done under physiological conditions by means of respiration whose functioning generates a membrane potential, positive outside and negative inside. Neutralization of charges can also be done by addition of an uncoupler, for example, FCCP which collapses the proton motive force;[194] however, in the assays with FCCP, nonsaturating concentrations of ATP were used; at saturating concentrations of ATP, the maximal rate of ATP transport was not significantly altered by FCCP.[75] On the other hand, FCCP markedly decreased the K_M value for ATP[75] and increased the K_M value for ADP.[141] As thoroughly discussed by Geck and Heinz,[195] three models of transport influenced by the electrical membrane potential can be proposed, depending on the effect of the membrane potential on the velocity or on the affinity or on both of them. Whether ADP/ATP transport belongs exclusively to the velocity type or the affinity type is questionable. The velocity type model of $ADP_{ex}^{3-}/ATP_{in}^{4-}$ transport reflects an electrophoretic process which is favored by the membrane potential poised negative inside. That the membrane potential is the main component of the proton motive force involved is demonstrated by the striking effect of thiocyanate, a membrane potential collapsing reagent which strongly alters the kinetics of ADP/ATP exchange, by the absence of effect of nigericin, a pH collapsing reagent,[72] and by the linear relationship between the membrane potential and the ratio of ADP/ATP outside to ADP/ATP inside.[196] In the affinity type model, the energy state is supposed to influence the conformation of the ADP/ATP carrier protein, which results in modification of the affinity for ADP and ATP. This model is supported by experiments showing that K_M, and not V_{max}, is affected by the membrane potential; for example, in intact mitochondria FCCP increases the K_M value for ATP uptake, without altering the maximal rate of ATP uptake at a saturating concentration of ATP.[75] Further, in inside-out submitochondrial particles, FCCP affects mostly the affinity for ADP and ATP, increasing the K_M values in both cases. Finally, with the reconstituted ADP/ATP carrier, both K_M and V_{max} for ADP and ATP uptake were reported to be affected by the applied potential.[197] Interpretation of K_M modifications deserves, however, some comments. For any transport across a membrane, three steps can be considered: (1) an association-dissociation reaction of the external substrate S_{ex} with the substrate site on the outer face of the carrier

$$S_{ex} + C \underset{k_{-1}}{\overset{k_{+1}}{\rightleftharpoons}} S_{ex} - C$$

(2) a vectorial reaction by which S_{ex} is carried to the inner side of the carrier and becomes internal substrate, S_{in}

$$S_{ex} - C \underset{k_{-2}}{\overset{k_{+2}}{\rightleftharpoons}} S_{in} - C$$

(3) an association-dissociation reaction of the internal substrate S_{in} with the substrate site on the inner face of the carrier:

$$S_{in} - C \rightleftharpoons S_{in} + C$$

By analogy with an enzymatic reaction, the K_M for S_{ex} is $(k_{-1} + k_{+2})/k_{+1}$. This shows that a modification of K_M may be due either to a change in the dissociation constant k_{-1}/k_{+1}, possibly in relation with a change of conformation of the carrier, or to a change in the rate of the vectorial reaction.

VIII. MECHANISM

A. Relevant Facts
In the following, the ADP/ATP transport mechanism will be discussed in light of the experimental data that have been presented above.

1.   There are more than one nucleotide sites per functional carrier unit. This is based on titration of nucleotide binding sites in detergent-solubilized carrier protein. It may be recalled that one site out of four nucleotide binding sites in the isolated carrier is of high affinity, and the others of low affinity (Section V). The four binding sites could belong to a dimeric carrier or to an oligomeric carrier with a higher degree of organization.
2.   Any given ADP/ATP carrier unit in the mitochondrial membrane is suggested to assume either the CATR conformation (recognized by CATR and ATR) or the BA conformation (recognized by BA and isoBA). These conformations exist prior to addition of inhibitory ligands. In other words, they are not primarily produced by an induced-fit mechanism; this does not mean that binding of CATR or BA to the carrier in the respective CATR and BA conformations does not result in further modification by an induced-fit mechanism. That the CATR and BA conformations are two distinct states occurring during transport is strongly suggested by the fact that only transportable nucleotides including ADP and ATP facilitate the transition between the two conformations. The

BA conformation is characterized by a NEM-reactive −SH group; after binding of this −SH to NEM, the BA conformation becomes totally blocked (Section VI.E).

3. The CATR (ATR) and BA (isoBA) binding sites are distinct, as shown by inactivation of their binding by appropriate chemical modifiers. In this respect, it is noteworthy that the interpretation of inactivation data is not univoqual. Whereas the results obtained with phenylglyoxal and butanedione can reasonably be interpreted as indicating the existence of distinct binding sites for CATR (ATR) and BA (isoBA), the NEM-inactivation data are consistent with the stabilization of the BA conformation by NEM and the subsequent inhibition of ATR and CATR binding by the inability of the carrier to take the CATR conformation (Section VI.E).

4. The transport kinetics are consistent with the formation of a ternary complex during the transport process (Section VII).

5. One mole of CATR binds to one mole of carrier dimer.[198] This can be interpreted either by a half-site reactivity mechanism or as a trivial sterical consequence of the fact that the two CATR binding sites face each other in the carrier dimer; in this case, the binding of 1 mol of CATR sterically prevents the binding of a second mole to the adjacent site.[199] A more convincing evidence for half-site reactivity was afforded by chemical modifications of the carrier by phenylglyoxal which showed that modification of 1 mol of arginine per carrier dimer inhibited ATR binding fully.[189]

B. Models

1. Binding of Inhibitory Ligands

It is now largely accepted that the ADP/ATP carrier is a protein spanning the inner mitochondrial membrane. Reactivity towards CATR (ATR) and BA (isoBA) is currently interpreted to mean that the ligand specificity is different on the inner and outer surfaces of the inner mitochondrial membrane, and that, therefore, the ADP/ATP carrier is asymmetric, the outer face of the carrier interacting with CATR (ATR) and the inner face with BA (isoBA). It must be borne in mind, however, that assignment of BA binding to the inner face of the carrier is based essentially on the fact that BA has to penetrate the mitochondrial membrane to inhibit transport; it is not excluded, therefore, that BA binds to the hydrophobic region of carrier in contact with the lipid core of the membrane. For the sake of simplicity, however, we shall consider only the first possibility.

Two models for the topography of the CATR and BA sites have been discussed in the past and are illustrated in Figure 8. In the first model referred to as single alternating site and advocated by Klingenberg,[3] CATR and BA are envisaged to bind to the same site. The binding specificity is dictated by the conformation taken by the carrier, either the CATR conformation with the site directed to the outside or the BA conformation with the site directed to the inside. It must be added that the single-site theory supposes that the same site is used for binding not only CATR and BA, but also nucleotides. In the dual site model favored by our group,[182,189] it is proposed that the CATR and BA sites differ at least by a set of amino acids; it remains, however, that the CATR site is available to CATR only when the carrier is in the CATR conformation and the BA site similarly when the carrier is in the BA conformation. In other words, in scheme B as well as in scheme A, CATR cannot bind at the same time as BA on the same carrier. A variant of scheme B involves the half-site reactivity typical of CATR and ATR binding; it shows that only one subunit of the dimer can bind CATR or ATR with high affinity. Whatever the mechanism, the rapid transition between the CATR and BA conformations requires the presence of traces of external ADP or ATP. At this point, it is worth recalling the experiments that were performed with the fluorescent N-ADP to probe the binding of CATR and BA to mitochondria (Section VI.E). These experiments showed that the carrier units are distributed in two populations, one in the CATR conformation, the other in the BA conformation, and that external ADP facilitates

CATR and BA BINDING SITES

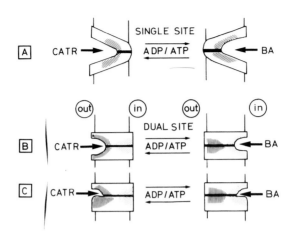

FIGURE 8.   Scheme illustrating the CATR(ATR) and BA (isoBA) binding data in terms of a single binding site (A) or a dual binding site (B and C). Scheme C illustrates the half-site reactivity of the carrier with respect to CATR(ATR) binding; a similar half-site reactivity is assumed for BA binding.

the transition between the two conformations. An important conclusion was that external ADP had to be recognized both by the carrier in the BA conformation and by the carrier in the CATR conformation, a conclusion which is not supported by the single-site mechanism.

2. ADP/ATP Transport

The following discussion is based on concepts of transport mechanism established long ago by Vidaver[201] and recently applied to glucose and choline transport in red cells.[202] For convenience, we shall assume that the functional ADP/ATP carrier is a dimer. It has to be noted, however, that although evidence has been provided that the CATR-carrier or BA-carrier complexes are dimers, the dimeric organization of the free ADP/ATP carrier has not been demonstrated. The ADP/ATP carrier models for transport fall in two classes depending on the number of nucleotide sites assigned to the carrier. The first type is the single-site model (scheme A, Figure 9) which involves a transition between two alternate conformations, i.e., outward-facing and inward-facing conformations, respectively. The second type (scheme B, Figure 9) is characterized by a two-site exposure. i.e., substrate sites simultaneously exposed on both sides of the membrane. In scheme B, each carrier dimer is supposed to possess four nucleotide sites; each subunit has two sites, one directed to the outside and the other to the inside. Assuming anticooperativity between subunits, when one subunit is actively transporting its bound substrate, the other is silent. Let us consider the active subunit *a*; its binding site is equivalent to a moveable crevice so that the substrate is transported from the outside to the inside by subunit *a*; the substrate carried by the moveable crevice is in contact with subunit *b*, but this subunit does not participate in transport. Once subunit *a* has delivered the transported substrate to the inside, subunit *b*, that was at rest before, now becomes active. How is the transmembrane nucleotide exchange in mitochondria triggered by the addition of ADP or ATP? Possibly, the rather firm binding of the externally added nucleotide is energetically favorable enough to elicit local and transient rearrangements of the peptide chain near the binding site; mobile crevices which were suggested to be involved in the movement of nucleotides through the carrier might originate from these transient rearrangements.

ADP/ATP SITES & TRANSPORT

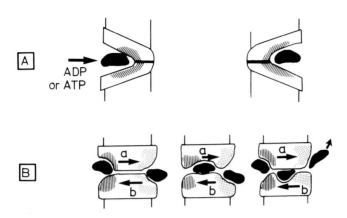

FIGURE 9.   Models of ADP/ATP transport corresponding to a single reorientable site (ping-pong mechanism) (A) or to multiple moveable sites (sequential mechanism) (B).

The moveable site carrier is supposed to contain separate CATR and BA sites (scheme C, Figure 8). This mechanism is consistent with a number of experimental data, namely, the presence of more than one nucleotide site per carrier unit, distinct CATR and BA sites, anticooperativity between subunits, and sequential kinetics. A criticism of the moveable site model concerns the difficulty in accommodating the large size of the substrates, ADP or ATP, with moveable sites on the carrier. In this respect, the single-site carrier would be more favorable since transport in that case corresponds to a "rocking" movement of the two subunits around the nucleotide molecule (scheme A, Figure 9). Another way to deal with the large size nucleotide in the case of the moveable site carrier is to consider an oligomeric carrier with more than two subunits.

The understanding of the transport mechanism of ADP and ATP is made still more complex by the unusual large size of the transported molecules, and by their high degree of asymmetry. This asymmetry is partly inherent in the nucleotide structure. In fact, the nucleotide molecule can be recognized by either of its two ends, namely, the adenine ring or the phosphate groups; possibly, the cytosolic face of the carrier in mitochondria may recognize one end of the nucleotide and the matrix face, the other end. It must be noted, however, that the free forms of ADP and ATP have a tendency to adopt, because of the *anti* gauche-gauche conformation, a bent shape (see Figure 1A2) in which the phosphate chain interacts with the adenine ring.[203] Also, the arrangement of the atom groups in each component of the nucleotide molecule, i.e., the adenine ring, the ribose, and the phosphate groups, is asymmetrical. The exquisite substrate specificity which is typical of the vectorial step of transport requires a strict steric complementarity over a large area between the nucleotide molecule and the carrier protein; how this steric complementarity is maintained during the course of transport of the nucleotide though the carrier protein remains enigmatic.

IX. SUMMARY

The mitochondrial ADP/ATP transport process is a transmembrane exchange involved in the cellular ADP/ATP cycle and catalyzed by a specific protein of intrinsic nature, with a minimal molecular weight close ot 32,000. ADP/ATP transport provides the cytosol with the mitochondrial ATP formed by oxidative phosphorylation, and allows ADP to enter mitochondria for phosphorylation. The ADP/ATP exchange is electrogenic; it is driven by the membrane potential which is developed in respiring mitochondria.

The knowledge that has accumulated concerning the structural features of the ADP/ATP transport protein has greatly benefited from the use of two groups of specific natural inhibitors, the atractylosides and the bongkrekic acids. The membrane-bound ADP/ATP carrier in mitochondria and in the reconstituted system is characterized by a binding asymmetry with respect to these inhibitors, ATR and CATR reacting with the carrier from the outside, and BA and isoBA from the inside. A number of ATR derivatives, spin-labeled ATR, photoactivable ATR, and fluorescent derivatives of ATR have been used to explore the topography of the membrane-bound carrier. Other topological approaches include immunochemistry and chemical modifications.

The beef heart ADP/ATP carrier protein has been isolated and purified in nonionic detergents. The predominance of positively charged amino acids explain its basic isoelectric point. The protein contains six hydrophobic stretches of amino acids, three of them being able to form α helices which are candidates to traverse the lipid core of the membrane. The inhibited forms of the carrier protein, CATR carrier and BA carrier, are dimers; the state of oligomerization of the nonliganded carrier is unknown. The isolated carrier has been incorporated into liposomes containing ATP; this reconstituted transport system catalyzes the exchange of internal ATP against added ADP at the same rate and with the same affinity as the membrane-bound carrier. A breakthrough in the attempt to explore the structure-function relationship in ADP/ATP transport was the finding that the carrier protein in detergent is still able to recognize substrate and inhibitory ligands, and to respond to their binding by conformational changes that are revealed, for example, by modification of the intrinsic fluorescence. The isolated protein in detergent is even able to discriminate between transportable and nontransportable nucleotides, and to respond to these two categories of ligands by different binding features.

The ADP/ATP carrier protein in the membrane-bound or the isolated form can assume two distinct conformations which differ by their reactivity to the specific inhibitors CATR (or ATR) and BA (or isoBA). For convenience, these native conformations are referred to as CATR and BA conformations. The sites recognized by CATR(ATR) and BA (isoBA) differ by a number of reactive amino acid residues as shown by chemical modification experiments; they are accessible to the ligands CATR(ATR) and BA(isoBA) in the CATR and BA conformations, respectively. A currently debated problem is how the two conformations are characterized in terms of amino acid residues exposed to chemical reagents; for example, the BA conformation, but not the CATR conformation, has an SH group available to alkylating reagents. In the absence of transportable nucleotides the CATR and BA conformations are not in rapid equilibrium; upon addition of micromolar concentrations of ADP or ATP, however, a rapid transition between the two conformations occurs. The ADP/ATP transport reaction appears to be of the sequential type, i.e., it involves the binding of two substrates to the same carrier unit. Although a large amount of information has accumulated about the conformational changes assumed by the carrier protein during transport, little information of the elementary steps of the transport is presently available.

A number of questions central to the transport mechanism are still unanswered. Is the functioning carrier organized as a dimer or does it possess a higher degree of oligomerization? How does binding of external ADP or ATP to the membrane-bound carrier in mitochondria initiate transport? What type of fluctuation occurs in the organization of the peptide chain of the carrier when transport is performed? What is the molecular basis of the cooperativity between the subunits of the carrier? Why is the vectorial step of nucleotide transport more specific than nucleotide binding? The final proof of structural features involved in carrier-substrate interactions must await X-ray diffraction analysis on crystallized preparations of the carrier protein. Evidence has been provided in this review that the free forms of ADP and ATP are the true substrates for the ADP/ATP carrier. However, it was recently suggested[204] that a carrier-bound metal, for example, Mg^{2+}, might be involved in transport; this hypothesis deserves consideration. Another unsolved question relative to the biogenesis of the carrier protein is that of the structural identity of the ADP/ATP carrier protein in the different organs of a same animal; in other words, do the different organs contain identical carrier proteins or structurally different, but functionally similar carrier proteins (isocarriers). The immunochemical approach to the organ specificity of the ADP has led, so far, to controversial results.[150,205,206,207]

ACKNOWLEDGMENTS

The authors are indebted to Ms. Jeannine Bournet who carefully and patiently typed the manuscript and to Mr. René Césarini who prepared the figures. The participation of R. Césarini and J. Doussière in the experimental part of the work performed in this laboratory is acknowledged.

REFERENCES

1. **Lindén, M., Gellerfors, P., and Nelson, B. D.,** Purification of a protein having pore formation activity from the rat liver mitochondrial outer membrane, *Biochem. J.,* 208, 77, 1982.
2. **Freitag, H., Genchi, Benz, R., Palmieri, F., and Neupert, W.,** Isolation of mitochondrial porin from *Neurospora crassa, FEBS Lett.,* 145, 72, 1982.
3. **Klingenberg, M.,** The ADP-ATP carrier in mitochondrial membranes, in *The Enzymes of Biological Membranes: Membrane Transport,* Vol. 3, Martonosi, A. N., Ed., John Wiley & Sons, New York, 1976, 383.
4. **Vignais, P. V.,** Molecular and physiological aspects of adenine nucleotide transport in mitochondria, *Biochim. Biophys. Acta,* 456, 1, 1976.
5. **Klingenberg, M., Hackenberg, H., Krämer, R., Lin, C. S., and Aquila, H.,** Two transport proteins from mitochondria. I. Mechanistic aspects of asymmetry of the ADP, ATP translocator. II. The uncoupling protein of brown adipose tissue mitochondria, *Ann. N.Y. Acad. Sci.,* 358, 83, 1980.
6. **Klingenberg, M.,** The ADP-ATP translocation in mitochondria, a membrane potential controlled transport, *J. Membrane Biol.,* 56, 97, 1980.
7. **Vignais, P. V. and Lauquin, G. J. M.,** Mitochondrial adenine nucleotide transport and its role in the economy of the cell, *Trends Biochem. Sci.,* 4, 90, 1979.
8. **Vignais, P. V., Block, M. R., Boulay, F., Brandolin, G., and Lauquin, G. J. M.,** Functional and topological aspects of the mitochondrial adenine nucleotide carrier, in *Membranes and Transport,* Vol. 1, Martonosi, A. N., Ed., Plenum Press, New York, 1982, 405.
9. **Vignais, P. M., Vignais, P. V., and Defaye, G.,** Structure-activity relationship of atractyloside and diterpenoid derivatives on oxidative phosphorylation and adenine nucleotide translocation in mitochondria, in *Atractyloside, Chemistry, Biochemistry and Toxicology,* Santi, R. and Luciani, S., Eds., Piccin Medical Books, Padua, Italy, 1978, 39.
10. **Stubbs, M.,** Inhibitors of adenine nucleotide translocase, *Pharmacol. Ther.,* 4, 329, 1979.
11. **Kováč, L., Lachowicz, T. M., and Slonimski, P. P.,** Biochemical genetics of oxidative phosphorylation, *Science,* 158, 1564, 1967.
12. **Kováč, L. and Hrušovskà, E.,** Oxidative phosphorylation in yeast. II. An oxidative phosphorylation-deficient mutant, *Biochim. Biophys. Acta,* 153, 43, 1968.
13. **Kolarov, J., Šubík, J., and Kováč, L.,** Oxidative phosphorylation in yeast. IX. Modification of the mitochondrial adenine nucleotide translocation system in the oxidative phosphorylation-deficient mutant op_1, *Biochim, Biophys. Acta,* 267, 465, 1972.
14. **Beck, J. C., Mattoon, J. R., Hawthorne, D. C., and Sherman, F.,** Genetic modification of energy-conserving systems in yeast mitochondria, *Proc. Natl. Acad. Sci. U.S.A.,* 60, 186, 1968.
15. **Kolarov, J. and Klingenberg, M.,** The adenine nucleotide translocator in genetically and physiologically modified yeast mitochondria, *FEBS Lett.,* 45, 320, 1974.

16. **Lauquin, G. J. M.,** Approche Moléculaire de la Structure et de la Fonction du Transporteur d'Adénine-Nucléotides des Mitochondries, D. Sc. Thesis, University of Grenoble, 1978.

17. **Lauquin, G., Lunardi, J., and Vignais, P. V.,** Effect of genetic and physiological manipulations on the kinetic and binding parameters of the adenine-nucleotide translocator in *Saccharomyces cerevisiae* and *Candida utilis, Biochimie,* 58, 1213, 1976.

18. **Groot, G. S. P., Out, T. A., and Souverijn, J. H. M.,** The presence of the adenine-nucleotide translocator in ρ⁻ yeast mitochondria, *FEBS Lett.,* 49, 314, 1975.

19. **Lauquin, G., Vignais, P. V., and Mattoon, J. R.,** Yeast mutants resistant to bongkrekic acid, an inhibitor of mitochondrial adenine-nucleotide translocation, *FEBS Lett.,* 35, 198, 1973.

20. **O'Malley, K., Pratt, P., Robertson, J., Lilly, M., and Douglas, M. G.,** Selection of the nuclear gene for the mitochondrial adenine nucleotide translocator by genetic complementation of the op₁ mutation in yeast, *J. Biol. Chem.,* 257, 2097, 1982.

21. **Lauquin, G. J. M., Block, M. R., Boulay, F., Brandolin, G., and Vignais, P. V.,** Molecular and genetic aspects of the mitochondrial ADP/ATP transport, in *2nd Bioenerg. Conf.,* LBTM-CNRS Edition, Lyon/Villeurbanne, 1982, 449.

22. **Šubík, J., Kolarov, J., and Kováč, L.,** Bongkrekic acid sensitivity of respiration-deficient mutants and of petite-negative species of yeasts, *Biochim. Biophys. Acta,* 357, 453, 1974.

23. **Hackenberg, H., Riccio, P., and Klingenberg, M.,** The biosynthesis of the mitochondrial ADP, ATP translocator, *Eur. J. Biochem.,* 88, 373, 1978.

24. **Zimmerman, R. and Neupert, W.,** Transport of proteins into mitochondria. Postranslational transfer of ADP/ATP carrier into mitochondria in vitro, *Eur. J. Biochem.,* 109, 217, 1980.

25. **Zwizinski, C., Schleyer, M., and Neupert, W.,** Transfer of proteins into mitochondria. Precursor to the ADP/ATP carrier binds to receptor sites on isolated mitochondria, *J. Biol. Chem.,* 258, 4071, 1983.

26. **Hatalová, I. and Kolarov, J.,** Synthesis and intracellular transport of cytochrome oxidase subunit IV and ADP/ATP translocator protein in intact hepatoma cells, *Biochem. Biophys. Res. Commun.,* 110, 132, 1983.

27. **Schleyer, M., Schmidt, B., and Neupert, W.,** Requirement of a membrane potential for the post translational transfer of proteins into mitochondria, *Eur. J. Biochem.,* 125, 109, 1982.

28. **Pfaff, E., Klingenberg, M., and Heldt, H. W.,** Unspecific permeation and specific exchange of adenine nucleotides in liver mitochondria, *Biochim. Biophys. Acta,* 104, 312, 1965.

29. **Duée, E. D. and Vignais, P. V.,** Echange entre adénine-nucléotides extra- et intramitochondriaux, *Biochim. Biophys. Acta,* 107, 184, 1965.

30. **D'Souza, M. P. and Wilson, D. F.,** Adenine nucleotide efflux in mitochondria induced by inorganic pyrophosphate, *Biochim. Biophys. Acta,* 680, 28, 1982.

31. **Vignais, P. V., Duée, E. D., and Huet, J.,** Discrimination between efflux and exchange of mitochondrial adenine nucleotides revealed by studies with atractyloside, *Life Sci.,* 7, 641, 1968.

32. **Meisner, H. and Klingenberg, M.,** Efflux of adenine nucleotides from rat liver mitochondria, *J. Biol. Chem.,* 243, 3631, 1968.

33. **Out, T. A., Kemp, A., Jr., and Souverijn, J. H. M.,** The effect of bongkrekic acid on the Ca²⁺-stimulated oxidation in rat liver mitochondria and its relation to the efflux of intramitochondrial adenine nucleotides, *Biochim. Biophys. Acta,* 245, 299, 1971.

34. **Zoccarato, F., Rugolo, M., Siliprandi, D., and Siliprandi, N.,** Correlated effluxes of adenine nucleotides, Mg²⁺ and Ca²⁺ induced in rat liver mitochondrial by external Ca²⁺ and phosphate, *Eur. J. Biochem.,* 114, 195, 1981.

35. **Harris, E. J. and Chen, M.-S.,** The losses of adenine nucleotide accompanying efflux of Ca²⁺ from heart, liver and kidney mitochondria, *Biochem. Biophys. Res. Commun.,* 104, 1264, 1982.

36. **LaNoue, K. F., Watts, J. A., and Koch, C. D.,** Adenine nucleotide transport during cardiac ischemia, *Am. J. Physiol.,* 241, H663, 1981.

36a. **Knowles, A. E.,** Characteristics of adenine nucleotide fluxes and transport in human tumor mitochondria, *Biochim. Biophys. Acta,* 764, 203, 1984.

37. **Asimakis, G. K. and Aprille, J. R.,** Net uptake of adenine nucleotides in isolated rat liver mitochondria, *FEBS Lett.,* 117, 157, 1980.

38. **Abou-Khalil, S. and Hanson, J. B.,** Net adenosine diphosphate accumulation in mitochondria, *Arch. Biochem. Biophys.,* 183, 581, 1977.

39. **Pollak, J. K. and Sutton, R.,** The transport and accumulation of adenine nucleotides during mitochondrial biogenesis, *Biochem. J.,* 192, 75, 1980.

40. **Aprille, J. R.,** Net uptake of adenine nucleotides by newborn rat liver mitochondria, *Arch. Biochem. Biophys.,* 207, 157, 1981.

41. **Rulfs, J. and Aprille, J. R.,** Adenine nucleotide pool size, adenine nucleotide translocase activity, and respiratory activity in newborn rabbit liver mitochondria, *Biochim. Biophys. Acta,* 681, 300, 1982.

42. **Reynafarje, B. and Lehninger, A. L.,** An alternative membrane transport pathway for phosphate and adenine nucleotides in mitochondria and its possible function, *Proc. Natl. Acad. Sci. U.S.A.,* 75, 4788, 1978.

43. **Fonyó, A. and Vignais, P. V.,** Phosphate retention and release during ATP hydrolysis in liver mitochondria, *FEBS Lett.,* 102, 301, 1979.
44. **Tyler, D. D.,** The pathway of inorganic-phosphate efflux from isolated liver mitochondria during adenosine triphosphate hydrolysis, *Biochem. J.,* 192, 821, 1980.
45. **Brandolin, G., Doussière, J., Gulik, A., Gulik-Krzywicki, T., Lauquin, G. J. M., and Vignais, P. V.,** Kinetic, binding and ultrastructural properties of the beef heart adenine nucleotide carrier protein after incorporation into phospholipid vesicles, *Biochim. Biophys. Acta,* 592, 592, 1980.
46. **Vignais, P. V., Vignais, P. M., and Doussière, J.,** Functional relationship between the ADP/ATP carrier and the F_1-ATPase in mitochondria, *Biochim. Biophys. Acta,* 376, 219, 1975.
47. **Out, T. A., Valeton, E., and Kemp, A., Jr.,** Role of the intramitochondrial adenine nucleotides as intermediates in the uncoupler-induced hydrolysis of extramitochondrial ATP, *Biochim. Biophys. Acta,* 440, 697, 1976.
48. **Duée, E. D. and Vignais, P. V.,** Kinetics and specificity of the adenine nucleotide translocation in rat liver mitochondria, *J. Biol. Chem.,* 244, 3920, 1969.
49. **Srere, P. A.,** The structure of the mitochondrial inner membrane-matrix compartment, *Trends Biochem. Sci.,* 7, 375, 1982.
50. **Hartung, K. J., Böhme, G., and Kunz, W.,** Involvement of intramitochondrial adenine nucleotides and inorganic phosphate in oxidative phosphorylation of extramitochondrially added adenosine-5′-diphosphate, *Biomed. Biochem. Acta,* 42, 15, 1983.
51. **Scholte, H. R., Weijers, P. J., and Witt-Peeters, E. M.,** The localization of mitochondrial creatine kinase, and its use for the determination of the sidedness of submitochondrial particles, *Biochim. Biophys. Acta,* 291, 764, 1973.
52. **Erickson-Viitanen, S., Viitanen, P., Geiger, P. J., Yang, W. C. T., and Bessman, S. P.,** Compartmentation of mitochondrial creatine phosphokinase. Direct demonstration of compartmentation with the use of labeled precursors, *J. Biol. Chem.,* 257, 14395, 1982.
53. **Moreadith, R. W. and Jacobus, W. E.,** Creatine kinase of heart mitochondria. Functional coupling of ADP transfer to the adenine nucleotide translocase, *J. Biol. Chem.,* 257, 899, 1982.
54. **Chance, B.,** Quantitative aspects of the control of oxygen utilization, in *Regulation of Cell Metabolism, Ciba Foundation Symp.,* Wolstenholme, G. E. W. and O'Connor, C. M., Eds., J. A. Churchill, London, 1959, 91.
55. **Wilson, D. F., Stubbs, M., Veech, R. L., Erecińska, M., and Krebs, H. A.,** Equilibrium relations between the oxidation-reduction reactions and the adenosine triphosphate synthesis in suspensions of isolated liver cells, *Biochem. J.,* 140, 57, 1974.
56. **Erecińska, M., Veech, R. L., and Wilson, D. F.,** Thermodynamic relationships between the oxidation-reduction reactions and ATP synthesis in suspensions of isolated pigeon heart mitochondria, *Arch. Biochem. Biophys.,* 160, 412, 1974.
57. **Forman, N. G. and Wilson, D. F.,** Energetics and stoichiometry of oxidative phosphorylation from NADH to cytochrome *c* in isolated rat liver mitochondria, *J. Biol. Chem.,* 257, 12908, 1982.
58. **Forman, N. G. and Wilson, D. F.,** Dependence of mitochondrial oxidative phosporylation on activity of the adenine nucleotide translocase, *J. Biol. Chem.,* 258, 8649, 1983.
59. **Davis, E. J. and Lumeng, L.,** Relationships between the phosphorylation potentials generated by liver mitochondria and respiratory state under conditions of adenosine diphosphate control *J. Biol. Chem.,* 250, 2275, 1975.
60. **Davis, E. J. and Davies-Van Thienen, W. I. A.,** Control of mitochondrial metabolism by the ATP/ADP ratio, *Biochem. Biophys. Res. Commun.,* 83, 1260, 1978.
61. **Küster, U., Bohnensack, R., and Kunz, W.,** Control of oxidative phosphorylation by the extramitochondrial ATP/ADP ratio, *Biochim. Biophys. Acta,* 440, 391, 1976.
62. **Kunz, W., Bohnensack, R., Böhme, G., Küster, U., Letko, G., and Schönfeld, P.,** Relations between extramitochondrial and intramitochondrial adenine-nucleotide systems, *Arch. Biochem. Biophys.,* 209, 219, 1981.
63. **Wanders, R. J. A., Groen, A. K., Meijer, A. J., and Tager, J. M.,** Determination of the free-energy difference of the adenine nucleotide translocator reaction in rat liver mitochondria using intra- and extramitochondrial ATP-utilizing reactions, *FEBS Lett.,* 132, 201, 1981.
64. **Tager, J. M., Wanders, R. J. A., Groen, A. K., Kunz, W., Bohnensack, R., Küster, U., Letko, G., Böhme, G., Duszynski, J., and Wojtczak, L.,** Control of mitochondrial respiration, *FEBS Lett.,* 151, 1, 1983.
65. **Stubbs, M., Vignais, P. V., and Krebs, H. A.,** Is the adenine nucleotide translocator rate limiting for oxidative phosphorylation?, *Biochem. J.,* 172, 333, 1978.
66. **Kacser, H. and Burns, J. A.,** Molecular democracy: who shares the controls?, *Biochem. Soc. Trans.,* 7, 1149, 1979.

67. **Heinrich, R. and Rapoport, T. A.,** A linear steady-state treatment of enzymatic chains. Critique of the crossover theorem and a general procedure to identify interaction sites with an effector, *Eur. J. Biochem.,* 42, 97, 1974.

68. **Groen, A. K., Wanders, R. J. A., Wersterhoff, H. V., Van der Meer, R., and Tager, J. M.,** Quantification of the contribution of various steps to the control of mitochondrial respiration, *J. Biol. Chem.,* 257, 2754, 1982.

68a. **Wanders, R. J. A., Groen, A. K., Van Roermund, C. W. T., and Tager, J. M.,** Factors determining the relative contribution of the adenine-nucleotide translocator and the ADP-regenerating system to the control of oxidative phosphorylation in isolated rat liver mitochondria, *Eur. J. Biochem.,* 142, 417, 1984.

69. **Gellerich, F. N., Bohnensack, R., and Kunz, W.,** Control of mitochondrial respiration. The contribution of the adenine nucleotides translocator depends of the ATP- and ADP-consuming enzymes, *Biochim. Biophys. Acta,* 722, 381, 1983.

69a. **Forman, N. G. and Wilson, D. F.,** Dependence of mitochondrial oxidative phosphorylation on activity of the adenine nucleotide translocase, *J. Biol. Chem.,* 258, 8649, 1983.

69b. **Doussière, J., Ligeti, E., Brandolin, G., and Vignais, P. V.,** Control of oxidative phosphorylation in rat heart mitochondria. The role of the adenine nucleotide carrier, *Biochim. Biophys. Acta,* 766, 492, 1984.

69c. **Hale, D. E. and Williamson, J. R.,** Developmental changes in the adenine nucleotide translocase in the guinea-pig, *J. Biol. Chem.,* 259, 8737, 1984.

70. **Laris, P. C.,** Evidence for the electrogenic nature of the ADP/ATP exchange in rat liver mitochondria, *Biochim. Biophys. Acta,* 722, 381, 1983.

71. **LaNoue, K., Mizani, S. M., and Klingenberg, M.,** Electrical unbalance of adenine nucleotide transport across the mitochondrial membrane, *J. Biol. Chem.,* 253, 191, 1978.

72. **Villiers, C., Michejda, J. W., Block, M., Lauquin, G. J. M., and Vignais, P. V.,** The electrogenic nature of ADP/ATP transport in inside-out submitochondrial particles, *Biochim. Biophys. Acta,* 546, 157, 1979.

72a. **Wilson, D. F., Erecińska, M., and Schramm, V. L.,** Evaluation of the relationship between the intra- and extramitochondrial ADP/ATP ratios using phosphoenolpyruvate kinase, *J. Biol. Chem.,* 258, 10464, 1983.

73. **Heldt, H. W., Klingenberg, M., and Milovancev, M.,** Differences between the ATP/ADP ratios in the mitochondrial matrix and in the extramitochondrial space, *Eur. J. Biochem.,* 30, 434, 1972.

74. **Duszyński, J., Bogucka, K., Letko, G., Küster, U., and Wojtczak, L.,** Relationship between the energy cost of ATP transport and ATP synthesis in mitochondria, *Biochim. Biophys. Acta,* 637, 217, 1981.

75. **Souverijn, J. H. M., Huisman, L. A., Rosing, J., and Kemp, A., Jr.,** Comparison of ADP and ATP as substrates for the adenine nucleotide translocator in rat liver mitochondria, *Biochim. Biophys. Acta,* 305, 185, 1973.

76. **Brandolin, G., Dupont, Y., and Vignais, P. V.,** Substrate-induced fluorescence changes of the isolated ADP/ATP carrier protein in solution, *Biochem. Biophys. Res. Commun.,* 98, 28, 1981.

77. **Akerboom, T. P. M., Bookelman, H., Zuurendonk, P. F., Van der Meer, R., and Tager, J. M.,** Intramitochondrial and extramitochondrial concentrations of adenine nucleotides and inorganic phosphate in isolated hepatocytes from fasted rats, *Eur. J. Biochem.,* 84, 413, 1978.

78. **Wilson, D. F., Nelson, D., and Erecińska, M.,** Binding of the intramitochondrial ADP and its relationship to adenine nucleotide translocation, *FEBS Lett.,* 143, 228, 1982.

79. **Gankema, H. S., Groen, A. K., Wanders, R. J. A., and Tager, J. M.,** Measurement of binding of adenine nucleotides and phosphate cytosolic proteins in permeabilized rat-liver cells, *Eur. J. Biochem.,* 131, 447, 1983.

80. **Lauquin, G. and Vignais, P. V.,** Adenine nucleotide translocation in yeast mitochondria. Effect of inhibitors of mitochondrial biogenesis on the ADP translocase, *Biochim. Biophys. Acta,* 305, 534, 1973.

81. **Mak, I. T., Shrago, E., and Elson, C. E.,** Modification of liver mitochondrial lipids and of adenine nucleotide translocase and oxidative phosphorylation by cold adaptation, *Biochim. Biophys. Acta,* 722, 302, 1983.

82. **Baldassare, J. J.,** Enhancement of the reconstituted glucose transport activity from LM cells by phosphatidylethanolamine, *Biochim. Biophys. Acta,* 258, 10223, 1983.

83. **Cullis, P. R. and De Kruijff, B.,** Lipid polymorphism and the functional roles of lipids in biological membranes, *Biochim. Biophys. Acta,* 559, 399, 1979.

84. **Devaux, P. F., Bienvenüe, A., Lauquin, G., Brisson, A. D., Vignais, P. M., and Vignais, P. V.,** Interaction between spin-labeled acyl-coenzyme A and the mitochondrial adenosine diphosphate carrier, *Biochemistry,* 14, 1272, 1975.

85. **Krämer, R.,** Cholesterol as activator of ADP-ATP exchange in reconstituted liposomes and in mitochondria, *Biochim. Biophys. Acta,* 693, 296, 1982.

86. **Colbeau, A., Nachbaur, J., and Vignais, P. V.,** Enzymic characerization and lipid composition of rat liver subcellular membranes, *Biochim. Biophys. Acta,* 249, 462, 1971.

87. **Pande, S. V. and Blanchaer, M. C.,** Reversible inhibitior of mitochondrial adenosine diphosphate phosphorylation by long chain acyl CoA esters, *J. Biol. Chem.,* 246, 402, 1971.

88. **Shug, A., Lerner, E., Elson, C., and Shrago, E.,** The inhibition of adenine nucleotides translocase activity by oleyl CoA and its reversal in rat liver mitochondria,, *Biochem. Biophys. Res. Commun.,* 43, 557, 1971.

89. **Morel, F., Lauquin, G., Lunardi, J., Duszynski, J., and Vignais, P. V.,** An appraisal of the functional significance of the inhibitory effect of long chain acyl-CoAs on mitochondrial transports, *FEBS Lett.,* 39, 133, 1974.

90. **Lauquin, G. J. M., Villiers, C., Michejda, J. W., Hryniewiecka, L. V., and Vignais, P. V.,** Adenine nucleotide transport in sonic submitochondrial particles. Kinetics properties and binding of specific inhibitors, *Biochim. Biophys. Acta,* 460, 331, 1977.

91. **Idell-Wenger, J. A., Grotyohann, L. W., and Neely, J. R.,** Coenzyme A and carnitine distribution in normal and ischemic hearts, *J. Biol Chem.,* 253, 4310, 1978.

92. **Woldegiorgis, G., Yousufzai, S. Y. K., and Shrago, E.,** Studies on the interaction of palmityl-CoA with the adenine nucleotide translocase, *J. Biol Chem.,* 257, 14783, 1982.

93. **Akerboom, T. P. M., Bookelman, H., and Tager, J. M.,** Control of ATP transport across the mitochondrial membrane in isolated rat-liver cells, *FEBS Lett.,* 74, 50, 1977.

93a. **Soboll, S., Seitz, H. J., Sies, H., Ziegler, B., and Scholz, R.,** Effect of long chain fatty acyl CoA on mitochondrial and cytosolic ATP/ADP ratios in the intact liver cell, *Eur. J. Biochem.,* 220, 371, 1984.

94. **Shug, A. L., Shrago, E., Bittar, N., Folts, J. D., and Koke, J. R.,** Acyl CoA inhibition of adenine nucleotide translocation in ischemic myocardium *Am J. Physiol.,* 228, 689, 1975.

94a. **Paulson, D. J. and Shug, A. L.,** Inhibition of the adenine nucleotide translocator by matrix-localized palmityl CoA in rat heart mitochondria, *Biochim. Biophys. Acta,* 766, 70, 1984.

95. **Brecher, P.,** The interaction of long chain acylCoA with membranes, *Mol. Cell Biochem.,* 57, 3, 1983.

95a. **Holness, M., Crespo-Armas, A., and Mowbray, J.,** The influence of thyroid hormone on the degree of control of oxidative phosphorylation exerted by the adenine nucleotide translocator, *FEBS Lett.,* 177, 231, 1984.

95b. **Mak, I. T., Shrago, E., and Nelson, C. E.,** Effects of thyroidectomy on the kinetics of ADP-ATP translocation in liver mitochondria, *Arch. Biochem. Biophys.,* 226, 317, 1983.

96. **Pfaff, E. and Klingenberg, M.,** Adenine nucleotide translocation of mitochondria. I. Specificity and control, *Eur. J. Biochem.,* 6, 66, 1968.

97. **Duée, E. D. and Vignais, P. V.,** Atractyloside-sensitive translocation of phosphonic acid analogues of adenine nucleotides in mitochondria, *Biochem. Biophys. Res. Commun.,* 30, 420, 1968.

98. **Vignais, P. V., Setondji, J., and Ebel, J. P.,** Translocation intramitochondriale d'un analogue structural de l'ADP, l'adénosine 5'-hypophosphate, *Biochimie,* 53, 127, 1971.

99. **Klingenberg, M.,** cited as personal communication in Yount, R. G., ATP analogs, *Adv. Enzymol.,* 43, 1, 1975.

100. **Schlimme, E. and Stahl, K. W.,** Activity of 8-C-bromo-derivatives of ATP and ADP in the mitochondrial adenine nucleotide translocation system, *Hoppe-Seyler's Z. Physiol. Chem.,* 355, 1139, 1974.

101. **Lauquin, G. J. M., Brandolin, G., Lunardi, J., and Vignais, P. V.,** Photoaffinity labeling of the adenine nucleotide carrier in heart and yeast mitochondria by an arylazido ADP analog, *Biochim. Biophys. Acta,* 501, 10, 1978.

102. **Block, M. R., Lauquin, G. J. M., and Vignais, P. V.,** Interaction of 3'-0(1-napthoyl) adenosine 5'-diphosphate, a fluorescent adenosine 5'-diphosphate analogue, with the adenosine 5'-diphosphate/adenosine 5'-triphosphate carrier protein in the mitochondrial membrane, *Biochemistry,* 21, 5451, 1982.

103. **Block, M. R., Lauquin, G. J. M., and Vignais, P. V.,** Use of 3'-O-naphtholy adenosine 5'-diphosphate to probe distinct conformational states of membrane-bound adenosine 5'-diphosphate/adenosine 5'-triphosphate carrier, *Biochemistry,* 22, 2202, 1983.

104. **Boos, K.-S. and Schlimme, E.,** Mitochondrial adenine nucleotide carrier. Investigation of principal structural, steric, and contact requirements for substrate binding and transport by means of ribose-modified substrate analogues, *Biochemistry,* 18, 5304, 1979.

105. **Ts'o, P. O. P.,** Bases, nucleosides and nucleotides, in *Basic Principles in Nucleic Acid Chemistry,* Vol. 1, Ts'o, P. O. P., Ed., Academic Press, New York, 1974, 453.

106. **Donohue, J. and Trueblood, K. N.,** Base pairing in DNA, *J. Mol. Biol.,* 2, 363, 1960.

107. **Haschemeyer, A. E. V. and Rich, A.,** Nucleoside conformations: an analysis of steric barriers to rotation about the glycosidic bond, *J. Mol. Biol.,* 27, 369, 1967.

108. **Viswamitra, M. A., Hosur, M. V., Shakked, Z., and Kennard, O.,** X-ray study of the rubidium salt of ADP, *Nature (London),* 262, 234, 1976.

109. **Davies, D. B. and Danyluk, S. S.,** Nuclear magnetic resonance studies of 5'-ribo- and deoxyribonucleotide structures in solution, *Biochemistry,* 13, 4417, 1974.

110. **Sarma, R. H., Lee, C. H., Evans, F. E., Yathindra, N., and Sundaralingam, M.,** Sugar pucker and the backbone conformation in C(8) substituted adenine nucleotides by ^{1}H and ^{1}H-^{31}P fast Fourier transform nuclear magnetic resonance methods and conformational energy calculations, *J. Am. Chem. Soc.*, 96, 7338, 1974.

111. **Schäfer, G. and Penades, S.,** Photolabeling of the adenine nucleotides carrier by 8-azido ADP, *Biochem Biophys. Res. Commun.*, 78, 811, 1977.

112. **Schlimme, E. and Schäfer, G.,** Properties of ADP- and ATP-1-N-oxide in the adenine nucleotides translocation in rat liver mitochondria, *FEBS Lett.*, 20, 359, 1972.

113. **Schlimme, E., Boos, K., and de Groot, E. J.,** Adenosine di- and triphosphate transport in mitochondria. Role of the amidine region for substrate binding and transport, *Biochemistry*, 19, 5569, 1980.

114. **Petrescu, I., Lascu, I., Goia, I., Markert, M., Schmidt, F. H., Deaciuc, I. V., Kezdi, M., and Bârzu, O.,** Phosphorylation and hydrolysis of 7-deazaadenine nucleotides by rat liver and beef heart mitochondria, *Biochemistry*, 21, 886, 1982.

115. **Graue, C. and Klingenberg, M.,** Studies of the ADP/ATP carrier of mitochondria with fluorescent ADP analogue, formycin diphosphate, *Biochim. Biophys. Acta*, 546, 539, 1979.

116. **Remin, M., Darzynkiewicz, E., Dworak, A., and Shugar, D.,** Proton magnetic resonance studies of the effects of sugar hydroxyl dissociation on nucleoside conformation. Arabinosyl nucleosides with an intramolecular hydrogen bond between the pentose O(5') and O(2'), *J. Am. Chem. Soc.*, 98, 367, 1976.

117. **Schäfer, G. and Onur, G.,** 3' Esters of ADP as energy-transfer inhibitors and probes of the catalytic site of oxidative phosphorylation, *Eur. J. Biochem.*, 97, 415, 1979.

118. **Skorka, G., Shuker, P., Gill, D., Zabicky, J., and Parola, H.,** Fluorescent substrate analogue for adenosine deaminase: 3'-0-5-(dimethyl-amino) naphtalene-1-sulfonyl adenosine, *Biochemistry*, 20, 3103, 1981.

119. **Schlimme, E., Lamprecht, W., Eckstein, F., and Goody, R. S.,** Thiophosphate-analogues and 1-N-oxides of ATP and ADP in mitochondrial translocation and phosphoryl-transfer reactions, *Eur. J. Biochem.*, 40, 485, 1973.

120. **Larsen, M., Willet, R., and Yount, R. G.,** Imidodiphosphate and pyrophosphate: possible biological significance of similar structures, *Science*, 166, 1510, 1969.

121. **Martin, R. B. and Myriam, Y. H.,** Interactions between metal ions and nucleic bases, nucleosides and nucleotides in solution, in *Metal Ions in Biological Systems*, Vol. 8, Sigel, H., Ed., Marcel Dekker, New York, 1979, 57.

122. **Bruni, A. and Contessa, A. R.,** Inhibition of phosphorylation by atractyloside, *Nature (London)*, 191, 818, 1961.

123. **Vignais, P. V., Vignais, P. M., and Stanislas, E.,** Inhibition of adenosine triphosphate-adenosine diphosphate exchange and adenosine triphosphatase activity by potassium atractylate, *Biochim. Biophys. Acta*, 51, 394, 1961.

124. **Bruni, A., Contessa, A. R., and Luciani, S.,** Atractyloside as inhibitor of energy-transfer reactions in liver mitochondria, *Biochim. Biophys. Acta*, 60, 301, 1962.

125. **Vignais, P. V., Vignais, P. M., and Stanislas, E.,** Action of potassium atractylate on oxidative phosphorylation in mitochondria and submitochondrial particles, *Biochim. Biophys. Acta*, 60, 284, 1962.

126. **Bruni, A. Luciani, S., and Contessa, A. R.,** Inhibition by atractyloside of the binding of adenosine nucleotides to rat liver mitochondria, *Nature (London)*, 201, 1219, 1964.

127. **Kemp, A., Jr. and Slater, E. C.,** The site of action of atractyloside, *Biochim. Biophys. Acta*, 92, 178, 1964.

128. **Chappell, J. B. and Crofts, A. R.,** The effect of atractylate and oligomycin on the behavior of mitochondria towards adenine nucleotides, *Biochem. J.*, 95, 707, 1965.

129. **Heldt, H. W., Jacobs, H., and Klingenberg, M.,** Endogenous ADP of mitochondria, an early phosphate acceptor of oxidative phosphorylation as disclosed by kinetic studies with ^{14}C-labeled ADP and ATP and with atractyloside, *Biochem. Biophys. Res. Commun.*, 18, 174, 1965.

130. **Vignais, P. V., Vignais, P. M., and Defaye, G.,** Adenosine diphosphate translocation in mitochondria. Nature of the receptor site for carboxyatractyloside (gummiferin), *Biochemistry*, 12, 1508, 1973.

131. **Stanislas, E. and Vignais, P. M.,** Sur les principes toxiques d'Atractylis gummifera, *C. R. Acad. Sci. Paris*, 259, 4872, 1964.

132. **Defaye, G., Vignais, P. M., and Vignais, P. V.,** Evidence expérimentale pour l'identification del la gummiférine au carboxyatractyloside, *C. R. Acad. Sci. Paris*, 273, 2672, 1971.

133. **Vignais, P. V., Duée, E. D., Vignais, P. M., and Huet, J.,** Effects of atractyligenin and its structural analogues on oxidative phosphorylation and on the translocation of adenine nucleotides in mitochondria, *Biochim. Biophys. Acta*, 118, 465, 1966.

134. **Chávez, E. and Klapp, M.,** A new inhibitor of adenine nucleotide translocase in mitochondria: agaric acid, *Biochem. Biophys. Res. Commun.*, 67, 272, 1975.

135. **Stipani, I., Francia, F., Palmieri, F., and Quagliariello, E.,** A new powerful inhibitor of the tricarboxylate carrier in rat liver mitochondria, *Bull. Mol. Biol. Med.*, 2, 72, 1977.

136. **Kemp, A., Jr., Souverijn, J. H. M., and Out, T. A.,** Effects of bongkrekic acid and hydrobongkrekic acid on oxidation and adenine nucleotide transport in mitochondria, in *Energy Transduction in Respiration and Photosynthesis,* Quagliariello, E., Papa, S., and Rossi, C. S., Eds., Adriatica Editrice, Bari, 1971, 959.

137. **Lauquin, G. J. M. and Vignais, P. V.,** Interaction of (^{3}H)bongkrekic acid with the mitochondrial adenine nucleotide translocator, *Biochemistry,* 15, 2316, 1976.

138. **Lauquin, G. J. M., Duplaa, A., Klein, G., Rousseau, A., and Vignais, P. V.,** Isobongkrekic acid, a new inhibitor of mitochondrial ADP/ATP transport: radioactive labeling and chemical and biological properties, *Biochemistry,* 15, 2323, 1976.

139. **Vignais, P. V., Douce, R., Lauquin, G. J. M., and Vignais, P. M.,** Binding of radioactively labeled carboxyatractyloside, atractyloside and bongkrekic acid to the ADP translocator of potato mitochondria, *Biochem. Biophys. Acta,* 440, 688, 1976.

140. **Vignais, P. V., Vignais, P. M., and Colomb, M. G.,** ^{35}S-Atractyloside binding affinity to the inner mitochondrial membrane, *FEBS Lett.,* 8, 328, 1970.

141. **Vignais, P. V., Vignais, P. M., Lauquin, G., and Morel, F.,** Binding of adenosine diphosphate and antagonistic ligands to the mitochondrial ADP carrier, *Biochimie,* 55, 763, 1973.

142. **Brandolin, G., Meyer, C., Defaye, G., Vignais, P. M., and Vignais, P. V.,** Partial purification of an atractyloside binding protein from mitochondria, *FEBS Lett.,* 46, 149, 1974.

143. **Block, M. R., Pougeois, R., and Vignais, P. V.,** Chemical radiolabeling of carboxyatractyloside by (^{14}C)acetic anhydride. Binding properties of (^{14}C)acetylcarboxyatractyloside to the mitochondrial ADP/ATP carrier, *FEBS Lett.,* 117, 335, 1980.

143a. **Ishiyama, S., Hiraga, K., and Tuboi, S.,** An improved method for labelling carboxyatractyloside by (^{3}H) KB$_4$, *Biochem. Int.,* 8, 305, 1984.

144. **Krämer, R. and Klingenberg, M.,** Reconstitution of inhibitor binding properties of the isolated adenosine 5'-triphosphate carrier-linked binding protein, *Biochemistry,* 16, 4954, 1977.

145. **Lauquin, G., Brandolin, G., and Vignais, P. V.,** Aryl-azido atractylosides as photoaffinity labels for the mitochondrial adenine nucleotide carrier, *FEBS Lett.,* 67, 306, 1976.

146. **Lauquin, G. J. M., Devaux, P. F., Bienvenüe, A., Villiers, C., and Vignais, P. V.,** Spin-labeled acyl atractyloside as a probe of the mitochondrial adenosine diphosphate carrier. Asymmetry of the carrier and direct lipid environment, *Biochemistry,* 16, 1202, 1977.

147. **Boulay, F., Brandolin, G., Lauquin, G. J. M., and Vignais, P. V.,** Synthesis and properties of fluorescent derivatives of atractyloside as potential probes of the mitochondrial ADP/ATP carrier protein, *Anal. Biochem.,* 128, 323, 1983.

148. **Boulay, F., Brandolin, G., Lauquin, G. J. M., Jolles, J., Jolles, P., and Vignais, P. V.,** An ADP- and atractyloside-binding protein involved in ADP/ATP transport in yeast mitochondria, *FEBS Lett.,* 98, 161, 1979.

149. **Brandolin, G., Lauquin, G. J. M., Silva Lima, M., and Vignais, P. V.,** Characterization by photoaffinity labeling of the adenine nucleotide carrier in plant mitochondria, *Biochim. Biophys. Acta,* 548, 30, 1979.

150. **Boulay, F.,** Etude Topographique de la Protéine de Transport ADP/ATP. Localisation des Sites de Fixation de l'ADP et de l'atractyloside. Identification des Domaines Antigéniques, D. Sc. thesis, University of Grenoble, 1983.

150a. **Boulay, F. and Vignais, P. V.,** Localization of the N-ethylmaleimide reactive cysteine in beef heart mitochondrial ADP/ATP carrier protein, *Biochemistry,* 23, 4807, 1984.

151. **Boos, K.-S. and Schlimme, E.,** Anthraquinone dyes: a new class of potent inhibitors of mitochondrial adenine nucleotide translocation and oxidative phosphorylation, *FEBS Lett.,* 127, 40, 1981.

152. **Riccio, P., Aquila, H., and Klingenberg, M.,** Solubilization of the carboxyatractylate binding protein from mitochondria, *FEBS Lett.,* 56, 129, 1975.

153. **Riccio, P., Aquila, H., and Klingenberg, M.,** Purification of the carboxyatractylate binding protein from mitochondria, *FEBS Lett.,* 56, 133, 1975.

154. **Klingenberg, M.,** The use of detergents for the isolation of intact carrier proteins, exemplified by the ADP/ATP carrier of mitochondria, in *Membranes and Transport,* Vol. 1, Martonosi, A., Ed., Plenum Press, New York, 1982, chap. 26.

155. **Dupont, Y., Brandolin, G., and Vignais, P. V.,** Exploration of the nucleotide binding sites of the isolated ADP/ATP carrier protein from beef heart mitochondria. I. Probing of the nucleotide sites by naphthoyl-ATP, a fluorescent non transportable analogue of ATP, *Biochemistry,* 21, 6343, 1982.

156. **Brandolin, G., Dupont, Y., and Vignais, P. V.,** Exploration of the nucleotide binding sites of the isolated ADP/ATP carrier protein from beef heart mitochondria. II. Probing of the nucleotide sites by formycin triphosphate, a fluorescent transportable analogue of ATP, *Biochemistry,* 21, 6348, 1982.

157. **Boulay, F., Lauquin, G. J. M., and Vignais, P. V.,** Fragmentation of the ADP/ATP carrier protein from beef heart mitochondria. Localization of the atractyloside binding site in a peptide obtained by cyanogen bromide cleavage, *FEBS Lett.,* 108, 390, 1979.

158. **Boulay, F., Lauquin, G. J. M., Tsugita, A., and Vignais, P. V.,** Photolabeling approach to the study of the topography of the atractyloside binding site in mitochondrial adenosine 5'-diphosphate/adenosine 5'-triphosphate carrier protein, *Biochemistry,* 22, 477, 1983.

159. **Hackenberg, H. and Klingenberg, M.,** Molecular weight and hydrodynamic parameters of the adenosine 5'-diphosphate — adenosine 5'-triphosphate carrier in Triton X-100, *Biochemistry,* 19, 548, 1980.

160. **Block, M. R., Zaccaï, G., Lauquin, G. J. M., and Vignais, P. V.,** Small angle neutron scattering of the mitochondrial ADP/ATP carrier protein in detergent, *Biochem. Biophys. Res. Commun.,* 109, 471, 1982.

161. **Aquila, H., Misra, D., Eulitz, M., and Klingenberg, M.,** Complete amino acid sequence of the ADP/ATP carrier protein from beef heart mitochondria, *Hoppe-Seyler's Z. Physiol. Chem.,* 363, 345, 1982.

162. **Saraste, M. and Walker, J. E.,** Internal sequence repeats and the path of polypeptide in mitochondrial ADP/ATP translocase, *FEBS Lett.,* 144, 250, 1982.

163. **Walker, J. E., Saraste, M., Runswick, M. J., and Gay, N. J.,** Distantly related sequences in the α and β subunits of ATP synthase, myosin, kinases and other ATP-requiring enzymes and a common nucleotide binding fold, *EMBO J.,* 1, 945, 1982.

163a. **Arends, H. and Sebald, W.,** Nucleotide sequence of the cloned mRNA and gene of the ADP/ATP carrier from *Neurospora crassa, EMBO J.,* 3, 377, 1984.

164. **Krämer, R. and Klingenberg, M.,** Reconstitution of adenine nucleotide transport from beef heart mitochondria, *Biochemistry,* 18, 4209, 1979.

165. **Bogner, W., Aquila, H., and Klingenberg, M.,** Surface labeling of membrane bound ADP/ATP carrier by pyridoxal phosphate, *FEBS Lett.,* 146, 259, 1982.

166. **Leblanc, P. and Clauser, H.,** ADP-dependent inhibition of sarcosomal adenine nucleotide translocase by N-ethylmaleimide, *FEBS Lett.,* 23, 107, 1972.

167. **Vignais, P. V. and Vignais, P. M.,** Effect of SH reagents on atractyloside binding to mitochondria and ADP translocation. Potentiation by ADP and its prevention by uncoupler FCCP, *FEBS Lett.,* 26, 27, 1972.

168. **Vignais, P. M., Chabert, J., and Vignais, P. V.,** The use of sulfhydryl reagents to identify proteins undergoing ADP-dependent conformational changes in mitochondrial membrane, in *Biomembrane, Structure and Function, 9th FEBS Meet.,* Gardos, G. and Szasz, J., Eds., North-Holland, Amsterdam, 1975, 307.

169. **Aquila, H., Eiermann, W., and Klingenberg, M.,** Incorporation of N-ethylmaleimide into the membrane-bound ADP/ATP translocator. Isolation of the protein labeled with N-(^{3}H)ethylmaleimide, *Eur. J. Biochem.,* 122, 133, 1982.

170. **Vignais, P. M.,** Fluidity of mitochondrial lipids, in *Mitochondria, Bioenergetics, Biogenesis and Membrane Structure,* Packer, L. and Gómez-Puyou, A., Eds., Academic Press, New York, 1976, 367.

171. **Munding, A., Beyer, K., and Klingenberg, M.,** Binding of spin-labeled carboxyatractylate to mitochondrial adenosine 5'-diphosphate/adenosine 5'-triphosphate carrier as studied by electron spin resonance, *Biochemistry,* 22, 1941, 1983.

172. **Müller, M., Krebs, J. R. J., Cherry, R. J., and Kawato, S.,** Selective labeling and rotational diffusion of the ADP/ATP translocator in the inner mitochondrial membrane, *J. Biol. Chem.,* 257, 1117, 1982.

173. **Weidemann, M. J., Erdelt, H., and Klingenberg, M.,** Adenine nucleotide translocation of mitochondria. Identification of carrier sites, *Eur. J. Biochem.,* 16, 313, 1970.

174. **Block, M. R.,** Protéine de Transport ADP/ATP liée à la Membrane Mitochondriale. Sites de Fixation des Inhibiteurs, Atractyloside et Acide Bongkrékique, et de Nucleotides. Stoechiométrie des Sites et États Conformationnels de la Protéine, D. Sc. thesis, University of Grenoble, 1983.

175. **Michejda, J. and Vignais, P. V.,** The energy-dependent unmasking of -SH groups in the mitochondrial ADP/ATP carrier, and its prevention by nigericin, *FEBS Lett.,* 132, 129, 1981.

176. **Vignais, P. V., Michejda, J., and Doussière, J.,** Inhibition by valinomycin of atractyloside binding to the membrane-bound ADP/ATP carrier: counteracting effect of cations, *J. Bioenerg. Biomembr.,* 15, 237, 1983.

177. **Klingenberg, M., Grebe, K., and Appel, M.,** Temperature dependence of ADP/ATP translocation in mitochondria, *Eur. J. Biochem.,* 126, 263, 1982.

178. **Vignais, P. V., Vignais, P. M., and Defaye, G.,** Gummiferin, an inhibitor of the adenine-nucleotide translocation. Study of its binding properties to mitochondria, *FEBS Lett.,* 17, 281, 1971.

179. **Vignais, P. V., Vignais, P. M., Defaye, G., Chabert, J., Doussière, J., and Brandolin, G.,** (^{35}S)Atractyloside and (^{35}S)atractyloside-derivatives as environmental probes of the adenine-nucleotide carrier in mitochondria, in *Biochemistry and Biophysics of Mitochondrial Membranes,* Azzone, G. F., Carafoli, E., Lehninger, A. L., Quagliariello, E., and Siliprandi, N., Eds., Academic Press, New York, 1972, 447.

180. **Buchanan, B. B., Eiermann, W., Riccio, P., Aquila, H., and Klingenberg, M.,** Antibody evidence for different conformational states of ADP, ATP translocator protein isolated from mitochondria, *Proc. Natl. Acad. Sci. U.S.A.* 73, 2280, 1976.

181. **Brdiczka, D. and Schumacher, D.,** Iodination of peripheral mitochondrial membrane proteins in correlation to the functional state of the ADP/ATP carrier, *Biochem. Biophys. Res. Commun.,* 73, 823, 1976.

182. **Block, M. R., Lauquin, G. J. M., and Vignais, P. V.,** Atractyloside and bongkrekic acid sites in the mitochondrial ADP/ATP carrier protein. An appraisal of their unicity by chemical modifications, *FEBS Lett.*, 131, 213, 1981.

183. **Brandolin, G.,** Propriétés du Transporteur Mitochondrial d'Adénine-Nucléotides Après Purification. Etude de la Protéine de Transport sous Forme Isolée et sous Forme Réincorporée dans des Vésicules Phospholipidiques, D. Sc. thesis, University of Grenoble, 1983.

184. **Silva Lima, M. and Denslow, N. D.,** The effect of atractyloside and carboxyatractyloside on adenine nucleotide translocation in mitochondria of *Vigna sinensis* (L.) Savi cv. Serido, *Arch. Biochem. Biophys.* 193, 368, 1979.

185. **Stoner, C. D. and Sirak, H. D.,** Adenine nucleotide-induced contraction of the inner mitochondrial membrane. I. General characterization, *J. Cell Biol.*, 56, 51, 1973.

186. **Stoner, C. D. and Sirak, H. D.,** Adenine nucleotide-induced contraction of the inner mitochondrial membrane. II. Effect of bongkrekic acid, *J. Cell Biol.*, 56, 65, 1973.

187. **Scherer, B. and Klingenberg, M.,** Demonstration of the relationship between the adenine nucleotide carrier and the structural changes of mitochondria as induced by adenosine 5'-diphosphate, *Biochemistry*, 13, 161, 1974.

188. **Klingenberg, M., Appel, M., Babel, W., and Aquila, H.,** The binding of bongkrekate to mitochondria, *Eur. J. Biochem.*, 131, 647, 1983.

188a. **Block, M. R. and Vignais, P. V.,** Substrate-site interactions in the membrane-bound adenine nucleotide carrier as disclosed by ADP and ATP analogs, *Biochim. Biophys. Acta*, 767, 369, 1984.

189. **Block, M. R., Lauquin, G. J. M., and Vignais, P. V.,** Chemical modifications of atractyloside and bongkrekic acid binding sites of the mitochondrial adenine nucleotide carrier. Are there distinct binding sites?, *Biochemistry*, 20, 2692, 1981.

190. **Block, M. R., Lauquin, G. J. M., and Vignais, P. V.,** Differential inactivation of atractyloside and bongkrekic acid binding sites on the adenine nucleotide carrier by ultraviolet light. Its implication for the carrier mechanism, *FEBS Lett.*, 104, 425, 1979.

191. **Vallee, B. L. and Riordan, J. F.,** Chemical approaches to the properties of active sites of enzymes, *Ann. Rev. Biochem.*, 38, 733, 1969.

192. **Duyckaerts, C., Sluse-Goffart, C. M., Fux, J. P., Sluse, F. E., and Liébecq, C.,** Kinetic mechanism of the exchanges catalyzed by the adenine nucleotide carrier, *Eur. J. Biochem.*, 106, 1, 1980.

193. **Barbour, R. L. and Chan, S. H. P.,** Characterization of the kinetics and mechanism of the mitochondrial ADP/ATP carrier, *J. Biol. Chem.*, 256, 1940, 1981.

194. **Klingenberg, M., Wulf, R., Heldt, H. W., and Pfaff, E.,** Control of adenine nucleotide translocation, in *5th FEBS Meet.*, Vol 17, Ernster, L. and Drahota, Z., Eds., Academic Press, New York, 1969, 59.

195. **Geck, P. and Heinz, E.,** Coupling in secondary transport. Effect of electrical potentials on the kinetics of ion linked co-transport, *Biochim. Biophys. Acta*, 443, 49, 1976.

196. **Klingenberg, M. and Rottenberg, H.,** Relation between the gradient of the ADP/ATP ratio and the membrane potential across the mitochondrial membrane, *Eur. J. Biochem.*, 73, 125, 1977.

197. **Krämer, R. and Klingenberg, M.,** Electrophoretic control of reconstituted adenine nucleotide translocation, *Biochemistry*, 21, 1082, 1982.

198. **Klingenberg, M., Riccio, P. and Aquila, H.,** Isolation of the ADP, ATP carrier as the carboxyatractylate protein complex from mitochondria, *Biochim. Biophys. Acta*, 503, 193, 1978.

199. **Klingenberg, M.,** Membrane protein oligomeric structure and transport function, *Nature (London)*, 290, 449, 1981.

200. **Klingenberg, M. and Appel, M.,** Is there a common binding center in the ADP, ATP carrier for substrate and inhibitors?, *FEBS Lett.*, 119, 195, 1980.

201. **Vidaver, G. A.,** Inhibition of parallel flux and augmentation of counter flux shown by transport models not involving a mobile carrier, *J. Theor. Biol.*, 10, 301, 1966.

202. **Krupka, R. M. and Deves, R.,** An experimental test for cyclic versus linear transport models. The mechanism of glucose and choline transport in erythrocytes, *J. Biol. Chem.*, 256, 5410, 1981.

203. **Schlimme, E., Boos, K. S., Bojanovski, D., and Lüstorff, J.,** Investigations of mitochondrial adenine nucleotide translocation by means of nucleotide analogs, *Angew. Chem. Int. Ed. Engl.*, 16, 695, 1977.

204. **Boos, K.-S. and Schlimme, E.,** A model for metalloprotein-catalyzed ADP, ATP transport in mitochondria, *FEBS Lett.*, 160, 11, 1983.

205. **Eiermann, W., Aquila, H., and Klingenberg, M.,** Immunological characterization of the ADP, ATP translocator protein isolated from mitochondria of liver, heart, and other organs, *FEBS Lett.*, 74, 209, 1977.

206. **Kolarov, J., Kužela, S., Krempaský, V., Lakota, J., and Ujházy, V.,** ADP, ATP translocator protein of rat heart, liver and hepatoma mitochondria exhibits cross-reactivity, *FEBS Lett.*, 96, 373, 1978.

207. **Schultheiss, H. P. and Klingenberg, M.,** Immunochemical characterization of the adenine nucleotide translocator. Organ specificity and conformation specificity, *Eur. J. Biochem*, 143, 599, 1984.

Chapter 8

EXOCYTOSIS, ENDOCYTOSIS, AND RECYCLING OF MEMBRANES*

D. James Morré

TABLE OF CONTENTS

* Work supported in part by grants from the National Institutes of Health HD 11508 and CA 18801.

I. INTRODUCTION

Exocytosis, outbound membrane flow or traffic, and endocytosis, the inbound movement of membranes (Figure 1), are responsible for the exit (secretion) and entry (phagocytosis/pinocytosis) of various macromolecular complexes, solutes, and fluids as well as among the primary processes whereby cell surfaces are formed, renewed, and remodeled. To the extent that the same membrane components or compartments participate sequentially in one or more series of exocytotic/endocytotic events, recycling occurs. Recycling has been envisioned as involving essentially the same cell compartments moving in both directions in the manner of a shuttle,[1—5] or as a more subtle reutilization of membrane constituents where the bulk of the inbound traffic and the bulk of the outbound traffic occur by separate pathways involving different organelle systems (Figure 2).[6,7] This latter view is the more traditional where traffic from the plasma membrane to the lysosomes (endocytosis) and from the Golgi apparatus to the plasma membrane (exocytosis) are unidirectional. It does not, however, contradict growing evidence that membrane recycling provides a major mechanism for the conservation of cell surface and cell surface receptors during both receptor-mediated and other forms of endocytosis.[5]

Mechanistically, exocytosis, endocytosis, and recycling of membranes appear to occur as distinct, albeit overlapping, cellular processes. Each has unique characteristics and fulfills different subcellular functions. Recycling is not to be regarded as synonymous with either exocytosis or endocytosis per se, nor can it lead to net membrane biogenesis or internalization. The term recycling might well be restricted to processes of rapid return of membrane. These would include (1) rapid transport back to the cell surface of cell surface constituents from inbound material following its endocytosis and (2) the corresponding process of rapid return back to the Golgi apparatus of internal membranes first delivered to the cell surface by exocytosis. By definition, the term recycling should be reserved for those aspects of membrane flow and circulation that are both bidirectional and reversible.

II. EXOCYTOSIS

Exocytosis is the major membrane pathway to the cell surface whereby vesicles resulting from products of synthesis are carried to the plasma membrane (Figure 3). Here the vesicle membrane fuses with and becomes incorporated into the plasma membrane and any vesicle content is discharged to the cell surface.[1—4,7—15] This type of pathway is considered to be the primary route through which various intrinsic membrane constituents (glycoproteins, steroids, phospholipids, glycolipids) are delivered to the plasma membrane.[7,8,16]

Exocytosis may occur in multiple and/or divergent pathways within the same cell. For example, Rodriquez Boulan et al.[17,18] have shown that envelope glycoproteins of different viruses may be selectively delivered to different regions of the cell's surface. Glycoproteins of Sendai virus are inserted into the apical domain and vesicular stomatitis virus glycoprotein is delivered to the basolateral domain. Presumably, both glycoproteins pass through the Golgi apparatus, but, thereafter, the pathways must diverge. The mechanisms whereby sorting and delivery to the different domains of the plasma membrane are accomplished are unknown.[1]

The Golgi apparatus has manifold and complex roles in exocytosis involving not only the sorting and concentration of products destined for secretion but the processing and sorting of membrane constituents, some of which arrive at the plasma membrane, as well as delivery of hydrolytic enzymes to lysosomes.[1—4,6—13] The complexity of Golgi apparatus operation is compounded by the fact that the Golgi apparatus is a center where multiple routes of vesicular traffic may both converge and diverge (Figure 1).[1]

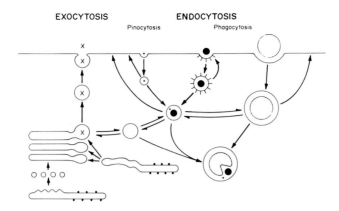

FIGURE 1. Diagrammatic representation of exocytosis and endocytosis. Uptake of large particles by phagocytosis, including receptor mediated via coated vesicles, and uptake of fluids and solutes by pinocytosis are depicted as pathways separate and distinct but interacting with the biosynthetic and secretory pathways of exocytosis to the cell's exterior.

The passage of secretory materials through the Golgi apparatus has been amply demonstrated by electron microscope autoradiography using tritiated precursors as labels.[1,19—24] As examples, for the guinea pig pancreas[21] and the rabbit parotid gland[23] following a 3- to 5-min pulse label with [³H] amino acid, the half-time for exit of labeled secretory proteins from the rough endoplasmic reticulum is 7 to 22 min and the half-time of entry into mature granules is greater than 60 min (Figure 4).

However, Golgi apparatus activities associated with packaging and secretion may extend beyond those of segregation, processing, and sorting. Certain marine algae carry out a complex process of stepwise assembly and secretion of complex wall units known as scales entirely within the cisternae of the Golgi apparatus.[25,26] In *Pleurochrysis scherffeli*, the completed cellulosic scales are discharged at a rate of one about every 2 min, each scale surrounded completely by the membrane of a single Golgi apparatus cisterna. The process has been viewed and recorded in living cells,[27] on the one hand, to provide an opportunity to monitor precisely the kinetics of cisternal formation and discharge by Golgi apparatus.[28] Additionally, from a knowledge of the complex architecture of the scales,[26] it provides an impressive example of a complex assembly process mediated soley within the confines of a single Golgi apparatus cisterna.

The rate of scale release to the cell surface in the scale-forming algae is in accord with measurements of cisternal progression in other systems[7] that suggest a proximal constitution of a cisterna every 2 to 4 min. For example, in carrot cells, when a cisternal discharge is blocked by the ionophore, monensin, a new cisterna per Golgi apparatus stack is formed within 3 to 4 min and a second new cisterna is formed within 6 to 8 min.[29]

Despite the plausibility of the notion of cisternal progression (Figures 2A and 5), there is no formal demonstration that secretory products do pass obligatorily and successively through all cisternae of the stack. Some evidence favors a peripheral route (Figure 5).[30-32] Implicit in certain models is the notion that cisternae are immobile and that migration through the system is accomplished by shuttle vesicles that move material from one cisterna to the next (Figure 2C).[8,33]

For convenience, biogenesis and exocytotic transport to the cell surface may be divided into six steps.[21] Step 1 is a synthesis[16] and step 2 is insertion into a membrane or luminal space enclosed by a membrane.[34] For proteins, synthesis occurs on polyribosomes in communication with the cytosolic environment.[34] For membrane-attached polyribosomes, the

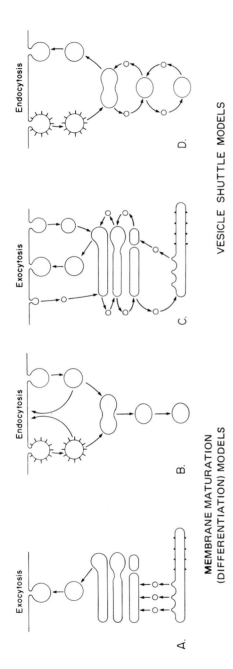

MEMBRANE MATURATION (DIFFERENTIATION) MODELS

VESICLE SHUTTLE MODELS

FIGURE 2.   Comparison of membrane maturation (differentiation)[7] and vesicle shuttle[1,5,8] models of exocytosis and endocytosis. In the vesicle shuttle model (C) the cisternae of the Golgi apparatus would be considered as immobile and that migration through the system would be accomplished by shuttle vesicles that move material from ER to the Golgi apparatus and back and from one cisternae to the next. Vesicle membrane discharged to the plasma membrane would be returned by compensatory endocytosis. In the membrane differentiation model (A) cell surface membranes would be synthesized at the ER with cisternal formation at an input (cis) face, followed by membrane modification and processing of membrane constituents at successive levels of the Golgi apparatus stack, and eventual formation of secretory vesicles and their discharge at an exit (trans) face for incorporation into the plasma membrane. The endocytotic counterparts are diagrammed at the right of the figure (B and D). By definition shuttle mechanisms accomplish net transfer of content materials under conditions where net transfer of membrane would be minimal. Membrane maturation models allow net transfer of both content materials and membrane.

EXOCYTOSIS

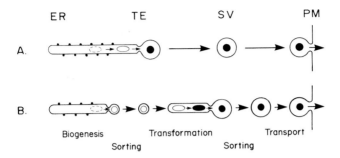

FIGURE 3. Generalized interpretation of membrane compartments involved in exocytosis. The simplest pathway involves the formation of transport (secretory) vesicles from part rough, part smooth transitional cisternae of ER (A) as encountered, for example, in many species of fungi.[14,15] In other cells there may be intermediate structures including small transition vesicles and specialized secretory vesicle-producing transitional elements (B). In most cells, the latter occurs in stacked configurations as part of the Golgi apparatus complex (Figure 2).

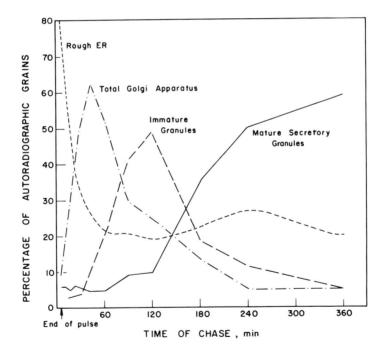

FIGURE 4. Kinetics of transport of labeled secretory proteins by acinar cells of the rabbit parotid gland.[23] Results are based on counts of autoradiographic grains at successive intervals after a labeling period. More than 75% of the incorporated label moves as a wave as it passes from rough ER to the Golgi apparatus to condensing vacuoles and, finally, to mature zymogen granules.

polypeptide is directed vectorially into the membrane of the rough endoplasmic reticulum, (ER) during chain elongation. Carbohydrate chains are added by glycosyltransferases with active sites located at the luminal surface of the membranes so that synthesis and segregation

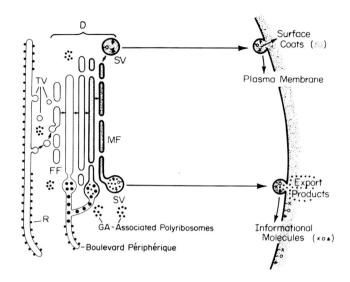

FIGURE 5.   Diagrammatic representation of endomembrane functioning in flow-differentiation of membranes as a mechanism of cell surface biogenesis during growth or for membrane renewal. Golgi apparatus (GA)-associated polyribosomes may serve as a source of membrane proteins for posttranslational insertion at various points in the flow pathway. R = ribosome; RER = rough endoplasmic reticulum; TV = transition vesicles; SV = secretory vesicle; D = dictyosome (Golgi apparatus stack); FF = forming (cis or input) face; MF = maturing (trans or exit) face.

occur simultaneously.[35] These transferases are located largely within the Golgi apparatus as determined from autoradiographic[19,20] and biochemical[35—37] investigations. The activated sugar nucleotide precursors are generated in the cytosol and transported across the membrane.[38,39] Step 3 involves processing in which both membrane and luminal contents are further modified.[8] Membrane composition changes from that of ER to one more closely resembling plasma membrane.[7,24]

Step 4 involves intracellular transport from the initial site of segregation to the eventual site of packaging for export to the cell surface.[40,41] With proteins, transport to the Golgi apparatus is considered to occur via either small transition vesicles[2,42,43] or via direct tubular connections[31] or both (Figure 5). This is a step most often considered to be the most sensitive to metabolic inhibitors[41,43] but one about which we know very little.

Step 5 is intracellular storage in those cells and tissues where it occurs.[13] In continuously secreting cells, vesicles are discharged from the Golgi apparatus and migrate to the plasma membrane as they mature. For the most part, discharge in such cells is free of hormonal control and independent of extracellular calcium.[44,45] In contrast, in discontinuously secreting cells, product accumulates as granules, exocytosis is slow or infrequent in the absence of stimulation, and the rate-limiting step in the process becomes the discharge or coalescence of granules with each other or with the plasma membrane.[13] In these cells, exocytosis requires extracellular calcium and an appropriate stimulus. Examples of the latter include the classical endocrine and exocrine gland cells such as the pancreatic beta cells, pancreatic acinar cells, adrenal medulla, and neurohypophysis. Gumbier and Kelley[46] report a pituitary cell line that exhibits both storage of mature adrenocorticotropic hormone (ACTH) as well as constituitive synthesis and secretion of ACTH precursors.

Step 6 is exocytosis.[2,41] Here the final transport vehicle moves vectorially from its site of formation or storage to undergo lateral (side-to-side) fusion with the plasma membrane.

As a result of this final fusion event, the vesicle membrane is incorporated into the plasma membrane and the vesicle content is discharged to the cell's exterior.[1,2,41] This last step is perhaps the best understood in terms of its regulation.[21] It depends on intracellular elevation of Ca^{++},[43] requires ATP,[40] and, in discontinuously secreting cells, is initiated by appropriate neural or hormonal stimuli via interactions with receptors on the plasma membrane.[21] A second messenger, Ca^{++} and/or cyclic nucleotides generated as a result of the effector-receptor interaction, is involved,[47,48] but details remain to be investigated.

The exocytotic processes of membrane biogenesis, flow, and secretion are summarized diagrammatically in Figure 3. Once the membranes of the secretory vesicles are delivered to the plasma membrane, they become available for retrieval by various recycling mechanisms (see Section III.F). While details of the process are beyond the scope of this article, some amplification of current concepts of Golgi apparatus structure and function are provided in the sections that follow. This information, hopefully, will aid in the comparison and interpretation of the membrane maturation (differentiation) and vesicle shuttle models of both exocytosis and endocytosis contrasted in Figure 2.

A. Studies of Golgi Apparatus Structure and Function

Early electron micrographs indicated that the Golgi apparatus was a system of saccules or cisternae flattened and stacked together in a characteristic structure, the Golgi apparatus stack or dictyosome.[14,24] The stack of saccules and associated vesicles appeared as an oriented structure with one pole or "face" adjacent to ER. The opposite pole was associated with the secretory vesicles that carry sequestered or elaborated products to various intra- or extracellular compartments.[14,24] This oriented appearance of the Golgi apparatus, with ER at one pole and mature secretory vesicles at the opposite pole, led to the concept of an input face (cis face, forming face, or proximal pole) and an exit face (trans face, maturing face, or distal pole). A morphological basis for ER-Golgi apparatus continuity was provided by small (about 50 nm) transition vesicles usually aligned between a part rough, part smooth cisterna of ER and the Golgi apparatus.[14,42,49]

Additional opportunities for Golgi apparatus-ER communications are provided by direct tubular connections at the Golgi apparatus periphery.[31] These connections have been suggested to function as a "boulevard périphérique" in rat liver to deliver very low density lipoprotein particles[31] and albumin[32] to the forming secretory vesicles, but a similar involvement in other cells remains to be investigated.

An anastomotic tubular network of the Golgi apparatus periphery was first shown by negative staining of isolated plant[50] and animal[51] dictyosomes in the 1960s and has since been found to be a feature of at least some cisternae of all Golgi apparatus.[52] These tubules may serve to connect adjacent cisternal stacks of the Golgi apparatus, to connect Golgi apparatus cisternae and secretory vesicles, and, as transition elements, to connect the Golgi apparatus and ER. The extent of the tubular interconnections that typify the Golgi apparatus has been elegantly demonstrated by high voltage electron microscopy after impregnation with osmium tetroxide, especially by the detailed studies of Rambourg et al.[53]

B. Membrane Differentiation

Different types of endomembranes were reported by Sjöstrand and others in the early 1960s.[24,54] That Golgi apparatus membranes were differentiated across the stacked cisternae was first demonstrated in the fungus *Pythium ultimum* by Grove et al.[55] Changes in both membrane thickness and in staining intensity were observed. Membranes of cisternae at the forming face appeared to be similar to ER and the nuclear envelope, while membranes at the maturing face (including vesicle membranes) were similar to plasma membranes. Membranes of intercalary cisternae were intermediate in appearance so that each successive cisterna was more like the plasma membrane from the forming to the maturing face (i.e., denser, thicker, and having a clearer dark-light-dark pattern). In general, membrane thickness progressed from thin to thick along the ER-Golgi apparatus-secretory vesicle-plasma membrane secretory route. Staining intensity, however, can vary. Yet, regardless of the specific staining pattern, membranes of the forming face of the Golgi apparatus resembled ER, while those at the maturing face (or of secretory vesicles) resembled plasma membrane.[56] Cytochemical evidence for membrane differentiation in rat IgM myeloma cells has been provided from studies of lectin binding suggestive of proximal to distal maturation of oligosaccharides[57] and in Golgi apparatus of rat liver from the appearance toward the mature face of a NADH-ferricyanide reductase resistant to glutaraldehyde.[58]

Considerable biochemical evidence for membrane differentiation within the Golgi apparatus has accrued from studies that compared ER, Golgi apparatus, and plasma membrane fractions isolated from rodent liver and rodent and bovine mammary gland.[24,59,60] Phospholipid and fatty acids of the major lipid classes of Golgi apparatus were intermediate between those of the ER and plasma membrane, and all membrane fractions (ER, Golgi apparatus, and plasma membrane) had major protein bands in common. Some enzymatic activities characteristic of plasma membranes, however, appeared to be acquired at the Golgi apparatus, whereas enzymatic activities characteristic of ER membranes appeared to be lost.[24,60]

C. Membrane Flow

Membrane flow or, more correctly, flow differentiation has emerged as a major biogenetic mechanism contributing to the origin and specificity of surface membranes in addition to providing the vehicles for discharge of secretory products during exocytosis.[7,60]

The basic premise of the membrane flow hypothesis[7,61] states that cell surface membranes are synthesized beginning at the ER, continuing with their modification and processing within certain transitional membrane elements, chiefly the Golgi apparatus, and their final delivery to the plasma membrane as preformed plasma membrane units. Evidence in support of the membrane flow concept applied to exocytosis was reported first by Franke et al.[61] from kinetics of incorporation of radioactivity from L-[guanido-^{14}C] arginine. The radioactive precursor was administered as a pulse after which the animals were sacrificed at intervals, and the various cell components were isolated, purified, and analyzed. The order of labeling of membrane proteins was ER, Golgi apparatus, and plasma membrane as predicted from morphological considerations.[14]

More recently, H-2 determinants of mouse liver were utilized as flow markers.[62] They are glycoproteins firmly bound to membranes and are present in ER and Golgi apparatus as well as plasma membranes.[63] In the basic experiment, membranes of ER, Golgi apparatus, and plasma membrane were isolated at different times after injection of [^{35}S] methionine during a pulse-chase experiment. Analysis of the fractions by immunoprecipitation with allospecific antisera (and conformed by PAGE-fluorography) demonstrated rapid accumulation and disappearance of label in ER, followed by appearance of label in Golgi apparatus and, finally, accumulation of label in plasma membrane (Figure 6). These results agree with the previous findings obtained for mixed membrane proteins[61] and for the migration and maturation of viral membrane proteins.[64,65] Additionally, they provide a specific example of transfer of a class of intrinsic membrane glycoproteins from ER to Golgi apparatus to plasma membrane via a flow-differentiation mechanism of membrane biogenesis.[62]

The glycosylation of gangliosides takes place also in the Golgi apparatus (and/or ER)[66—69] from where they are transported to a major site of deposition, the plasma membrane. Results of Miller-Podraza and Fishman[70] indicate that transfer of gangliosides from sites of synthesis to the cell surface requires approximately 20 min in cultured N18 and rat glioma C6 cells. Landa et al.[71] reported blockage of transport of gangliosides by colchicine and vinblastine in chick optic nerve suggestive of vesicular transport in keeping with the kinetics of transfer. Alternatively, Sonnino et al.[72,73] proposed that soluble gangliosides of the cytosol,

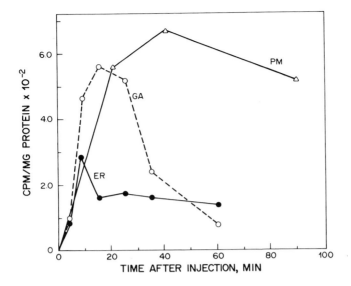

FIGURE 6.   Kinetics of incorporation of radioactivity from [^{35}S] methionine into H-2 antigens of ER, Golgi apparatus (GA), and plasma membranes (PM) of mouse liver. Specific activity refers to cpm in
immunoprecipitates per milligram protein of pooled starting fractions from
eight mice.[62]

as protein-ganglioside complexes, may represent a transient pool en route to the plasma
membrane.

D. Mechanisms of Membrane Differentiation (Processing)

An important feature of the flow differentiation concept of membrane biogenesis and
exocytosis is that of selectivity.[7] Membrane flow and membrane differentiation cooperate
to ensure selective removals or additions of membrane components so that functionally
appropriate constituents are transferred while functionally inappropriate constituents are
excluded. It is clear that ER and plasma membranes are different, structurally, functionally,
and their chemical and enzymatic make-up.[24] Golgi apparatus, while showing many characteristics intermediate between those of ER and plasma membrane, have unique identifying
characteristics.[1,24,67] Therefore, selection must occur. While some membrane constituents,
those common to all endomembranes, may be transferred directly, other constituents characteristic of plasma membranes must be added. Still other constituents characteristic only
of ER must be removed.

The major function of the Golgi apparatus is to elaborate (presumably from precursor
materials derived in part, from ER) a secretory vesicle having a membrane that is plasma
membrane-like and capable of fusing with the plasma membrane. To acquire these characteristics and to function properly in the sequestration of secretory products, secretory
vesicle membranes are somehow modified during their elaboration at the Golgi apparatus.[1—4,7—15] While mechanisms for various forms of protein modification[8] and carbohydrate
addition and removal[35—37] are well understood, evidence for protein additions or deletions
is more fragmentary.

1. Golgi Apparatus-Associated Polyribosomes, an Opportunity for Protein Additions during Membrane Differentiation

A mechanism whereby membrane proteins may be added during flow differentiation of

Table 1
PROPERTIES OF BOVINE
GALACTOSYL TRANSFERASE
(LACTOSE SYNTHETASE)[81]

Isolated from	Molecular weight (daltons)
Golgi apparatus	63,000
Milk fat globule membranes	63,000
Milk fat globule supernatant	51,000
Milk	51,000

membranes has been suggested by work of Elder and Morré.[74] Discrete clusters of free polyribosomes are associated with Golgi apparatus of rat liver and various other animal and plant species (Figure 5).[75] These polyribosomes are not membrane-bound like those attached to ER but are best described as a class of ''free'' polyribosomes located within a distinct ''zone of exclusion'' or specialized cytoplasmic zone surrounding the Golgi apparatus.[75] On an RNA basis, the specific incorporation of radioactive amino acids in a cell-free system for protein synthesis by Golgi apparatus-associated polyribosomes of rat liver is ten times that of free polyribosomes of the cytosol.[74] The peptides synthesized correspond to electrophoretic bands absent from ER but present in plasma membranes or correspond to unique Golgi apparatus-associated peptides.[74] One possibility under investigation is that these polyribosomes are programmed to translate messenger RNAs of proteins inserted into Golgi apparatus membranes during their maturation.

The existence of Golgi apparatus-associated polyribosomes helps to explain why early incorporation of radioactive amino acids into Golgi apparatus in vivo proceeds without appreciable lag and parallels, for a time incorporation into ER (Figure 6).[61] It also may help to explain why maximum specific radioactivities of the total membrane proteins of the Golgi apparatus and plasma membrane may exceed those of the total membrane proteins of ER (Figure 5).[61,62] Input from two sources, membrane-bound polyribosomes of the ER and Golgi apparatus-associated polyribosomes (and possibly other classes of free polyribosomes[75]), all are indicated. Borgese et al.[76] found a parallel incorporation of NADH-cytochrome b$_5$ reductase, an integral membrane protein common to ER, Golgi apparatus, and plasma membrane.[58,77,78] newly synthesized on free polyribosomes, into several intracellular membranes including Golgi apparatus.

To what extent other classes of cytoplasmic polyribosomes may contribute to membrane biogenesis during exocytosis is not known. Mollenhauer and Morré[79] reported polyribosomes associated with forming acrosome membranes which were similar in appearance to the Golgi apparatus-associated polyribosomes and were found only in certain developmental stages of spermiogenesis.

2. Lactose Synthetase and Glycosyltransferases, an Opportunity for Protein Deletion during Membrane Differentiation

Glycosyltransferases, with acceptor specificity for exogenous substrates, while present in ER, are most concentrated in Golgi apparatus and are frequently absent from plasma membranes.[80] In mammary gland, the galactosyltransferase of lactose formation is an enzyme of relatively high molecular weight closely associated with membranes of the Golgi apparatus (Table 1). The soluble form of the enzyme in milk is of lower molecular weight.[81] At the apical plasma membrane, the enzyme is no longer active and is apparently released into the milk as the lower molecular weight form of the enzyme, probably through proteolysis.

Experimental evidence for this concept has come from studies where derivatives of the

Table 2
LACTOSE SYNTHETASE-SPECIFIC
ACTIVITIES OF MAMMARY GLAND
FRACTIONS[81,82]

	Protein (nmol/min/mg)	
Fraction analyzed	**Pellet**	**Supernatant**
Whole gland	0.5	
Golgi apparatus	8.6	
Milk fat globule membrane	1.9	
Freshly isolated, 0—4°C	1.9	
1 *M* sucrose wash, 38°C, 1 hr	0.6	5.1
2nd sucrose wash, 38°C, 1 hr	0.3	1.5
Milk		0.2

apical plasma membrane, the membranes surrounding globules of milk fat, have been isolated carefully under conditions unfavorable to proteolysis (low temperature, rapid isolation). Here some of the membrane-associated high molecular weight form of the lactose synthetase is retained associated with the globule membrane (Table 2). This apparent loss through proteolytic action provides one potential example of selective removal of a protein during flow differentiation of membranes. Galactosyltransferase in HeLa cells may have a different fate.[83] Here transport from the ER to the Golgi apparatus and eventually to the cell surface was observed but with a very long transit time in the Golgi apparatus. Results were with a galactosyltransferase antibody and pulse-chase methodology.[83] Other examples where proteins may be deleted during flow differentiation of membranes include degradation of cytochrome b_5 and P-450 as partially studied examples.[84]

3. The Role of Transition Vesicles, an Opportunity for Selective Transfer

While selective removal provides one mechanism for loss of ER characteristics from Golgi apparatus membranes, the same result could be accomplished by selective transfer either in the manner of shuttle between ER and Golgi apparatus (Figure 2C),[1—5] and from one cisterna to the next,[8] or by unidirectional transfer of some constituents and the exclusion of others (Figure 2A).[7] ER-associated transition vesicles, as opposed to clathrin-coated vesicles of the mature Golgi apparatus face,[85,86] have, thus far, eluded isolation and biochemical characterization. Thus, it is not clear even that transition vesicles participate in the delivery of secretory products to the Golgi apparatus much less as to what role they may play in the selective transfer of membrane between ER and forming Golgi apparatus cisternae. With myeloma cells, where immunocytochemistry with peroxidase-conjugated antimouse FAB has been used on serial section, IgG is present in rough ER, including perinuclear cisternae, and to some degree in all cisternae of the Golgi apparatus.[87] Yet the micrographs reveal no clear localization of IgG in the free transition vesicles that are abundant in these cells. In contrast, some small vesicular profiles of the maturing or trans face did show the presence of IgG.

We know almost nothing about the molecular details whereby the exocytotic process per se is regulated. This is despite rapid progress in understanding, for example, the mechanisms whereby animal cells synthesize and process proteins for export.[1—4,16,34] There remain, as well, important controversies about the nature of the specific compartments involved and the precise routes followed. Major questions concern whether or not synthesis and delivery of membrane proteins along the flow route are coordinated, or do individual proteins or glycoproteins exhibit independent rates and routes of transfer and/or processing? Croze and

ENDOCYTOSIS

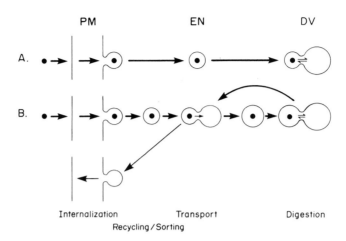

FIGURE 7.   Generalized interpretation of the compartments involved in endocytosis and intracellular digestion. In its simplest configuration (A), the internalized material is engulfed in a plasma membrane (PM)-derived endosome (EN) followed by fusion with a lysosome containing acid hydrolases to form a digestive vacuole (DV). In more complex recycling systems (B), transfer of internalized material to an intermediate series of endosomes may occur (multicompartment model).

Morré[62] observed different rates of migration of two H-2 antigenic determinants of different specificities. Similarly, Lodish et al[88] reported that in human hepatoma HepG2 cells, five secreted proteins mature from the rough ER to Golgi apparatus vesicles with characteristic rates that differ at least threefold, yet transit times through the Golgi apparatus appear to be similar.

III. ENDOCYTOSIS

The concept of membrane flow was applied originally to endocytosis,[89] i.e., the uptake of extracellular materials within membrane-bound vacuoles or vesicles. During endocytosis, evaginations of the plasma membrane pinch off from the cell surface and become cytoplasmic vesicles. Trapped within the vesicles are fluids, solutes, or solid extracellular materials (Figure 7).[5,90,91]

Classically, endocytosis has been divided into two types: phagocytosis (literally cell eating) and pinocytosis (cell drinking).[5,90,91] During these processes, a segment of the plasma membrane surrounds part of the extracellular fluid or material resulting in the formation of an intracellular vesicle. In addition to internalization of part of the environment, these processes also result in an internalization of part of the plasma membrane. Depending on the cell type, the amount of membrane internalized can carry between 1 to 20 times the total cell area per hour.[5,92] Pinocytosis describes the uptake of all smaller particles ranging from small, insoluble particles to low molecular solutes to the fluid itself. On the other hand, phagocytosis is operationally restricted to encompass the uptake of particles larger than about 0.2 μm diameter. Mechanistically, pinocytosis is further subdivided into *fluid-phase*, where substances enter by virtue of their presence in the fluid, and *absorptive*, where particles are bound to the membrane and subsequently internalized. Absorptive endocytosis can be specific (e.g., *receptor-mediated* where binding is to defined receptors) or unspecific (e.g., binding of cationic ferritin to anionic sites of the plasma membrane). *Compensatory endocytosis*, the retrieval of the membranes of secretory granules which have fused with the cell surface, will be discussed separately as a topic under membrane recycling (Section III.F).

Several different processes contribute to the internalization of the plasma membrane.[5,90] The first involves invagination of the cell surface and its entry into the cytoplasm. Following internalization, the vesicles (or endosomes) may be transported to different locations in the cell. Mostly, the content is delivered to lysosomes, although this passage can involve intermediate compartments (e.g. receptosomes) in which segregation of membrane constituents (e.g., receptors) from vesicle contents may occur (Figure 7).

Quantitation of pinocytosis, both fluid phase and absorptive, has been performed using a variety of markers and in vitro systems.[5,90—106] Content markers, while providing indications of volumes of materials taken up, may exhibit different fates from membrane markers once internalized. The latter may apply as well to noncovalently-bound membrane markers.

A. Endocytosis and Membrane Recycling

The elegant work of Steinman et al.[97] was fundamental in establishing new interest in the dynamic nature of endocytosis. For mouse fibroblasts, 0.5 cell surface equivalents per hour were estimated to be internalized from morphometric measurements (2/hr for macrophages). As an extreme, *Acanthamoeba* has been estimated to internalize an amount of membrane equal to 10 to 40 surface equivalents per hour.[95] In spite of this process, there is no significant reduction of the cell surface implying that the membrane is replaced at a corresponding rate. Yet, membrane replacement from *de novo* synthesis has been considered to be too slow a process to account for the total replacement. Therefore, it has been suggested that the internalized membrane is recycled to the cell surface.[5,97] In contrast to high rates of plasma membrane internalization, the half-life times of plasma membrane proteins based on enzymatic labeling with iodine,[107] acetylation,[108,109] incorporation of precursors,[61,110] or studies of the turnover of specific membrane receptors[111] suggest that the degradation of most membrane proteins occurs slowly with half-life times of 10 to 100 hr (see Morré et al.[7] for review). However, until recently, the evidence for recycling has been based largely on the ability of cells to take up ligands in excess of the numbers of binding sites on their surfaces and that such uptake can proceed in the absence of protein systhesis.

In a study of pinocytosis in *Dictostelium discoideum*, Thilo and Vogel[104] used [^{3}H] galactose as a marker of pinocytotic membranes and found that previously internalized label did reappear at the plasma membrane. Additionally, Adams et al.[112] used CHO cells and horseradish peroxidase (HRP) as a pinocytotic marker to show that after 10 min of pulse labeling followed by washing of the cells, 30 to 50% of the pinocytosed HRP reappeared in the medium within a 20-min period. Thereafter, little or no HRP was released from the cells and the HRP within the cells disappeared with a half-life corresponding to that of HRP in lysosomes of 6 to 8 hr.

Schneider et al.[113] reported that fibroblasts that had internalized fluorescein-labeled goat antirabbit IgG as a lysosomal content marker release some of this label to the medium when the cells were exposed subsequently for 24 hr to rabbit antiplasma membrane, IgG. They postulated that the rabbit antiplasma membrane IgG, bound to the plasma membrane, was internalized by absorptive pinocytosis and, within the lysosomes, complexed with the goat antirabbit IgG. The membranes coated with rabbit antiplasma membrane IgG apparently served as the vehicle both for the intracellular capture and the return to the cell surface of the previously internalized goat antirabbit IgG. Similarly, evidence was presented by Widnell et al.[114] for reappearance at the cell surface of internalized 5′-nucleotidase, an integral protein of the fibroblast plasma membrane.

Muller et al.[115,116] used lactoperoxidase coupled covalently to latex spheres to selectively iodinate the membranes of internalized phagolysosomes within living macrophages. Ap-

MEMBRANE RECYCLING

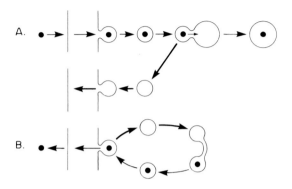

FIGURE 8.   Diagrammatic representation of membrane re-cycling. The more general situation would encompass recycling of endocytosed plasma membrane after fusion with an inter-mediate endosome (A) to effect transfer of endosomal content to a digestive vacuole but return of the bulk of the membrane to the cell surface. A special form of recycling is that of com-pensatory endocytosis (B) where membrane material added to the plasma membrane by exocytosis is returned intact by en-docytosis to be reutilized in subsequent exocytotic events.

proximately 24 polypeptides, similar to those available to iodination on the external surface of the plasma membrane, became labeled. When such cells were returned to culture, the label initially in the phagolysosome membrane returned to the cell surface and became randomly associated with the cell surface as determined by autoradiography. Analyses of isolated plasma membranes showed that all of the polypeptides initially iodinated within the phagolysosome took part in the flow. The movement of membranes from the phagolysosome compartment to the plasma membrane without loss or shrinkage of the phagolysosome implied that the membrane of the phagolysosome must be somehow replaced. The source of new membrane was apparently from the fusion of newly entering pinocytotic vesicles. Evidence for this pathway in *Dictyostelium* was provided by the redistribution of glycocon-jugates, enzymatically marked with radioactive galactose, between the membrane-enclosed and undigestible latex beads and the plasma membrane.[116a]

These studies demonstrated in a direct manner that living macrophages carried out a continuous, bidirectional flow of membrane polypeptides between the plasma membrane and the phagolysosome where flow was random and considerable recycling appeared to occur (Figure 8). Of the intracellularly iodinated proteins of the phagolysosome membrane, 20 to 50% were lost rapidly to digestion and the amount digested could be increased by increasing the phagocytic load. Again, all phagolysosome peptides were digested in parallel suggesting that membrane degradation may involve exposing whole segments of membranes to the digestive events such as may occur in the formation of multivesicular bodies. Both rapid and slow degradative turnover phases of phagolysosome membrane involved the same spectrum of polypeptides. Either some selective mechanism (e.g., specific for altered or denatured proteins) or random removal of some constant portion of the membrane would explain the data.

Storrie et al[117] used CHO cells to internalize lactoperoxidase through pinocytosis at 37°C followed by radioiodination of the proteins of those vesicles locked within the cell at 4°C.

Within 10 to 20 min after warming the intracellularly iodinated cells from 4 to 37°C, radioiodinated proteins appeared at the cell surface as determined from sensitivity to protease digestion. This also would imply that already internalized membranes may be returned to the cell surface as a unit rather than return of only one or a few specific types of receptors.

B. The Multicompartment Model to Explain Endocytosis and Membrane Recycling

A morphological approach was used by Steinman et al.[5,97] to establish that horseradish peroxidase labeled an internal membrane space within minutes of exposure to the enzyme, whereas the secondary lysosomes were labeled much more slowly (within an hour). They, therefore, provided the morphological correlation to kinetically derived models where the endosomes comprise the rapidly turning-over compartment (recycling) and lysosomes (secondary) are the slowly turning-over compartment.

A multicompartment model for endocytosis and membrane/receptor recycling that integrates both biochemical and morphological observations appears to represent the simplest and most economical way for the cell to maintain membrane balance (homeostasis), and a constant volume of the cell and vacuolar apparatus as well as for the efficient reutilization of surface receptors. In principle, such a model could operate without involving the Golgi apparatus or other elements of the apparatus involved in membrane biogenesis (Figure 8A).

Tietze et al.[118] provided evidence for two internal pools of mannose receptors in alveolar macrophages. One was sensitive to weak bases and receptor-ligand complexes could return from this pool intact. Chloroquine and ammonium chloride inhibited dissociation of ligand-receptor complexes entering the second pool and blocked endocytosis by preventing the return of free receptors.

Similarly, Besterman et al.[106] had previously modeled the process of fluid-phase pinocytosis and subsequent exocytosis in pulmonary alveolar macrophages and fetal lung fibroblasts using [14C] sucrose, a fluid phase marker. Cells are not freely permeable to the [14C] sucrose nor is the [14C] sucrose metabolized efficiently by these cells. Once taken up by the cells, it is released intact. By preloading the cells with [14C] sucrose for varying lengths of time and then monitoring the appearance of radioactivity into isotope-free medium, the following aspects were deduced. At least two intracellular compartments were required. One was small and turned over rapidly ($t_{1/2}$ of 5 min in macrophages, 6 to 8 min in fibroblasts). The other compartment was larger and turned over more slowly ($t_{1/2}$ of 180 min in macrophages, 430 to 620 min in fibroblasts).

A similar model was adopted by Burgert and Thilo[119] in their analysis of endocytotic internalization of membranes of a macrophage cell line P388D$_1$. The steady-state distribution of label between the plasma membrane and two consecutive intracellular membrane compartments was determined based on measurements of plasma membrane glycoconjugates labeled with [3H] galactose. Resistance to β-galactosidase was used as evidence for membrane internalization. Subsequent reappearance of β-galactosidase sensitivity was used as evidence for reappearance at the cell surface of previously internalized label. This was justified on the basis that the label remained membrane bound and that the composition of the labeled glycoconjugates, as determined from SDS-polyacrylamide gel electrophoresis patterns, was unchanged as a result of internalization.

Redistribution of label from the cell surface occurred with biphasic kinetics described as the sum of two exponential functions and consistent with the two-compartment model.[119] The relative surface areas were calculated to be 100:12.5:7.3 (plasma membrane-to-endosomes (pinosomes)-to-endosome-derived membrane of secondary lysosomes). In the macrophage line, the equivalent of the plasma membrane was internalized once every 20 min in the form of 0.24 μm endosomes. The residence time of membranes in the endosome compartment was calculated to be relatively short and in the order of 3 min. In contrast, the rate at which membranes entered the lysosome compartment was relatively slow (1/31 the rate of endocytosis). This provided the basis for their conclusion that only 3% of the total membrane surface internalized at any one time entered the secondary lysosome compartment. By inference the remainder (97%) must have been recycled back to the plasma membrane to participate in subsequent endocytotic events. Likewise, in cells showing a high rate of endocytosis, large amounts of membrane surface were internalized and rapid recycling of internalized membrane back to the cell surface emerges as an economical way of keeping the cell surface area (as well as that of the vacuolar or lysosomal compartment) constant. Additionally, recycling is a potentially important part of the general homeostatic mechanism operating in the cell to maintain membrane balance and cellular volume. Berlin et al.[102,120] and Quintart et al.[103] have established that endocytotic activities vary during the cell cycle and that pinocytosis is depressed during mitosis.

While a two-compartment model is consistent with the kinetic data of many endocytotic systems, additional compartments are not excluded, especially in different organisms. Ciliate protozoans, upon feeding, provide one of the most dramatic examples of the extensive membrane movements possible as a result of endocytosis as well as the potential involvement of multiple compartments.[121,122]

C. Receptor-Mediated Endocytosis

A special category of absorptive endocytosis is where ligands interact initially with defined receptors at the cell surface. The receptors become localized to or associated with clathrin-coated membrane surfaces or coated pits. Ligands internalized by this mechanism include low density lipoproteins,[123—126] peptide hormones,[113,127] and the transport protein of ferritin, transferrin.[128] Specificity is conferred by the receptor. The term receptor-mediated endocytosis, coined by Goldstein et al.,[124] has been applied to this process.

Electron-dense probes (ferritin, colloidal gold, etc.) conjugated to ligands enable studies at the electron microscope level. Like several other receptors that mediate the transport of protein ligands into cells, the majority of LDL receptors (about 70%) are located in coated pits. The LDL receptor has been purified to homogeneity from bovine adrenal cortex. The receptor is an acidic, single-chain glycoprotein with a molecular weight of 164,000.[129]

Coated membrane surfaces (vesicles) receive their name from the fact that they have on their cytoplasmic surfaces several proteins (33,000, 36,000, and 180,000 mol wt) organized into a cage-like structure composed of triskelions.[130] The predominant protein is the 180,000-mol wt peptide, clathrin, first described by Pearse.[131]

While fusion of some incoming pinosomes or phagosomes directly with primary or secondary lysosomes remains a possibility,[6] this is not always the route followed. Especially for receptor-mediated endocytosis, the more common route may be that of fusion of pinosomes first with an intermediate compartment (prelysosomal) which then functions to deliver the content to the lysosome. Various terms have now been applied to this intermediate compartment including receptosome[132] or sorting vesicle.[125]

Geuze and co-workers[133] described in rat liver cells a compartment of uncoupling receptor and ligand (CURL), a tubulo-vesicular structure of the sinusoidal cytoplasm revealed through immunocytochemistry. A similar model based on the demonstration of "tubules" or "channels" connecting the vacuoles directly with the cell surface has been suggested by Nichols.[134] All three terms (receptosome, sorting vesicle, CURL) imply only one aspect of what may turn out to be a complex set of functions carried out by this intermediate compartment or series of compartments. Efforts to isolate such intermediate compartments to permit biochemical characterization are underway in several laboratories.[135—137]

In formulating the receptosome concept, Pastan and Willingham[132,138] postulated the coated pit as a dynamic structure on the cell surface that served primarily to extend into the cytoplasm to deliver content to the intermediate endosome compartment or receptosome. One extreme of this possibility would be that the coated pits never completely dissociate from the plasma membrane but remain attached by narrow necks which are subsequently retracted following discharge to the intermediate endosome. Were this so, the bulk of the membrane plus receptors of the coated pits would return unaltered to the cell surface. There is considerable disagreement on this point since at least some clathrin-coated structures near the cell surface appear to exist as free vesicles. Others report that perhaps no more than 50% of the vesicles seen near the plasma membrane in single thin sections show attachments to the plasma membrane in serial sections.[139] This does not detract from the basic concept of the Pastan and Willingham proposal, however. Certainly, the essential features of the hypothesis could be accomplished by coated pits that enter the cell as coated vesicles and migrate short distances, fuse with the intermediate endosome, discharge their content, and then rebud from the intermediate endosome to return to the cell surface for reentry as a coated pit (Figure 1). One line of evidence in favor of this hypothesis is that despite the demonstration in vitro of an ATP-requiring dissociation of clathrin from coated vesicles,[140] the existence of free clathrin in the vicinity of the cell surface or in the cytoplasm in the region of putative coated vesicle-intermediate endosome associations has not been demonstrated using immunocytochemical procedures.[141,142] Wheland et al.[141] provided ultrastructural evidence that microinjected anticlathrin antibodies gained access to extramembranous cytosolic compartments and could bind clathrin microinjected into the cytoplasm. Yet, only preexisting coated pits were labeled. Ongoing receptor-mediated endocytosis was not inhibited nor were aggregates of endogenous clathrin induced to form. All of the clathrin detected by these procedures has been associated either with coated pits of the cell surface or with internalized coated pits/or vesicles. Only in rare instances have clathrin cages been observed free in the cytoplasm.[86,143] The major pitfall from this type of analysis is that the concentration of clathrin in the cytoplasm may be quite dilute, even locally, such that it would not be observed over the background staining with the cytochemical methods thus far employed for its localization.

In recent years, receptor-mediated endocytosis systems have been described for substances as diverse as plasma transport lipoproteins,[123—126,144] polypeptide hormones and growth factors,[145—148] circulating glycoproteins,[149] and hydrolytic enzymes.[150] The last two groups involve the recognition of specific carbohydrate residues.[151]

Ashwell and Morell[152] first demonstrated that sialic acid was essential for the viability of serum glycoproteins in the circulation. Upon injection into rabbits of a preparation of ceruloplasmin from which the nonreducing sialic acid had been removed enzymatically, the resulting asialoglycoprotein was cleared from the circulation within minutes. In contrast, the fully sialylated protein exhibited a normal half-life of several days. The penultimate galactosyl residues, exposed by the removal of sialic acid, were subsequently shown to be the critical determinants of clearance since modification of these galactoses by galactose oxidase treatment or by treatment with β-galactosidase prolonged the survival time of glycoproteins in the circulation.[153]

The clearance of asialoglycoproteins from the blood by liver parenchymal cells is via receptor-mediated endocytosis.[149] Upon binding, asialoglycoprotein ligand becomes internalized together with the receptor by means of clathrin-coated vesicles from which the ligand is transferred to larger smooth-surfaced vesicles within minutes.[154—156] The fate of the ligand is rapid lysosomal degradation, whereas the receptor is spared this fate and is reutilized.[154,157,158.] Thus, the receptor and ligand must dissociate following endocytosis.

The primary locus of binding of the asialoglycoproteins was the plasma membrane. The receptors for galactose-terminated glycoproteins were isolated and purified by affinity chromatography on Sepharose columns to which asialoorosomucoid was bound covalently.[159] The receptor also was a glycoprotein, oligomeric, with subunits of 48,000 and 40,000 mol wt. As such, the receptor was identified as the first lectin of mammalian origin.

The initial identification of the rat liver plasma membrane as the major locus of binding activity for galactose-terminated glycoproteins was subsequently expanded to include membranes of the Golgi apparatus, the smooth microsomes, and the lysosomes.[160] The receptor from each of these sources was shown to be identical in terms of specificity toward galactose, an absolute requirement for calcium for binding, sensitivity to neuraminidase, response to carrier dilution with added, nonradioactive asialoglycoprotein, and pattern of band formation on PAGE and formation of a single-fused precipitin line on double immunodiffusion. In contrast to plasma membrane, the asialoglycoprotein binding of Golgi apparatus and smooth microsomes was augmented greatly by the presence of detergents (or treatment with detergents). These findings have been interpreted to indicate that the receptors of Golgi apparatus and smooth ER are within the organelle and inaccessible to ligand in solution when present as closed vesicles or cisternae. The multiple intracellular loci of the binding protein have suggested the possibility of a biosynthetic pathway involving a progression via the Golgi apparatus, and, finally, to the plasma membrane.

In contrast to its location on the luminal surface of other organelles, the hepatic binding protein appears to be located on the cytosolic surface of the lysosomes.[111] On this basis, the marked differences in the survival rates of the receptor and ligand may be correlated with the location of the receptor on the external surface of the lysosome and that of the ligand on the interior of the lysosome.[111] This, however, would have to involve a spatial reorientation of the receptor upon fusion of receptosomes with the lysosomes different from that usually visualized for this type of fusion event.[151]

A receptor for *N*-acetylglucosamine-terminated glycoproteins also has been described for avian hepatocytes.[161] Again, the major locus for the binding protein is the liver.

The specific recognition of lysosomal enzymes by cellular receptors and the equation of mannose-6-phosphate with this recognition signal stemmed from the work of Neufeld et al.[150] on cultured fibroblasts from patients with genetic disorders to polysaccharide catabolism. These cells failed not only to segregate lysosomal enzymes properly but also failed to take up the lysosomal enzymes they produced by absorptive endocytosis. They would, however, take up lysosomal enzymes produced by normal cells to the extent that it was possible to correct the genetic disorder in fibroblasts by enzyme replacement.[151]

These findings can now be explained on the basis that the deficient cells lacked a recognition signal that normally directs the traffic of acid hydrolases to lysosomes. In the deficient cells, enzymes are secreted, instead, into the medium. Since the enzymes were defective for the recognition marker, they could not reenter the cell by absorptive endocytosis. Mannose-6-phosphate proved to be a potent competitive inhibitor of the uptake of lysosomal enzymes by normal fibroblasts.[162,163] There is now strong evidence from a variety of sources for mannose-6-phosphate as the recognition determinant for hydrolytic enzymes.[151,164]

It is no longer widely held that secretion-recapture is an obligatory pathway for the entry of hydrolytic enzymes into lysosomes.[164,165] Since they are glycoproteins, they pass through the Golgi apparatus where the recognition marker is normally added.[166,167] Following addition of the recognition marker, the enzymes, so modified are then segregated from normal secretory proteins and routed to the lysosomes. Thus, a defect in synthesis of the signal mannose-6-phosphate in I cell disease results in the extracellular accumulation of low uptake forms of lysosomal enzymes and accounts for the corresponding deficiency in intracellular digestive enzymes.[150]

Additional sugars that serve as recognition signals for receptor-mediated endocytosis and specific to certain cells include L-fucose (mammalian hepatocytes),[168] *N*-acetylglucosamine (avian hepatocytes),[161] and *N*-acetylglucosamine or mannose (mammalian reticuloendothelial cells).[169,170]

Receptor-mediated endocytosis in combination with membrane recycling allows for an efficient delivery of receptor-bound ligands to lysosomes without obligatory loss or degradation of the receptor itself. For example, recycling of fibroblast low density lipoprotein receptors would account for the extensive and continued uptake of lipoproteins in the absence of receptor synthesis.[123] This is in contrast to interactions with other ligands such as epidermal growth factor with their receptors which leads to removal of the receptors from the cell surface,[145] or so-called down regulation as exemplified by the insulin receptor.[170a] Receptors, along with the hormone, are destroyed in the lysosome.[171] The regulatory mechanisms that determine whether or not a ligand will be delivered to the lysosome for digestion (down-regulation) or returned to the cell surface as free receptor after discharge of the ligand now appears to be determined, at least in part, by the valency of the ligand. If multivalent, clustering of receptors, transfer to lysosomes, and degradation is the result. Alternatively, if monovalent, the ligands do not prevent recycling and the receptor may be returned to the cell surface.[172]

D. Endocytosis, Lysosomes, and Intracellular Digestion

According to the original lysosome concept, whatever entered the cell was assumed to be degraded although the fate of the entering membranes was unclear.[6] Recent developments[90] reaffirm that most content from endocytosed vesicles is delivered to lysosomes. Soluble proteins such as HRP and bovine serum albumin are delivered to lysosomes and completely digested. Also, there is little or no evidence for subsequent movement of fluid phase markers to compartments other than endosomes or lysosomes. Thus, the route followed by endosome content may be much less complicated than that of the pinocytic membrane. The majority of the content delivered to lysosomes is degraded to constituent molecules of dipeptides, amino acids, sugars, fatty acids, glycerol, phosphate, and purine or pyrimidine bases[6] which can be utilized by the cell in redirected synthesis of new complex molecules and membranes via the biosynthetic route.

When a particular vacuolar compartment becomes heavily loaded with internalized molecules, it is transformed into a secondary lysosome by receiving acid hydrolases by fusion either with "primary lysosomes" or fusion with preexisting secondary lysosomes or both. A low pH of this compartment is of importance for the release of specific ligands from their receptors prior to recycling of receptors back to the cell surface.[173,174]

Endosomes (the intermediate compartments) also may have a low pH (about 5.0) as shown by spectrofluorography at the microscope level.[135,174] This implies that the endosome membrane, like the lysosome membrane,[175—177] may contain a proton pump, driven by ATP and responsible for maintenance of a low internal pH. This low pH would facilitate dissociation of the ligand from its receptor. A number of ligands taken up by receptor-mediated endocytosis, LDL,[124—127] asialoglycoproteins,[156] epidermal growth factor,[178] and insulin[179] dissociate from their receptors as the pH falls below 5.5. This is important not only to promote dissociation of ligand and receptor before membrane and receptor are recycled back to the cell surface, but also perhaps to serve as an "acid bath" to "cleanse" the membrane of extraneous molecules taken up from the extracellular environment nonspecifically (by electrostatic interaction, for example).[180]

Nondegraded pinocytic substrates are retained within a class of secondary lysosomes termed residual bodies. These residual bodies function as a third compartment in series with endosomes and active secondary lysosomes (Figure 9).

Much of the work underlying the lysosome concept can be traced to the morphological and cytochemical studies of Novikoff and colleagues.[181—184] Their results, which have received confirmation by many other investigators, suggest that lysosomal enzymes are produced in the rough ER and that these enzymes are translocated to the Golgi apparatus where the lysosomal membrane is formed and where the final lysosome arises still in association with a specialized region of Golgi apparatus-ER-lysosome association known widely as the GERL system. In GERL, the acid phosphatase is localized in a cisterna at the distal face of the Golgi apparatus and in lysosomes and smooth tubules of ER associated with the apparatus (Figure 9).[184]

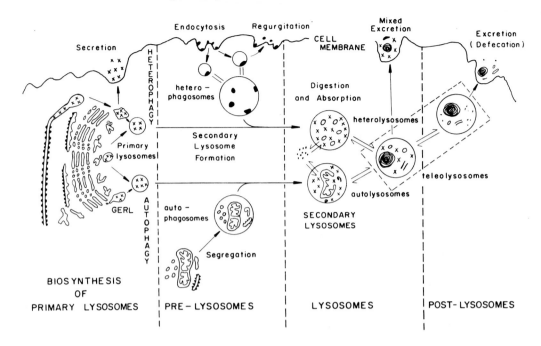

FIGURE 9. The vacuolar apparatus has been viewed traditionally as a system of structurally unique and well-defined compartments.[6,7] The route to the lysosomes was considered a unidirectional path taken by exogenous molecules and membrane destined for digestion. The concept of a vacuolar system that has emerged more recently is perhaps no less dynamic but places more emphasis on continuous and extensive interactions with the plasma membrane to effect rapid return of endocytosed membranes to the cell surface. In either model, intracellular digestion is initiated when the contents of lysosomes (either primary or secondary) are discharged into phagosomes or, conversely, the contents of phagosomes (endosomes) are discharged into lysosomes. When digestion is complete, the cell may be left with a compartment containing the residue that can either participate in subsequent rounds of digestion, remain in the cell as a residual body, or be discharged to the pericellular space. The terms designating the various lysosomal compartments are those originally proposed by de Duve and co-workers.[6]

While Golgi apparatus-associated vesicles (or, more correctly, GERL) appear to represent a source of lysosomes in many organisms, lysosomes may also originate directly from ER without apparent involvement of the Golgi apparatus.[185] Yet, the more usual route of lysosome formation seems to be the one involving GERL or some functional equivalent through cooperative action of ER and Golgi apparatus.

Despite extensive evidence for recycling, it is evident that the classic concept of degradation is still a major fate of internalized membranes. Membrane constituents or even intact portions of the plasma membrane can be degraded in lysosomes. This may be reflected in the formation of multivesicular bodies which invaginate small portions of their membrane into the matrix. Also, as suggested by Muller er al.,[115,116] it may be that internalization affords the cell an opportunity to rapidly and continuously survey the integrity of the plasma membrane or of plasma membrane constituents such that some are returned, apparently unaltered, whereas others may be degraded.

Those plasma membrane proteins internalized through endocytosis and not recycled are transported to lysosomes within 10 to 30 min. Some endocytosed membrane constituents that may be degraded in lysosomes include the receptor for epidermal growth factor[145] or for acetylcholine.[186] Others, such as the phosphomannose receptor in fibroblasts,[187] α-macroglobulin receptors in macrophages,[188] and asialoglycoprotein receptors in hepatocytes,[157] are conserved, perhaps, in a manner related to the valency of the ligand.[172]

A role of the vacuolar apparatus in membrane biogenesis is yet another possibility, especially in nonsecretory cells, since the polypeptide composition of phagolysosomes are virtually indistinguishable from that of the plasma membrane. New membranes from portions of the vacuolar apparatus that trace their origins to the Golgi apparatus or GERL would eventually reach the cell surface by this process. The unique finding of the Muller et al.[115,116] study is that membrane proteins introduced into phagolysosomes can escape degradation and return to the cell surface within 15 to 30 min after their internal iodination and that the process occurs by a mechanism involving the bulk movement of assembled membranes. However, the nature of the transport vehicle for the return trip has not yet been identified.

Small vacuoles have a high surface-to-volume ratio compared to large internal vesicular compartments. It has been suggested that the returning membrane may pass through the Golgi apparatus.[180,189,190] It is plausible, moreover, that the efflux of membrane to the cell surface is tightly coupled to and balanced by the rate of pinocytosis. The phagolysosome compartment, like the plasma membrane and other compartments involved in these dynamic processes, seldom swell with excess membrane so that an appreciation for the magnitude of membrane flux must come from approaches other than those based solely on stereology. Suitable methods to quantitate these aspects of recycling remain to be developed.

E. Transcellular Transport

One of the more extreme examples where internalized membrane taken into the cell does not fuse with lysosomes is that of the transcellular route of membrane trafficking.[191] Here, membrane incorporated by endocytosis moves across the cell and fuses with the opposite cell surface by exocytosis. The classic example is that of vesicular transport across endothelial cells in capillaries of skeletal and cardiac muscle. Here material picked up on the luminal front of the endothelium is released on the adventitial front. Rodewald[192,193] demonstrated that maternal IgG in milk binds to Fc receptors on the membranes of the brush borders at the luminal front of the cell. From there, it is internalized by receptor-mediated endocytosis into coated vesicles within which it is presumed to move across the cell to the lateral surface for release into the intercellular spaces of the vascular front of the cell. Similar routes may operate in the reverse direction for the transepithelial transport of IgA[194—196] in the liver. Here polymeric IgA binds to its receptor protein (secretory component) at the vascular front and is transported across the cell to the bile capillary where IgA is discharged into the bile along with a proteolytic fragment of the secretory component.[194—198] Similar findings have been reported by Sztul et al.[199] for secretory component and albumin.

The entire transport, however, may not take place in coated vesicles. Abrahamson and Rodewald[200] provide evidence comparing HRP, a fluid phase marker, and ferritin-conjugated IgG that binds to the membrane. The content marker and the membrane marker, while entering the cell in the same compartment, eventually become separated with the content going to lysosomes and the IgG being transported transcellularly.

An intriguing possibility suggested by Muller et al.[116] is that vesicular membrane flow might be used by the cell to deliver membrane from one area to another. Although our present level of resolution suggests that the process is a random one, it is possible that such a mechanism could be used to direct membrane from one region of the plasma membrane to another region where it could be utilized, for example, during cell movement or spreading.

F. Compensatory Endocytosis

Excess membrane delivered to the cell surface by fusion of secretory granules or vesicles or an equivalent amount of preexisting membrane must somehow be retrieved, removed, or degraded in nongrowing but actively secreting cells to account for the maintenance of a relatively constant cell surface area. From morphological studies aided by evidence from endocytosed tracers, an endocytosis-like retrieval mechanism has been envisioned that returns the surface to its original area.[11] It is sometimes assumed, although still not proven, that the membrane retrieved is reutilized in subsequent rounds of secretion (Figure 8B).[9,180,189,201]

In the parotid gland of the rabbit, isoproterenol initiates a massive burst of exocytosis through stimulation of the fusion of secretory granules with the plasma membrane of the gland lumen. The fused membrane should add about 1340 μm^2 of membrane but the actual increase is only 185 μm^2.[202] Thus, it is apparent that at least a portion of the membrane is removed, presumably through internalization. In the rat parotid, freeze etch analyses have indicated that the granule membranes upon fusion with the cell surface lack an opportunity to intermix and that large portions are deleted selectively from the exocytotic lacuna.[3,4,23,203] However, evidence that this membrane returns to the Golgi apparatus for use in subsequent secretory cycles without extensive modification is lacking.

The concept of compensatory endocytosis has been applied to nerve terminals where Golgi apparatus do not appear involved directly.[204,205] In secretory systems such as synapses large amounts of membranes are added to the cell surface during exocytosis, and membrane retrieval appears necessary if continuous increases in the area of the cell surface are to be avoided.[11] In certain neurosecretory terminals, retrieved membrane is incorporated into large vacuoles, apparently for transport up the axon and subsequent degradation or reuse in the cell body rather than in the production of new synaptic vesicles.[206] An important question raised by Holtzman[11] is how might the intensity of secretion influence the route.? With cells of adrenal medulla stimulated to secrete epinephrine rapidly, lysosomes seemed to play the major role in receiving the retrieved membranes.[207,208]

Other mechanisms to account for reductions in plasma membrane surface include shedding to the exterior[209,210] or withdrawal of molecules rather than intact membrane units.[3,4] The latter was suggested early for plant cells [56,211,212] where obvious evidence for endocytotic vesicles involved in membrane internalization were not seen.[213,214] Yet the plasma membranes of secretory cells retained their contours even in nongrowing cells during periods of active secretion.[214]

G. Pathway from the Plasma Membrane to the Golgi Apparatus (Deep Recycling)

Among the sites of delivery of endocytotic membranes indicated, in addition to lysosomes, are GERL,[215,216] Golgi apparatus,[9,189,217—220] condensing vacuoles,[189] and the ER.[219] Especially significant, in this regard, is that small amounts of membrane-associated tracers eventually reach some peripheral elements of the Golgi apparatus in addition to entering the lysosomes.[1,87,180,189,201,217,218—222a] Tracers used in these studies have been largely dextran (uncharged and relatively inert) and cationized ferritin (charged and capable of binding electrostatically to membranes). Farquhar[189] tagged the surface of rat anterior pituitary cells with cationized ferritin and demonstrated that this electron-dense marker was internalized and found subsequently in lysosomes, Golgi apparatus, and condensing granules. When dextran was infused into the gland, the tracer appeared first in coated pits and smooth apical vesicles.[218,222] After about 15 min, some of the dextran particles appeared in mature secretion granules close to the Golgi apparatus.[222] In this study, tracer was not found with cisternae of rough ER, transition elements, or peripheral Golgi apparatus elements.[207] More recently, Schwarz and Thilo,[222a] in a quantitative autoradiographic study of *Dictyostelium*, reported that after a lag of about 20 min a small but significant fraction (3%) of the total silver grains was found in the region of the Golgi apparatus during fluid phase pinocytosis with plasma membrane glycoconjugates surface labeled enzymatically with [^{3}H] galactose.

In thyroid follicle cells, with a variety of tracers, lysosomes were reached within about 5 min.[221,223] Beginning about 90 min, a few ferritin particles become associated with the Golgi apparatus peripheries. If cationized ferritin was attached first to latex beads, the complexes were internalized into lysosomes. The latex spheres remained with the lysosomes but some of the ferritin particles appeared near the Golgi apparatus to suggest that small portions of lysosomal membranes reach the Golgi apparatus.[217] Thus, Herzog and Miller[217,223] present data indicative of an indirect route from plasma membrane to the Golgi apparatus via lysosomes in thyroid cells. On the other hand, there is little or no evidence for a route to the Golgi apparatus via lysosomes in other cell types.[87]

Thus, while qualitative electron microscopic studies with cationized ferritin have suggested that membrane recycling occurs via the Golgi apparatus, either directly or indirectly through a lysosome compartment for cells of the pituitary,[189] thyroid follicle,[223] macrophages,[224,225] and plasma and myeloma cells,[87] other compartments also were labeled. As pointed out by Tartakoff,[13] the tracers employed were nonideal and the data did not prove that delivery to the Golgi apparatus either was direct or quantitatively a dominant pathway.

In studies of internalization of cationized ferritin into choroid plexus epithelial cells,[226] human skin fibroblasts,[227,228] and mouse fibroblasts,[229] no labeling of stacked Golgi apparatus was observed. Similarly, the cationic HRP poly(lysine)-conjugate used by Ryser et al.[228] did not label any stacked Golgi apparatus cisternae in L-cells. Comparing cationized ferritin as a marker in the presence of monensin, an obligatory role of the Golgi apparatus in membrane recycling has been questioned also for plasma cells.[230]

There are data that indicate that membranes of secretory vesicles turn over at a much slower rate than those of the vesicle content.[3,4] This would be predicted if secretory proteins exit from the cell while membrane proteins of the secretory vesicles were conserved and reutilized. However, earlier findings that membrane proteins of Golgi apparatus were more slowly labeled or not labeled at all have not been borne out by subsequent experimentation in other systems.[61,62] Continuous addition of newly synthesized membrane constituents into the plasma membrane by exocytosis most likely involves the Golgi apparatus complex.[34,60] Morré et al.[7] previously suggested that direct membrane recycling via the Golgi apparatus by return of intact, undegraded membranes (as opposed to membrane recycling via breakdown and reutilization in redirected synthesis or assembly) is probably of minor importance as a source of membrane exocytosis.

Even if quantitatively minor, the suggestion of bulk membrane recycling to the Golgi apparatus for utilization in subsequent secretion cycles poses new questions. Evidence favors, in those examples where it has been shown to occur, entry of endocytosed materials at the mature Golgi apparatus face. This face normally exhibits plasma membrane characteristics so that fusion compatibility with incoming plasma membrane-derived vesicles might be expected. This face is also that most frequently associated with lysosomes and with the location of GERL. A possibility, perhaps worthy of consideration, is that GERL may represent an input face adapted for membrane recycling and reutilization rather than a region entirely devoted to export activities as previously assumed. This would then necessitate postulation of an exit face for secretory vesicles proximal to the GERL-membrane utilization apparatus. A morphological organization consistent with such a suggestion is observed in hepatocytes but is less obvious in pancreas or parotid gland where extensive membrane recycling and reutilization has been postulated.

One consequence of deep recycling of plasma membrane, especially to the mature face of the Golgi apparatus or to the GERL regions, may reside in effecting a form of intracellular communication. Transport of even small portions of internalized plasma membranes to the Golgi apparatus would provide an opportunity for the internal membrane systems to acquire information on cell surface composition and topography as well as on ligand binding that may be important to the regulation of exocytosis and/or membrane biogenesis. In this manner, the biogenetic pathway might be guided to more appropriately renew altered portions of the plasma membrane or to alter membrane composition in response to various regulatory and/ or developmental signals.

IV. CONCLUDING COMMENTS

The concept that membranes move from one cell compartment to another is based primarily on microscopic studies using surface labeling of membranes (for endocytosis), short-term labeling and turnover studies (for exocytosis), and lactoperoxidase labeling of internal membrane compartments (for recycling). As a result of numerous correlative structure-function investigations spanning nearly three decades, transfer of membrane constituents to the plasma membrane from ER via the Golgi apparatus, and endocytotic internalization of cell surface membranes are well established cellular processes common to most, if not all, eukaryotes. Evidence for recycling of membranes is more recent and was first predicted on the basis of findings that actively pinocytosing cells, such as macrophages, internalize the equivalent of membrane corresponding to the entire cell surface in 1 hr or less. Since it was regarded as unlikely that such membrane losses could be counterbalanced completely by new synthesis, the alternative of membrane recycling was proposed within the context of a continuous endocytotic-exocytotic shuttle. The precise pathway by which recycling occurs and the identity of the cell components involved remain to be determined, but a substantial degree of separation of endocytotic and exocytotic pathways remains most likely. While it becomes clearer that membrane-bound compartments interact rapidly and extensively in dynamic transport processes, the identity and key topological features of major interacting components are preserved and membranes transferred by one process are most likely to be retrieved by another process. A major question concerns the role of the Golgi apparatus, an active site of processing of membrane constituents destined for export to the cell surface. To what extent is the Golgi apparatus involved in membrane recycling and reprocessing of retrieved membrane constituents? In most cell types, its role in the latter appears to be relatively minor. However, if only a small fraction of the total membrane picked at random passes through the Golgi apparatus, ultimately all of the retrieved membranes would pass through this cell component under steady-state conditions of membrane recycling.

Another major question concerns the identity and nature of the major compartment or compartments responsible for rapid recycling of receptors and other cell surface consitituents following endocytosis and mechanisms involved. Do, in fact, the internalized membranes even dissociate completely from the plasma membrane to form a receptosome free in the cytoplasm?

Once internalized, what controls the specificity of subsequent fusion events? We know very little concerning recognition signals for lysosomal enzymes and uptake signals for the variety of endocytosed ligands. Signals that may regulate membrane fusions are virtually unknown. The actual sorting mechanism and details through which directionality of sorting and membrane flow are determined and maintained remain enigmatic. Clathrin coats and interactions of membranes with components of the cytoskeleton have been implicated, but definitive information is lacking. Because of the high degree of interaction among the various cellular components during exocytotic and endocytotic events, the mechanisms whereby specificity and directionality are initiated, coordinated, and maintained emerge as, perhaps, one of the most challenging and potentially rewarding study areas of contemporary cell biology.

REFERENCES

1. **Farquhar, M. G. and Palade, G. E.,** The Golgi apparatus (complex) — (1954—1981) — from artifact to center stage, *J. Cell Biol.,* 91, 77s, 1981.
2. **Palade, G. E.,** Intracellular aspects of the process of protein secretion, *Science,* 189, 347, 1975.
3. **Meldolesi, J.,** Membranes and membrane surfaces. Dynamics of cytoplasmic membranes in pancreatic acinar cells, *Philos. Trans. R. Soc. London, Ser. B,* 268, 39, 1974.

4. **Meldolesi, J., Borgese, N., De Camilli, P., and Ceccarelli, B.,** Cytoplasmic membranes and the secretory process, in *Membrane Fusion,* Poste, G., and Nicolson, G. L., Eds., Elsevier/North-Holland, Amsterdam, 1978, 509.
5. **Steinman, R. M., Mellman, I. S., Muller, W. A., and Cohn, Z. A.,** Endocytosis and the recycling of plasma membrane, *J. Cell Biol.,* 96, 1, 1983.
6. **de Duve, C. and Wattiaux, R.,** Functions of lysosomes, *Ann. Rev. Physiol.,* 28, 435, 1966.
7. **Morré, D. J., Kartenbeck, J., and Franke, W. W.,** Membrane flow and interconversions among membranes, *Biochim. Biophys. Acta,* 449, 71, 1979.
8. **Rothman, J. E.,** The Golgi apparatus: two organelles in tandem, *Science,* 213, 1212, 1981.
9. **Farquhar, M. G.,** Membrane recycling in secretory cells: implications for traffic of products and specialized membranes within the Golgi complex, *Methods Cell Biol.,* 23, 399, 1981.
10. **Holtzman, E.,** The origin and fate of secretory packages, especially synaptic vesicles, *Neuroscience,* 2, 327, 1977.
11. **Holtzman, E.,** Membrane circulation: an overview, *Methods Cell Biol.,* 23, 379, 1981.
12. **Winkler, H.,** The biogenesis of adrenal chromaffin granule, *Neuroscience,* 2, 657, 1977.
13. **Tartakoff, A. M.,** The Golgi complex: crossroads for vesicular traffic, *Intern. Rev. Pathol.,* 22, 227, 1980.
14. **Morré, D. J., Mollenhauer, H. H., and Bracker, C. E.,** The origin and continuity of Golgi apparatus, in *Results and Problems in Cell Differentiation,* Vol. 2, Reinert, T. and Ursprung, H., Eds., Springer-Verlag, Berlin, 1971, 82.
15. **Bracker, C. E.,** Ultrastructure of fungi, *Ann. Rev. Phytopathol.,* 5, 343, 1967.
16. **Palade, G. E.,** Membrane biogenesis: an overview, *Methods Enzymol.,* 69, xxxix, 1983.
17. **Rodriquez Boulan, E. and Sabatini, D. D.,** Asymmetric budding of viruses in epithelial monolayers: a model system for the study of epithelial polarity, *Proc. Natl. Acad. Sci. U.S.A.,* 75, 5071, 1978.
18. **Rodriquez Boulan, E. and Pendergast, M.,** Polarized distribution of viral envelope proteins in the plasma membrane of infected epithelial cells, *Cell,* 20, 45, 1980.
19. **Leblond, C. P. and Bennett, G.,** Elaboration and turnover of cell coat glycoproteins, in *The Cell Surface and Development,* Moscona, A. A., Ed., John Wiley & Sons, New York, 1974, 24.
20. **Michaels, J. E. and Leblond, C. P.,** Transport of glycoprotein from Golgi apparatus to cell surface by means of "carrier" vesicles, as shown by radioautography of mouse colonic epithelium after injection of ^{3}H-fucose, *J. Microsc. Biol. Cell,* 25, 243, 1976.
21. **Jamieson, J. D. and Palade, G. E.,** Production of secretory proteins in animal cells, in *International Cell Biology 1976—1977,* Brinkley, B. R. and Porter, K. R., Eds., Rockefeller University Press, New York, 1977, 308.
22. **Flickinger, C. J.,** The relation between the Golgi apparatus, cell surface, and cytoplasmic vesicles in amoebae studied by electron microscope radioautography, *Exp. Cell Res.,* 96, 189, 1975.
23. **Castle, J., Jamieson, J., and Palade, G. E.,** Radioautographic analysis of the secretory process in the parotid acinar cell of the rabbit, *J. Cell Biol.,* 53, 290, 1972.
24. **Morré, D. J. and Ovtracht, L.,** The dynamics of Golgi apparatus: membrane differentiation and membrane flow, *Intern. Rev. Cytol. Suppl.,* 5, 61, 1977.
25. **Manton, I.,** Observations on scale production in *Pyramimonas amylifera* Conrad, *J. Cell Sci.,* 1, 375, 1966.
26. **Brown, R. M., Franke, W. W., Kleinig, H., Falk, H., and Sitte, P.,** Scale formation in Chrysophycean algae. I. Cellulosic and noncellulosic wall components made by the Golgi apparatus, *J. Cell Biol.,* 45, 246, 1970.
27. **Brown, R. M.,** Observations on the relationship of the Golgi apparatus to wall formation in the marine Chrysophycean algae, *Pleurochrysis scherffelii* Pringsheim, *J. Cell Biol.,* 41, 109, 1969.
28. **Williams, D. C.,** Studies of protistan mineralization. I. Kinetics of coccolith secretion in *Hymenomonas carterae, Calcif. Tissue Res.,* 16, 227, 1974.
29. **Morré, D. J., Boss, W. F., Grimes, H., and Mollenhauer, H. H.,** Kinetics of Golgi apparatus membrane flux following monensin treatment of embryogenic carrot cells, *Eur. J. Cell Biol.,* 30, 25, 1983.
30. **Jamieson, J. D. and Palade, G. E.,** Intracellular transport of secretory proteins in the pancreatic exocrine cell. I. Role of the peripheral elements of the Golgi complex, *J. Cell Biol.,* 34, 577, 1967.
31. **Morré, D. J. and Ovtracht, L.,** Structure of rat liver Golgi apparatus. Relationship to lipoprotein secretion, *J. Ultrastruct. Res.,* 74, 284, 1981.
32. **Franz, C. P., Croze, E. M., and Morré, D. J.,** Albumin secreted by rat liver bypasses Golgi apparatus cisternae, *Biochim. Biophys. Acta,* 678, 395, 1981.
33. **Rothman, J. E., Pettegrew, H. C., and Fine, R. E.,** Transport of the membrane glycoprotein of the vesicular stomatitis virus to the cell surface in two stages by clathrin-coated vesicles, *J. Cell Biol.,* 86, 162, 1980.
34. **Sabatini, D. D., Kreibich, G., Morimoto, T., and Adesnik, M.,** Mechanisms for the incorporation of proteins in membranes and organelles, *J. Cell Biol.,* 92, 1, 1982.

35. **Kornfeld, R. and Kornfeld, S.,** Structure of glycoproteins and their oligosaccharide units, in *The Bio-chemistry of Glycoproteins and Proteoglycans,* Lennarz, W. J., Ed., Plenum Press, New York, 1980, 1.

36. **Schachter, H.,** The subcellular sites of glycosylation, *Biochem. Soc. Symp.,* 40, 57, 1974.

37. **Schachter, H. and Roseman, S.,** Mammalian glycosyltransferases: their role in the synthesis and function of complex carbohydrates and glycolipids, in *The Biochemistry of Glycoproteins and Proteoglycans,* Lennarz, W. J., Ed., Plenum Press, New York, 1980, 85.

38. **Carey, D. J. and Hirschberg, C. B.,** Kinetics of glycosylation and intracellular transfer of sialoglyco-proteins in mouse liver, *J. Biol. Chem.,* 255, 4348, 1980.

39. **Creek, K. E. and Morré, D. J.,** Translocation of cytidine-5'-monophosphosialic acid across Golgi apparatus membranes, *Biochim. Biophys. Acta,* 643, 292, 1981.

40. **Jamieson, J. D. and Palade, G. E.,** Condensing vacuole conversion and zymogen granule discharge, *J. Cell Biol.,* 48, 503, 1971.

41. **Palade, G. E.,** Problems in intracellular membrane traffic, *Ciba Found. Symp.,* 92, 1, 1982.

42. **Friend, D. S.,** The fine structure of Brunner's gland in the mouse, *J. Cell Biol.,* 25, 563, 1965.

43. **Jamieson, J. D. and Palade, G. E.,** Intracellular transport of secretory proteins in the pancreatic exocrine cell. IV. Metabolic requirements, *J. Cell Biol.,* 39, 589, 1968.

44. **Tartakoff, A. M. and Vassalli, P.,** Plasma cell immunoglobulin secretion, *J. Exp. Med.,* 146, 1332, 1977.

45. **Tartakoff, A. M. and Vassalli, P.,** Comparative studies of intracellular transport of secretory proteins, *J. Cell Biol.,* 79, 694, 1978.

46. **Gumbier, B. and Kelley, R. B.,** Two distinct intracellular pathways transport secretory and membrane glycoproteins to the surface of pituitary cells, *Cell,* 28, 51, 1982.

47. **Schramm, M. and Selinger, Z.,** The functions of cyclic AMP and calcium as alternative second messengers in rat parotid gland and pancreas, *J. Cyclic Nucleotide Res.,* 1, 181, 1974.

48. **Rasmussen, H., Jensen, P., Lake, W., Friedmann, N., and Goodman, D. B. P.,** Cyclic nucleotides and cellular calcium metabolism, *Adv. Cyclic Nucleotide Res.,* 5, 375, 1975.

49. **Zeigel, R. F. and Dalton, A. J.,** Speculations based on the morphology of the Golgi apparatus in several types of protein secreting cells, *J. Cell Biol.,* 15, 45, 1962.

50. **Cunningham, W. P., Morré, D. J., and Mollenhauer, H. H.,** Structure of isolated plant Golgi apparatus revealed by negative staining, *J. Cell Biol.,* 28, 169, 1966.

51. **Mollenhauer, H. H., Morré, D. J., and Bergman, L.,** Homology of form in plant and animal Golgi apparatus, *Anat. Rec.,* 158, 313, 1967.

52. **Tandler, B. and Morré, D. J.,** The Golgi apparatus of ciliated cells in the cat trachea negatively stained *in situ* and in cell fractions, *Protoplasma,* 115, 193, 1983.

53. **Rambourg, A., Clermont, V., and Hermo, L.,** Three dimensional architecture of the Golgi apparatus in Sertoli cells of the rat, *Am. J. Anat.,* 154, 455, 1979.

54. **Sjöstrand, F. S.,** A comparison of plasma membrane, cytomembranes, and mitochondrial membranes with respect to ultrastructural features, *J. Ultrastruct. Res.,* 9, 561, 1963.

55. **Grove, S. N., Bracker, C. E., and Morré, D. J.,** Cytomembrane differentiation in the endoplasmic reticulum-Golgi apparatus-vesicle complex, *Science,* 161, 171, 1968.

56. **Morré, D. J. and Mollenhauer, H. H.,** The endomembrane concept: a functional integration of endoplasmic reticulum and Golgi apparatus, in *Dynamic Aspects of Plant Ultrastructure* , Robards, A. W., Ed., McGraw-Hill, New York, 1974, 84.

57. **Tartakoff, A. M. and Vassalli, P.,** Lectin binding sites as markers of Golgi subcompartments: proximal-to-distal maturation of oligosaccharides, *J. Cell Biol.,* 97, 1243, 1983.

58. **Morré, D. J. and Vigil, E. L.,** Membrane differentiation within Golgi apparatus of rat hepatocytes, *J. Ultrastruct. Res.,* 68, 317, 1974.

59. **Morré, D. J., Keenan, T. W., and Huang, C. M.,** Membrane flow and differentiation: origin of Golgi apparatus membranes from endoplasmic reticulum, in *Advances in Cytopharmacology,* Vol. 2, Ceccaralli, B., Clementi, F., and Meldolesi, J., Eds., Raven Press, New York, 1974, 107.

60. **Morré, D. J.,** The Golgi apparatus and membrane biogenesis, in *Cell Surface Reviews,* Vol. 4, Poste, G. and Nicolson, G. L., Eds., North-Holland, Amsterdam, 1977, 1.

61. **Franke, W. W., Morré, D. J., Deumling, B., Cheetham, R. D., Kartenbeck, J., Jarasch, E.-D., and Zentgraf, H.-W.,** Synthesis and turnover of membrane proteins of rat liver: an examination of the membrane flow hypothesis, *Z. Naturforsch.,* 26b, 1031, 1971.

62. **Croze, E. M. and Morré, D. J.,** Flow kinetics of mouse histocompatibility antigens, *Proc. Natl. Acad. Sci. U.S.A.,* 78, 1547, 1981.

63. **Morré, D. J., Schirrmacher, V., Robinson, P., Hess, K., and Franke, W. W.,** H-2 histocompatibility antigens of subcellular membranes of mouse liver, *Exp. Cell Res.,* 119, 265, 1979.

64. **Green, J., Griffiths, G., Louvard, D., Quinn, P., and Warran, G.,** Passage of viral membrane proteins through the Golgi complex, *J. Mol. Biol.,* 152, 663, 1981.

65. **Bergmann, J. E., Tokuyasu, K. T., and Singer, S. J.,** Passage of an integral membrane protein, the vesicular stomatitis virus glycoprotein, through the Golgi apparatus en route to the plasma membrane, *Proc. Natl. Acad. Sci. U.S.A.,* 78, 1746, 1981.

66. **Keenan, T. W., Morré, D. J., and Basu, S.,** Ganglioside biosynthesis: concentration of glycolipid glycosyltransferases in Golgi apparatus from rat liver, *J. Biol. Chem.,* 249, 310, 1974.

67. **Fleischer, B.,** Localization of some glycolipid glycosylating enzymes in the Golgi apparatus of rat kidney, *J. Supramol. Struct.,* 7, 79, 1977.

68. **Pachuszka, T., Duffard, R. O., Nishimura, R., N., Brady, R. O., and Fishman, P.,** Biosynthesis of bovine thyroid gangliosides, *J. Biol. Chem.,* 253, 5839, 1978.

69. **Eppler, C. M., Morré, D. J., and Keenan, T. W.,** Ganglioside biosynthesis in rat liver. Characterization of cytidine-5'-monophosphate-N-acetylneuraminic acid: hematoside (G_{M3}) sialyltransferase, *Biochim. Biophys. Acta,* 619, 318, 1980.

70. **Miller-Podraza, H. and Fishman, P. H.,** Translocation of newly synthesized gangliosides to the cell surface, *Biochemistry,* 21, 3265, 1982.

71. **Landa, C. A., Maccioni, H. J. F., and Caputto, R.,** The site of synthesis of gangliosides in the chick optic system, *J. Neurochem.,* 33, 825, 1979.

72. **Sonnino, S., Ghidoni, R., Marchesini, S., and Tettamanti, G.,** Cytosolic gangliosides: occurrence in calf brain as ganglioside-protein complexes, *J. Neurochem.,* 33, 117, 1979.

73. **Sonnino, S., Ghidoni, R., Masserini, M., Aporti, F., and Tettamanti, G.,** Changes in rabbit brain cytosolic and membrane-bound gangliosides during prenatal life, *J. Neurochem.,* 36, 227, 1981.

74. **Elder, J. H. and Morré, D. J.,** Synthesis *in vitro* of intrinsic membrane proteins by free, membrane-bound, and Golgi apparatus-associated polyribosomes from rat liver, *J. Biol. Chem.,* 251, 5054, 1976.

75. **Mollenhauer, H. H. and Morré, D. J.,** Structural compartmentation of the cytosol: zones of exclusion, zones of adhesion, cytoskeletal and intercisternal elements, *Subcell. Biochem.,* 5, 327, 1978.

76. **Borgese, N., Pietrini, G., and Meldolesi, J.,** Localization and biosynthesis of NADH-cytochrome b₅ reductase, an integral membrane protein, in rat liver cells. III. Evidence for the independent insertion and turnover of the enzyme in various subcellular compartments, *J. Cell Biol.,* 86, 38, 1980.

77. **Meldolesi, J., Corte, G., Pietrini, G., and Borgese, N.,** Localization and biosynthesis of NADH-cytochrome b₅ reductase, an integral membrane protein of rat liver cells. II. Evidence that a single enzyme accounts for the activity in various subcellular locations, *J. Cell Biol.,* 88, 516, 1980.

78. **Huang, C. M., Goldenberg, H., Frantz, C., Morré, D. J., Keenan, T. W., and Crane, F. L.,** Comparison of NADH-liked cytochrome c reductases of endoplasmic reticulum, Golgi apparatus and plasma membrane, *Int. J. Biochem.,* 10, 723, 1979.

79. **Mollenhauer, H. H. and Morré, D. J.,** Polyribosomes associated with forming acrosome membranes in guinea pig spermatids, *Science,* 200, 85, 1978.

80. **Merritt, W. D., Morré, D. J., Franke, W. W., and Keenan, T. W.,** Glycosyltransferases with endogenous acceptor activity in plasma membranes isolated from rat liver, *Biochim. Biophys. Acta,* 497, 820, 1977.

81. **Powell, J. T., Järlfors, U., and Brew, K.,** Enzymatic characteristics of fat globule membranes from bovine colostrum and bovine milk, *J. Cell Biol.,* 72, 617, 1977.

82. **Keenan, T. W., Morré, D. J., and Cheetham, R. D.,** Lactose synthetase in a Golgi apparatus fraction from rat mammary gland, *Nature (London),* 228, 1105, 1970.

83. **Strous, G. J. A. M. and Berger, E. G.,** Biosynthesis, intracellular transport and release of the Golgi enzyme galactosyltransferase (lactose synthetase A protein) in HeLa cells, *J. Biol. Chem.,* 257, 7623, 1982.

84. **Jarasch, E.-D., Kartenbeck, J., Bruder, G., Fink, A., Morré, D. J., and Franke, W. W.,** B-type cytochromes in plasma membranes isolated from rat liver, in comparison with those of endomembranes, *J. Cell Biol.,* 80, 37, 1979.

85. **Mollenhauer, H. H., Hass, B. S., and Morré, D. J.,** Membrane transformations in Golgi apparatus of rat spermatids. A role for thick cisternae and two classes of coated vesicles in acrosome formation, *J. Microsc. Biol. Cell. (Paris),* 27, 33, 1976.

86. **Croze, E. M., Morré, D. J., Morré, D. M., Kartenbeck, J., and Franke, W. W.,** Distribution of clathrin and spiny-coated vesicles on membranes within mature Golgi apparatus elements of mouse liver, *Eur. J. Cell Biol.,* 28, 130, 1982.

87. **Ottosen, P. D., Courtoy, P. J., and Farquhar, M. G.,** Pathways followed by membrane recovered from the surface of plasma cells and myeloma cells, *J. Exp. Med.,* 152, 1, 1980.

88. **Lodish, H. F., Kong, N., Snide, M., and Strous, G. E. A. M.,** Hepatoma secretory proteins migrate from rough endoplasmic reticulum to Golgi at characteristic rates, *Nature (London),* 304, 80, 1983.

89. **Bennett, H. S.,** The concepts of membrane flow and membrane vesiculation as mechanisms for active transport and ion pumping, *J. Biophys. Biochem. Cytol.,* 2 (Suppl.), 99, 1956.

90. **Silverstein, S. C., Steinman, R. M., and Cohn, Z. A.,** Endocytosis, *Ann. Rev. Biochem.,* 46, 649, 1977.

91. **Besterman, J. M. and Low, R. B.,** Endocytosis: a review of mechanisms and plasma membrane dynamics, *Biochem. J.,* 210, 1, 1983.

92. **Vogel, G., Thilo, L., Schwarz, H., and Steinhart, R.,** Mechanism of phagocytosis in *Dictyostelium discoideum*: phagocytosis is mediated by different recognition sites as disclosed by mutants with altered phagocytic properties, *J. Cell Biol.,* 86, 456, 1980.

93. **Van Furth, R.,** Origin and kinetics of monocytes and macrophages, *Semin. Hematol.,* 7, 125, 1970.

94. **Steinman, R. M. and Cohn, Z. A.,** The interaction of soluble horseradish peroxidase with mouse peritoneal macrophages *in vitro, J. Cell Biol.,* 55, 186, 1972.

95. **Bowers, B. and Olsezewski, T. E.,** Pinocytosis in *Acanthamoeba castellanii, J. Cell Biol.,* 53, 681, 1972.

96. **Steinman, R. M., Silver, J. M., and Cohn, Z. A.,** Pinocytosis in fibroblasts, *J. Cell Biol.,* 63, 949, 1974.

97. **Steinman, R. M., Brodie, S. F., and Cohn, Z. A.,** Membrane flow during pinocytosis. A stereologic analysis, *J. Cell Biol.,* 68, 665, 1976.

98. **Roberts, A. V. S., Williams, K. E., and Lloyd, J. B.,** The pinocytosis of ^{125}I-labeled poly (vinylpyrrolidone), ^{14}C sucrose and colloidal [198]Au gold by rat yolk sac cultured *in vitro, Biochem. J.,* 168, 239, 1977.

99. **Pratten, M. K., Williams, K. E., and Lloyd, J. B.,** A quantitative study of pinocytosis and intracellular proteolysis in rat peritoneal macrophages, *Biochem. J.,* 168, 365, 1977.

100. **Steinman, R. M., Silver, J. M., and Cohen, Z. A.,** Fluid phase pinocytosis, in *Transport of Molecules in Cellular Systems,* Life Sciences Research Report, Vol. 11, Silverstein, S. C., Ed., Dahlem Konferenzen, Berlin, 1978, 167.

101. **Kaplan, J. and Nielsen, M.,** Pinocytotic activity of rabbit alveolar macrophages *in vitro, J. Reticuloendothel. Soc.,* 24, 673, 1978.

102. **Berlin, R. D., Oliver, J. M., and Walter, R. J.,** Surface formation during mitosis. I. Phagocytosis, pinocytosis and mobility of surface-bound Con A, *Cell,* 15, 327, 1978.

103. **Quintart, J. M., Leroy-Houyet, A., Trouet, A., and Baudhuin, P.,** Endocytosis and chloroquine accumulation during the cell cycle of hepatoma cells in culture, *J. Cell Biol.,* 82, 644, 1979.

104. **Thilo, L. and Vogel, G.,** Kinetics of membrane internalization and recycling during pinocytosis in *Dictyostelium discoideum, Proc. Natl. Acad. Sci. U.S.A.,* 77, 1015, 1980.

105. **Ose, L., Ose, T., Reinertsen, R., and Berg, T.,** Fluid endocytosis in isolated rat perenchymal and nonparenchymal liver cells, *Exp. Cell Res.,* 126, 109, 1980.

106. **Besterman, J. M., Airhart, J. A., Woodworth, R. C., and Low, R. E.,** Exocytosis of pinocytosed fluid in cultured cells: kinetic evidence for rapid turnover and compartmentation, *J. Cell Biol.,* 91, 716, 1981.

107. **Hubbard, A. L. and Cohn, Z. A.,** Externally disposed plasma membrane proteins. II. Metabolic fate of iodinated polypeptides of mouse L cells, *J. Cell Biol.,* 64, 461, 1975.

108. **Roberts, R. M. and Yan, B. O.-C.,** Chemical modification of the plasma membrane polypeptides of cultured mammalian cells as an aid to studying protein turnover, *Biochemistry,* 13, 4846, 1974.

109. **Roberts, R. M. and Yuan, B. O.-C.,** Turnover of plasma membrane polypeptides in nonproliferating cultures of Chinese hamster ovary cells and human skin fibroblasts, *Arch. Biochem. Biophys.,* 171, 234, 1975.

110. **Dehlinger, P. J. and Schimke, R. T.,** Size distribution of membrane proteins of rat liver and their relative rates of degradation, *J. Biol. Chem.,* 246, 2574, 1971.

111. **Tanabe, T., Pricer, W. E., and Ashwell, G.,** Subcellular membrane topology and turnover of a rat hepatic binding protein specific for asialoglycoproteins, *J. Biol. Chem.,* 254, 1038, 1979.

112. **Adams, C. J., Maurey, K. M., and Storrie, B.,** Endocytosis of pinocytotic contents by Chinese hamster ovary cells, *J. Cell Biol.,* 93, 632, 1982.

113. **Schneider, Y.-J., Tulkens, P., de Duve, C., and Trouet, A.,** Fate of plasma membrane during endocytosis. II. Evidence for recycling (shuttle) of plasma membrane constituents, *J. Cell Biol.,* 82, 466, 1979.

114. **Widnell, C. C., Schneider, Y.-J., Pierre, B., Baudhuin, P., and Trouet, A.,** Evidence for a continual exchange of 5'-nucleotidase between the cell surface and membranes of the cytoplasm in cultured rat fibroblasts, *Cell,* 28, 61, 1981.

115. **Muller, W. A., Steinman, R. M., and Cohn, Z. A.,** The membrane proteins of the vacuolar system. I. Analysis by a novel method of intralysosomal iodination, *J. Cell Biol.,* 86, 292, 1980.

116. **Muller, W. A., Steinman, R. M., and Cohn, Z. A.,** The membrane proteins of the vacuolar system. II. Bidirectional flow between secondary lysosomes and plasma membranes, *J. Cell Biol.,* 86, 304, 1980.

116a. **de Chastellier, C., Ryter, A., and Thilo, L.,** Membrane shuttle between plasma membrane, phagosomes, and pinosomes in *Dictyostelium discoideum* amoeboid cells, *Eur. J. Cell Biol.,* 30, 233, 1983.

117. **Storrie, B., Dreesen, T. D., and Maurey, K. M.,** Rapid cell surface appearance of endocytotic membrane proteins in Chinese hamster ovary cells, *Mol. Cell. Biol.,* 1, 261, 1981.

118. **Tietze, C., Schlesinger, P., and Stahl, P.,** Mannose-specific endocytosis receptor of alveolar macrophages: demonstration of two functionally distinct intracellular pools of receptor and their roles in receptor recycling, *J. Cell Biol.,* 92, 417, 1982.

119. **Burgert, H.-G. and Thilo, L.,** Internalization and recycling of plasma membrane glycoconjugates during pinocytosis in the macrophage cell line, P388D$_1$, *Exp. Cell Res.,* 144, 127, 1983.

120. **Berlin, R. D. and Oliver, J. M.,** Surface functions during mitosis. II. Quantitation of pinocytosis and kinetic characterization of the mitotic cycle with a new fluorescence technique, *J. Cell Biol.,* 85, 660, 1980.

121. **Allen, R. D.,** Membranes of ciliates: ultrastructure, biochemistry and fusion, *Cell Surf. Rev.,* 5, 657, 1978.

122. **McKanna, J. A.,** Membrane recycling: vesiculation of the *Amoeba* contractile vacuole at systole, *Science,* 179, 88, 1973.

123. **Anderson, R. G. W., Goldstein, J. L., and Brown, M. S.,** A mutation that impairs the ability of lipoprotein receptors to localize in coated pits on the cell surface of human fibroblasts, *Nature (London),* 270, 695, 1977.

124. **Goldstein, J. L., Anderson, R. G. W., and Brown, M. S.,** Coated pits, coated vesicles, and receptor-mediated endocytosis, *Nature (London),* 279, 679, 1979.

125. **Goldstein, J. L., Anderson, R. G. W., and Brown, M. S.,** Receptor-mediated endocytosis and the cellular uptake of low density lipoprotein, *Ciba Found. Symp.,* 92, 77, 1982.

126. **Anderson, R. G. W., Brown, M. S., Beisiegel, U., and Goldstein, J. L.,** Surface distribution and recycling of the LDL receptor as visualized with antireceptor antibodies, *J. Cell Biol.,* 93, 523, 1982.

127. **King, A. C. and Cuatrecasas, P.,** Peptide hormone-induced receptor mobility, aggregation, and internalization, *N. Engl. J. Med.,* 305, 77, 1981.

128. **Regoeczi, E., Chindemi, P. A., DeBanne, M. T., and Charlwood, P. A.,** Partial resialylation of human asialotransferrin type 3 in the rat, *Proc. Natl. Acad. Sci. U.S.A.,* 79, 2226, 1982.

129. **Schneider, W. J., Beisiegel, U., Goldstein, J. L., and Brown, M. S.,** Purification of the low density lipoprotein receptor, an acidic glycoprotein of 164,000 molecular weight, *J. Biol. Chem.,* 257, 2664, 1982.

130. **Ungewickell, E. and Branton, D.,** Assembly of units of clathrin coats, *Nature (London),* 289, 420, 1981.

131. **Pearse, B. M. F.,** On the structural and functional composition of coated vesicles, *J. Mol. Biol.,* 126, 803, 1978.

132. **Willingham, M. C. and Pastan, I.,** The receptosome: an intermediate organelle of receptor-mediated endocytosis in cultured fibroblasts, *Cell,* 21, 67, 1980.

133. **Geuze, H. J., Slot, J. W., Strous, G. J. A. M., Lodish, H. F., and Schwartz, A. L.,** Intracellular site of asialoglycoprotein receptor-ligand coupling: double label immunoelectron microscopy during receptor-mediated endocytosis, *Cell,* 32, 277, 1983.

134. **Nichols, B. A.,** Uptake and digestion of horseradish peroxidase in rabbit alveolar macrophages. Formation of a pathway connecting lysosomes to the cell surface, *Lab. Invest.,* 47, 235, 1982.

135. **Galloway, C. J., Dean, G. E., March, M., Rudnick, G., and Mellman, I.,** Acidification of macrophage and fibroblast endocytic vesicles *in vitro, Proc. Natl. Acad. Sci. U.S.A.,* 80, 3334, 1983.

136. **Smith, G. D. and Peters, T. J.,** The localization in rat liver of alkaline phosphodiesterase to a discrete organelle implicated in ligand internalization, *Biochim. Biophys. Acta,* 716, 24, 1982.

137. **Smith, G. D., Flint, N., Evans, W. H., and Peters, T. J.,** Hepatic processing of IgA: studies using analytical and preparative subcellular fractionation, *Biosci. Rep.,* 1, 921, 1981.

138. **Pastan, I. H. and Willingham, M. C.,** Journey to the center of the cell. Role of the receptosome, *Science,* 214, 504, 1981.

139. **Fan, J. Y., Carpentier, J. L., van Obberghen, E., Blackett, N. M., Grunfeld, C., Gorden, P., and Orchi, L.,** Receptor mediated endocytosis of insulin: the role of microvilli, coated pits and coated vesicles, *Proc. Natl. Acad. Sci. U.S.A.,* 79, 7788, 1982.

140. **Patzer, E. J., Schlossman, D. M., and Rothman, J. E.,** Release of clathrin from coated vesicles dependent upon a nucleoside triphosphatase and a cytosol fraction, *J. Cell Biol.,* 93, 230, 1982.

141. **Wheland, J., Willingham, M., Dickson, R., and Pastan, I.,** Microinjection of anticlathrin antibodies into fibroblasts does not interfere with the receptor-mediated endocytosis of α$_2$-macroglobulin, *Cell,* 25, 105, 1981.

142. **Kartenbeck, J., Schmid, E., Miller, H., and Franke, W. W.,** Immunological identification and localization of clathrin and coated vesicles in cultured cells and tissues, *Exp. Cell Res.,* 133, 191, 1981.

143. **Merisko, E. M., Farquhar, M. G., and Palade, G. E.,** Changes in clathrin distribution in pancreatic exocrine cells under anoxic conditions, *J. Cell Biol.,* 97, 174a, 1983.

144. **Brown, M. A. and Goldstein, J. L.,** Receptor-mediated endocytosis: insights from the lipoprotein receptor system, *Proc. Natl. Acad. Sci. U.S.A.,* 76, 3330, 1979.

145. **Carpenter, G. and Cohen, G.,** ^{125}I-labeled human epidermal growth factor. Binding, internalization and degradation in human fibroblasts, *J. Cell Biol.,* 71, 159, 1976.

146. **Maxfield, F. R., Schlessinger, J., Shechter, Y., Pastan, I., and Willingham, M. C.,** Collection of insulin, EGF, and α$_2$-macroglobulin in the same patches on the surface of cultured fibroblasts and common internalization, *Cell,* 14, 805, 1978.

147. **Willingham, M. C., Maxfield, F. R., and Pastan, I. H.,** α-2 Macroglobulin binding to the plasma membrane of cultured fibroblasts. Diffuse binding by clustering in coated regions, *J. Cell Biol.*, 82, 614, 1979.

148. **Cheng, S. Y., Maxfield, F. R., Robbins, J., Willingham, M. C., and Pastan, I.,** Receptor-mediated uptake of 3,3′,5-triiodo-L-thyronine by cultured fibroblasts, *Proc. Natl. Acad. Sci. U.S.A.*, 77, 3425, 1980.

149. **Ashwell, G. and Harford, J.,** Carbohydrate specific receptors of liver, *Ann. Rev. Biochem.*, 51, 531, 1982.

150. **Neufeld, E. F., Sando, G. N., Garvin, A. J., and Rome, L. H.,** The transport of lysosomal enzymes, *J. Supramol. Struct.*, 6, 95, 1977.

151. **Neufeld, E. F. and Ashwell, G.,** Carbohydrate recognition systems for receptor-mediated endocytosis, in *The Biochemistry of Glycoproteins and Proteoglycans,* Lennarz, W. J., Ed., Plenum Press, New York, 1980, 241.

152. **Ashwell, G. and Morell, A. G.,** The role of surface carbohydrates in the hepatic recognition and transport of circulating glycoproteins, *Adv. Enzymol.*, 41, 99, 1974.

153. **Morell, A. G., Irvine, R. A., Sternlieb, I., Scheinberg, I. M., and Ashwell, G.,** Physical and chemical studies on ceruloplasmin. V. Metabolic studies on sialic acid-free ceruloplasmin *in vivo, J. Biol. Chem.*, 243, 155, 1968.

154. **Schwartz, A. L., Fridovitch, S. E., and Lodish, H. F.,** Kinetics of internalization and recycling of the asialoglycoprotein receptor in a hepatoma cell line, *J. Biol. Chem.*, 257, 4230, 1982.

155. **Ciechannover, A., Schwartz, A. L., and Lodish, H. F.,** The asialoglycoprotein receptor internalizes and recycles independently of the transferrin and insulin receptors, *Cell*, 32, 267, 1983.

156. **Wall, D. A., Wilson, G., and Hubbard, A. L.,** The galactose-specific recognition system of mammalian liver. The route of ligand internalization in rat hepatocytes, *Cell*, 21, 79, 1980.

157. **Steer, C. J. and Ashwell, G.,** Studies on a mammalian hepatic binding protein specific for asialoglyco-proteins. Evidence for receptor recycling in isolated rat hepatocytes, *J. Biol. Chem.*, 255, 3008, 1980.

158. **Bridges, K., Harford, J., Ashwell, G., and Klausner, R. D.,** Fate of receptor and ligand during endocytosis of asialoglycoprotein by isolated hepatocytes, *Proc. Natl. Acad. Sci. U.S.A.*, 79, 350, 1982.

159. **Hudgin, R. L., Pricer, W. E., Ashwell, G., Stockert, R. J., and Morell, A. G.,** The isolation and properties of a rabbit liver binding protein specific for asialoglycoproteins, *J. Biol. Chem.*, 249, 5536, 1974.

160. **Pricer, W. E. and Ashwell, G.,** The binding of desialylated glycoproteins by plasma membranes of rat liver, *J. Biol. Chem.*, 246, 4825, 1971.

161. **Lunney, J. and Ashwell, G.,** A hepatic receptor of avian origin capable of binding specifically modified glycoproteins, *Proc. Natl. Acad. Sci. U.S.A.*, 73, 341, 1976.

162. **Sando, G. N. and Neufeld, E. F.,** Recognition and receptor-mediated uptake of a lysosomal enzyme, alpha-L-iduronidase, by cultured fibroblasts, *Cell*, 12, 619, 1977.

163. **Kaplan, A., Achord, D. T., and Sly, W. S.,** Phosphohexosyl recognition. A general characteristic of pinocytosis of lysosomal glycosidases by human fibroblasts, *Proc. Natl. Acad. Sci. U.S.A.*, 74, 2026, 1977.

164. **Sly, W. S. and Stahl, P.,** Receptor-mediated uptake of lysosomal enzymes, in *Transport of Macromolecules in Cellular Systems,* Silverstein, S. C., Ed., Dahlem Konferenzen, Berlin, 1978, 229.

165. **Gonzales-Noriega, A., Grubb, J. H., Talkad, U., and Sly, W. S.,** Chloroquine inhibits lysosomal enzyme pinocytosis and enhances lysosomal enzyme secretion by impairing receptor recycling, *J. Cell Biol.*, 85, 839, 1980.

166. **Varke, A. and Kornfeld, S.,** Identification of a rat liver α-N-acetylglucosaminyl phosphodiesterase capable of removing "blocking" α-N-acetylglucosamine residues from phosphorylated high mannose oligosaccha-rides of lysosomal enzymes, *J. Biol. Chem.*, 255, 8398, 1980.

167. **Waheed, A., Pohlman, R., Hasilik, A., and von Figura, K.,** Subcellular location of two enzymes involved in the synthesis of phosphorylated recognition markers in lysosomal enzymes, *J. Biol. Chem.*, 256, 4150, 1981.

168. **Prieels, J. P., Pizzo, S. V., Glascow, L. R., Paulson, J. C., and Hill, R. L.,** Hepatic receptor that specifically binds oligosaccharides containing fucosyl α1 → 3 N-acetylglucosamine linkages, *Proc. Natl. Acad. Sci. U.S.A.*, 75, 2115, 1978.

169. **Achord, D. T., Brot, F. E., Bell, C. E., and Sly, W. S.,** Human beta-glucuronidase. *In vivo* clearance and *in vitro* uptake by a glycoprotein recognition system on reticuloendothelial cells, *Cell*, 15, 269, 1978.

170. **Schlesinger, P. H., Doebber, T. W., Mandell, B. F., White, R., de Schriyver, C., Rodman, J. S., Miller, M. J., and Stahl, P.,** Plasma clearance of glycoprotein with terminal mannose and N-acetylglu-cosamine by liver non-parenchymal cells, *Biochem. J.*, 176, 103, 1978.

170a. **Carpentier, J.-L., Gorden, P., Amherdt, M., Van Obberghen, E., Kahn, C. R., and Orci, L.,** ^{125}I-insulin binding to cultured human lymphocytes: initial localization and fate of hormone determined by quantitative electron microscopic autoradiography, *J. Clin. Invest.*, 61, 1056, 1978.

171. **McKanna, J. A., Haigler, H. T., and Cohen, S.,** Hormone receptor topology and dynamics: morphological analysis using ferritin-labeled epidermal growth factor, *Proc. Natl. Acad. Sci. U.S.A.,* 76, 5689, 1979.

172. **Ukkonen, P., Helenius, A., and Mellman, I.,** Valency influences the transport of Fc receptor-bound ligands to lysosomes, *J. Cell Biol.,* 97, 168a, 1983.

173. **Tycko, B. and Maxfield, F. R.,** Rapid acidification of endocytic vesicles containing alpha$_2$-macroglobulin, *Cell,* 28, 643, 1982.

174. **Maxfield, F. R.,** Weak bases and ionophores rapidly and reversibly raise the pH of endocytic vesicles in cultured mouse fibroblasts, *J. Cell Biol.,* 95, 676, 1982.

175. **Schneider, D. L.,** ATP-dependent acidification of intact and disrupted lysosomes. Evidence for an ATP-driven proton pump, *J. Biol. Chem.,* 256, 3858, 1981.

176. **Ohkuma, S., Moryama, Y., and Takano, T.,** Identification and characterization of a proton pump on lysosomes by fluorescein isothiocyanate-dextran fluorescence, *Proc. Natl. Acad. Sci. U.S.A.,* 79, 2758, 1982.

177. **Geisow, M.,** Lysosome proton pump identified, *Nature (London),* 298, 515, 1982.

178. **Haigler, H. T., Maxfield, F. R., Willingham, M. C., and Pastan, I.,** Dansyl cadaverine inhibits ^{125}I-epidermal growth factor in Balb 3T3 cells, *J. Biol. Chem.,* 255, 1239, 1980.

179. **Posner, B. I., Bergeron, J. J. M., Josefsberg, Z., Khan, M. N., Khan, R. J., Patel, B. A., Sikstrom, A., and Verma, A. K.,** Polypeptide hormones: intracellular receptors and internalization, *Recent Prog. Horm. Res.,* 37, 539, 1981.

180. **Farquhar, M. G.,** Multiple pathways of exocytosis, endocytosis and membrane recycling: validation of a Golgi route, *Fed. Proc., Fed. Am. Soc. Exp. Biol.,* 42, 2407, 1983.

181. **Novikoff, A. B.,** GERL, its form and function in neurons of rat spinal ganglia, *Biol. Bull.,* 127, 358, 1964.

182. **Essner, E. and Novikoff, A.,** Cytological studies on two functional hepatomas: interrelations of endoplasmic reticulum, Golgi apparatus and lysosomes, *J. Cell Biol.,* 15, 298, 1962.

183. **Novikoff, A. B. and Shin, W. Y.,** The endoplasmic reticulum in the Golgi zone and its relations to microbodies, Golgi apparatus and autophagic vacuoles in rat liver cells, *J. Microsc.,* 3, 187, 1964.

184. **Novikoff, A. B. and Novikoff, P. M.,** Cytochemical contributions to differentiating GERL from the Golgi apparatus, *Histochem. J.,* 9, 525, 1977.

185. **Moe, H., Rostgaard, J., and Behnke, O.,** On the morphology and origin of virgin lysosomes in the intestinal epithelium of the rat, *J. Ultrastruct. Res.,* 12, 396, 1965.

186. **Gardner, J. M. and Fambrough, D. M.,** Acetylcholine receptor degradation measured by density labelling: effects of cholinergic ligands and evidence against recycling, *Cell,* 1, 16, 1979.

187. **Stahl, P., Schlesinger, H., Sigardson, E., Rodman, J. S., and Lee, Y. C.,** Receptor-mediated pincoytosis of mannose glycoconjugates by macrophages. Characterization and evidence for receptor recycling, *Cell,* 19, 207, 1980.

188. **Kaplan, J.,** Evidence for reutilization of surface receptors for α-macroglobulin-protease complexes in rabbit alveolar macrophages, *Cell,* 19, 197, 1980.

189. **Farquhar, M. G.,** Recovery of surface membrane in anterior pituitary cells. Variations in traffic detected with anionic and cationic ferritin, *J. Cell Biol.,* 77, R35, 1978.

190. **Gonatas, N. K., Steiber, A., Kim, S. U., Graham, D. I., and Avrameas, S.,** Internalization of neuronal plasma membrane ricin receptors into the Golgi apparatus, *Exp. Cell Res.,* 94, 426, 1975.

191. **Palade, G. E., Simionescu, M., and Simionescu, N.,** Structural aspects of the permeability of the microvascular endothelium, *Acta Physiol. Scand. Suppl.,* 463, 11, 1979.

192. **Rodewald, R.,** Intestinal transport of antibodies in the newborn rat, *J. Cell Biol.,* 58, 189, 1973.

193. **Rodewald, R.,** Distribution of immunoglobulin G receptors in the small intestine of the young rat, *J. Cell Biol.,* 85, 18, 1980.

194. **Nagura, H., Nakane, P. K., and Brown, W. R.,** Translocation of dimeric IgA through neoplastic colon cells *in vitro,* *J. Immunol.,* 123, 2359, 1979.

195. **Renston, R. H., Jones, A. L., Christiansen, W. D., and Hradek, G.,** Evidence for a vesicular transport mechanism in hepatocytes for biliary secretion of immunoglobulin A, *Science,* 208, 1276, 1980.

196. **Jones, A. L., Huling, S., and Hradek, G.,** Uptake and intracellular deposition of immunoglobulin A by rat hepatocytes in monolayer culture, *Hepatology,* 2, 769, 1982.

197. **Goldman, I. S., Jones, A. L., Hradek, G. T., and Huling, S.,** Hepatocyte handling of immunoglobulin A in the rat: the role of microtubules, *Gastroenterology,* 85, 130, 1983.

198. **Kloppel, T. M. and Brown, W. R.,** Rat liver membrane secretory component is larger than free secretory component in bile: evidence of proteolytic conversion of membrane form to free form, *J. Cell. Biochem.,* 24, 307, 1984.

199. **Sztul, E. S., Howell, K. E., and Palade, G. E.,** Intracellular and transcellular transport of secretory component and albumin in rat hepatocytes, *J. Cell Biol.,* 97, 1582, 1983.

200. **Abrahamson, D. R. and Rodewald, R.,** Evidence of the sorting of endocytic vesicle contents during the receptor-mediated transport of IgG across the newborn rat intesting, *J. Cell Biol.,* 91, 270, 1981.

201. **Farquhar, M. G.,** Membrane recycling in secretory cells: pathway to the Golgi complex, *Ciba Found. Symp.,* 92, 157, 1982.

202. **Cope, G. H. and Williams, M. H.,** Quantitative analyses of the constituent membranes of parotid acinar cells and of the changes evident after induced exocytosis, *Z. Zellforsch. Mikrosk. Anat.,* 145, 311, 1973.

203. **De Camilli, P., Pluchetti, D., and Meldolesi, J.,** Dynamic changes of the luminal plasmalemma in stimulated parotid acinar cells. A freeze-fracture study, *J. Cell Biol.,* 70, 59, 1976.

204. **Ceccarelli, B., Hurlbut, W. P., and Mauro, A.,** Turnover of transmitter and synaptic vesicles at the frog neuromuscular junction, *J. Cell Biol.,* 57, 499, 1973.

205. **Heuser, J. E. and Reese, T. S.,** Evidence for recycling of synaptic vesicle membranes during transmitter release at the frog neuromuscular junction, *J. Cell Biol.,* 57, 315, 1973.

206. **Morris, J. F. and Nordmann, J. J.,** Membrane recapture after hormonal release from nerve endings in the neural lobe of the rat pituitary gland, *Neuroscience,* 5, 639, 1980.

207. **Abraham, S. and Holtzman, E.,** Secretion and endocytosis in insulin-stimulated rat adrenal medulla cells, *J. Cell Biol.,* 56, 540, 1973.

208. **Holtzman, E., Schacher, S., Evans, J., and Teichberg, S.,** Origin and fate of the membranes of secretion granules and synaptic vesicles: membrane circulation in neurons, gland cells and retinal photoreceptors, *Cell Surf. Rev.,* 4, 116, 1977.

209. **Lawson, D., Raff, M. C., Gomperts, B., Fewtrell, C., and Gilula, N. B.,** Molecular events during membrane fusion. A study of exocytosis in rat peritoneal mast cells, *J. Cell Biol.,* 72, 242, 1977.

210. **Specian, R. L. and Neutra, M. R.,** Goblet cells: membrane loss during rapid secretion, *J. Cell Biol.,* 83, 429a, 1979.

211. **Schnepf, E.,** Gland cells, in *Dynamic Aspects of Plant Ultrastructure,* Robards, A. W., Ed., McGraw-Hill, New York, 1974, 331.

212. **Schnepf, E. and Busch, J.,** Morphology and kinetics of slime secretion in gland cells of *Mimulus tilingii,* *Z. Pflanzenphysiol.,* 79, 62, 1976.

213. **Fowke, L. C., Griffing, L. R., Mersey, B. G., and Van der Valk, P.,** Protoplasts for studies of plasma membrane and associated cell organelles, *Experientia Suppl.,* 46, 101, 1983.

214. **Morré, D. J. and Mollenhauer, H. H.,** Dictyosome polarity and membrane differentiation in outer cap cells of the maize root tip, *Eur. J. Cell Biol.,* 29, 126, 1983.

215. **Gonatas, N. K., Stiber, A., Kim, S. U., Graham, D. I., and Avrameas, S.,** Internalization of neuronal plasma membrane ricin receptors into the Golgi apparatus, *Exp. Cell Res.,* 94, 426, 1977.

216. **Gonatas, N. K.,** The role of the neuronal Golgi apparatus in a centripetal membrane vesicular traffic, *J. Neuropathol. Exp. Neurol.,* 41, 6, 1982.

217. **Herzog, V.,** Pathways of endocytosis, *Trends Biochem. Sci.,* 6, 319, 1981.

218. **Herzog, V. and Farquhar, M. G.,** Luminal membrane retrieved after exocytosis reaches most Golgi cisternae in secretory cells, *Proc. Natl. Acad. Sci. U.S.A.,* 74, 5073, 1977.

219. **Bergeron, J. J. M., Sikstrom, R., Hand, A. R., and Posner, B. I.,** Binding and uptake of ^{125}I-insulin into rat liver hepatocytes and endothelium. An *in vivo* radioautographic study, *J. Cell Biol.,* 80, 427, 1979.

220. **Varga, J. M., Moellmann, G., Fritsch, P., Godawska, E., and Lerner, A. B.,** Association of cell surface receptors for melanotropin with the Golgi region in mouse melanoma cells, *Proc. Natl. Acad. Sci. U.S.A.,* 73, 559, 1976.

221. **Herzog, V. and Miller, F.,** Structural and functional polarity of inside-out follicles prepared from pig thyroid gland, *Eur. J. Cell Biol.,* 24, 74, 1981.

222. **Herzog, V. and Reggio, H.,** Pathways of membranes retrieved from the luminal surface of exocrine cells of the pancreas, *Eur. J. Cell Biol.,* 21, 11, 1980.

222a. **Schwarz, H. and Thilo, L.,** Membrane traffic in *Dictyostelium discoideum*: plasma membrane glycoconjugates internalized and recycled during fluid phase pinocytosis enter the Golgi apparatus, *Eur. J. Cell Biol.,* 31, 212, 1983.

223. **Herzog, V. and Miller, F.,** Membrane retrieval in epithelial cells of isolated thyroid follicles, *Eur. J. Cell Biol.,* 19, 203, 1979.

224. **Thyberg, J.,** Internalization of cationized ferritin into the Golgi complex of cultured mouse peritoneal macrophages. Effects of colchicine and cytochalasin B, *Eur. J. Cell Biol.,* 23, 95, 1980.

225. **Thyberg, J., Nilsson, J., and Hellgrem, D.,** Recirculation of cationized ferritin in cultured mouse peritoneal macrophages. Electron microscopic and cytochemical studies with double labeling techniques, *Eur. J. Cell Biol.,* 23, 85, 1980.

226. **van Deurs, B., von Bülow, F., and Møller, M.,** Vesicular transport of cationized ferritin by the epithelium of the rat choroid plexus, *J. Cell Biol.,* 89, 131, 1981.

227. **van Deurs, B., Nilausen, K., Faegerman, O., and Meinertz, H.,** Coated pits and pinocytosis of cationized ferritin in human skin fibroblasts, *Eur. J. Cell Biol.,* 27, 270, 1982.

228. **Ryser, H. J.-P., Drummond, I., and Shen, W.-C.,** The cellular uptake of horseradish peroxidase and its poly (lysine) conjugate by cultured fibroblasts is qualitatively similar despite a 900-fold difference in rate, *J. Cell Physiol.,* 113, 167, 1982.

229. **van Deurs, B. and Nilausen, K.,** Pinocytosis in mouse L-fibroblasts: ultrastructural evidence for a direct membrane shuttle between the plasma membrane and the lysosomal compartment, *J. Cell Biol.,* 94, 279, 1982.

230. **Tartakoff, A., Vassalli, P., and Montesano, R.,** Plasma cell endocytosis: is it related to immunoglobulin secretion?, *Eur. J. Cell Biol.,* 26, 188, 1981.

Chapter 9

THE SURFACE POTENTIAL OF MEMBRANES: ITS EFFECT ON MEMBRANE-BOUND ENZYMES AND TRANSPORT PROCESSES

Lech Wojtczak and Maciej J. Nałęcz

TABLE OF CONTENTS

I. SURFACE CHARGE AND SURFACE POTENTIAL: GENERAL CONSIDERATIONS

A difference in the electric potential is usually formed at the boundary between two phases due to the transfer of ions and/or electrons from one phase to another. This occurs when ions or electrons dissociate from one phase and become associated with molecules of the other phase. A potential difference can also arise at neutral interface if there are mobile ions which have different solubility (or mobility) in either of the phases. The potential difference between the surface of phase A and a point infinitely distant from the interface within phase B is defined as the *surface potential* (more precisely: *potential at the surface*, ψ_s) of phase A.

This general definition applies, in particular, to the interface between a solid phase, or a liquid immiscible with water, and the aqueous solution of an electrolyte. The resulting surface potential at the solid phase produces an uneven distribution of ions in the aqueous phase. Ions of the charge which is opposite to that of the surface are concentrated in the vicinity of the surface due to electrostatic attraction, whereas ions of the same charge are repulsed. The concentration of ions in the aqueous phase at the surface of the solid phase (C_0) is a function of the surface potential and is described by the Boltzmann expression

$$C_0 = C_\infty \exp(-ze\psi_s/kT) \tag{1}$$

where C_∞ denotes the concentration in the bulk solution, e is the elementary charge ($1.60210 \cdot 10^{-19}$ C), k is the Boltzmann constant ($1.38054 \cdot 10^{-23}$ J$\cdot$°K^{-1}), T is the absolute temperature, and z is a positive or negative integer denoting the number of charges of the ion. The same dependence can also be expressed in another convention as

$$C_0 = C_\infty \exp(-zF\psi_s/RT) \tag{2}$$

where F is the Faraday constant (96,487 C$\cdot$mol^{-1} or 23,063 cal$\cdot$volt$^{-1}\cdot$mol^{-1}) and R is the gas constant (8.3143 J$\cdot$mol$^{-1}\cdot$°K^{-1} or 1.987 cal$\cdot$mol$^{-1}\cdot$°K^{-1}). From these equations it is obvious that $C_0 < C_\infty$ when z and ψ_s have the same sign (both are positive or negative) and $C_0 > C_\infty$ when z and ψ_s are of opposite signs.

Equations 1 and 2 describe the concentration of ions in the immediate vicinity of the surface, i.e., at the distance $x = 0$ from the interface. The separation of ions due to the surface potential is, however, counteracted by their thermal motion, resulting in the formation of a diffuse concentration gradient, or a *diffuse layer*, within the aqueous phase, whereas the surface together with the adjacent layer of counterions in the aqueous phase form what is called the *electrical double layer*. The structure and properties of these layers are described by the Gouy-Chapman theory.[1,2] A comprehensive approach to this theory is presented in numerous books and special articles.[3-10] In this chapter we shall limit our considerations only to consequences of the Gouy-Chapman theory which are important for the situation existing at the surface of biological membranes.

The Gouy-Chapman theory expresses the dependence between the surface potential and the density of the surface charge (σ) by the following equation:

$$\sigma = \pm \left| \left\{ 2 \, \epsilon_0 \epsilon_r RT \sum_n C_{n\infty} \left[\exp(-z_n F\psi_s/RT) - 1 \right] \right\} \right|^{1/2} \tag{3}$$

where ϵ_0 is the permittivity of the vacuum ($8.8542 \cdot 10^{-12}$ C$^2\cdot$J$^{-1}\cdot$m^{-1}), ϵ_r is the relative permittivity (dielectric constant) of the solution, C_n^∞ is the concentration of the ion n, and z_n is its valency. A relatively simple form of this equation requires two following assumptions:

(1) the surface is planar and (2) the relative permittivity remains constant throughout the diffuse layer and the bulk solution. For the surface of biological membranes, whose curvature is usually negligible as compared to the thickness of the diffuse layer and whose charge density is not too large, these two assumptions hold within limits of acceptable approximation.

When the aqueous phase is a solution of a symmetric electrolyte, i.e., one containing cations and anions of the same valency, e.g., KCl or $MgSO_4$, Equation (3) can be simplified to

$$\sigma = 2(2RT\ \epsilon_0\epsilon_r C_\infty)^{1/2}\sinh(zF\psi_s/2RT) \tag{4}$$

Numerical substitution for 25°C gives

$$\sigma = 0.1185\ C_\infty^{1/2}\ \sinh(z\psi_s/51.36) \tag{5}$$

where σ is expressed in C/m^2, C_∞ in mol/ℓ (*M*), and ψ_s in mV.

In biological membranes the surface charge density depends on their chemical composition and the orientation of molecules within the membrane structure. Since the surface charge density determines the magnitude of the surface potential, Equation (4) is often applicable in its reversed form

$$\psi_s = \frac{2RT}{z\ F}\ \text{arc sinh}\ [\sigma(8RT\ \epsilon_0\epsilon_r C_\infty)^{-1/2}] \tag{6}$$

and after numerical substitution for 25°C

$$\psi_s = \frac{51.36}{z}\ \text{arc sinh}\ (8.44\ \sigma\ C_\infty^{-1/2}) \tag{7}$$

Since arc sinh x for low values of x approaches a linear function of x, Equation (6) can be further simplified to

$$\psi_s = \frac{2RT}{z\ F}\ \sigma\ (8RT\ \epsilon_0\epsilon_r C_\infty)^{-1/2} = \frac{\sigma}{zF}\left(\frac{RT}{2\ \epsilon_0\epsilon_r C_\infty}\right)^{1/2} \tag{8}$$

and after numerical substitution

$$\psi_s = 433.4\ \frac{\sigma}{z}\ C_\infty^{-1/2} \tag{9}$$

This linear approximation holds for surface charge density of 0.01 C/m^2 or less, which is often the case for natural membranes, and C_∞ of 0.1 *M* or more.

It is thus evident that for low values of σ the surface potential is approximately proportional to the surface charge density and inversely proportional to the square root of the electrolyte concentration. It is also inversely proportional to the electrolyte valency (z). This is illustrated by Figure 1. It shows, for example, that the same reduction of ψ_s as produced by 100 m*M* monovalent electrolyte can be obtained with 18 m*M* divalent electrolyte and with approximately 5 m*M* trivalent electrolyte. At high electrolyte concentrations, i.e., 1 *M* or higher, the surface potential largely disappears.

The Gouy-Chapman theory also allows to plot the potential profiles in the vicinity of the surface. As shown in Figure 2, in the millimolar range of a monovalent electrolyte the potential rapidly declines practically to zero within a short distance from the interface. For example, at physiological electrolyte concentration, about 100 m*M*, the diffuse layer extends

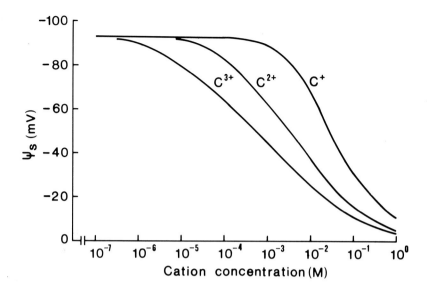

FIGURE 1.   Effectivness of mono-, di-, and trivalent cations in reducing the surface potential (ψ_s) of surface having a net negative charge density of -0.025 C/m² suspended in a medium having a background monovalent electrolyte concentration of 5 mM. The curves have been calculated assuming symmetrical electrolytes using Equation (10). (Adapted from Barber, J., *Biochim. Biophys. Acta*, 594, 253, 1980. With permission.)

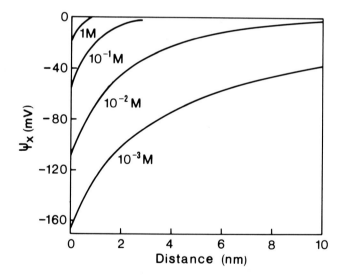

FIGURE 2.   Electrical potential profiles for a surface having a net negative charge density of -0.05 C/m² immersed in a solution containing monovalent electrolyte ranging from 1 mM to 1M. ψ_x Designates the potential at the distance indicated at the abscissa. (From Barber, J., *Biochim. Biophys. Acta*, 594, 253, 1980. With permission.)

over a few nanometers and even at 10 mM concentration it does not exceed 10 nm. A divalent electrolyte at 9 mM concentration results in almost the same potential profile as 100 mM monovalent electrolyte (not shown).

Combination of Equations (2) and (6) allows us to calculate the concentration of ions in

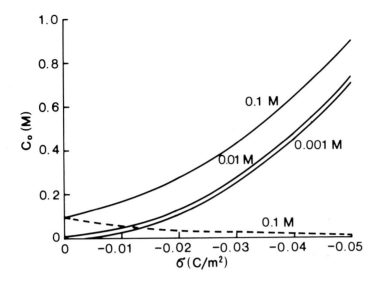

FIGURE 3. Dependence of the concentration of monovalent cation (solid lines) and anion (dashed line) at the surface (C_0) on the surface charge density (σ) and the bulk concentration ranging from 0.001 to 0.1 M.

the vicinity of the surface for a single symmetric electrolyte as a function of the surface charge density. From such calculation it comes out that large variations of the bulk concentration are accompanied by much smaller changes of the local concentration at the membrane. For example (Figure 3), a tenfold decrease of the bulk concentration of a cation results only in a twofold decrease at the surface when its charge density is -0.02 C/m^2. This is due to the fact that, with the decrease of the electrolyte concentration, the surface potential increases at a fixed value of σ (Equation 6).

For situations encountered in biological systems it is of particular importance to consider local concentrations at the membrane surface of minor ionic components present in the bulk solution of a relatively high ionic strength. This is exemplified by enzyme substrates and activators (e.g., Mg^{2+}) dissolved in cellular or extracellular fluids. To calculate this, Equation (3) is used. For a mixture of a monovalent and a divalent electrolyte it can be rearranged as follows:

$$2C''_\infty \cosh^2(F\psi_s/RT) + C'_\infty \cosh(F\psi_s/RT) - \left(2C''_\infty + C'_\infty + \frac{\sigma^2}{4RT\epsilon_0\epsilon_r}\right) = 0 \qquad (10)$$

where C'_∞ and C''_∞ are bulk concentrations of monovalent and divalent electrolyte, respectively. The dependence of local concentration, of mono- and divalent cations, present in the bulk solution at 1 mM concentration, on σ is illustrated in Figure 4. It clearly shows that divalent cations are concentrated to a much higher degree than monovalent cations.

These considerations also apply to hydrogen ions whose concentration in the microenvironment close to a charged surface depends in a similar way on the surface potential. Rearranging Equation (2), one obtains the following dependence of the local pH on the surface potential

$$pH_0 = pH_\infty + F\psi_s/2.3RT \qquad (11)$$

where pH_∞ and pH_0 are pH values in the bulk solution and at the surface, respectively.

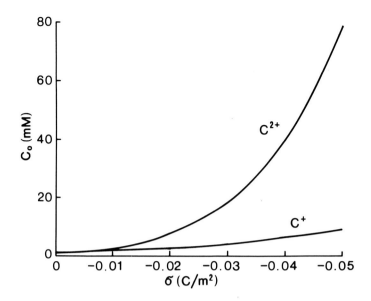

FIGURE 4.   Dependence of the concentration of mono- and divalent cations at the surface (C_0) on the surface charge density (σ). The medium contains 1 mM concentration of the cation in question (C^+ or C^{2+}) and a background monovalent electrolyte at 100 mM concentration.

The Gouy-Chapman theory assumes that ions are point charges and that their interaction with the solid phase is of purely electric nature. However, concentrations of counterions at the surface, as calculated from the assumptions of the Gouy-Chapman theory, are often unrealistically high. Moreover, ions may interact chemically or physically with the surface, so that a tiny layer of immobilized counterions is formed whose properties are not compatible with the Gouy-Chapman theory. This is the *Stern layer* (Figure 5). It also includes oriented dipolar molecules of water. The Stern layer results in a decrease of the surface potential and, in particular cases when "overloading" occurs, may even reverse its sign (Figure 6). The Stern layer usually extends over a few tenths of a nanometer from the interface.

When the liquid phase moves with respect to the solid phase, a thin layer of the liquid, adhering to the solid phase, remains immobile. It usually extends over a distance of a few tenths of a nanometer and is called the *shear layer*. The potential at the plane of the shear layer, designated as ζ potential, determines electrokinetic phenomena at the interface (electrophoresis and electroosmosis). The thickness of the shear layer is assumed to be generally not much different than that of the Stern layer (usually it is slightly larger, see Figures 5 and 6). Therefore, the ζ potential which, for biological membranes and colloidal particles, can be determined by measuring their electrophoretic mobility (Section III.A) can be, with some approximation, regarded as equal to the surface potential at the Stern layer (i.e., the Gouy-Chapman potential).

II. SURFACE CHARGE AND SURFACE POTENTIAL OF BIOLOGICAL MEMBRANES

Biological membranes can be regarded as a solid phase in spite of a certain fluidity of their lipid components and motion of integral proteins. Therefore, the interface between the membrane and the adjacent biological fluid has many properties of a solid-liquid interface. Biological membranes are composed of the lipid core, forming a bimolecular layer, and integral proteins. They are lined with peripheral proteins and glycoproteins. Because of this

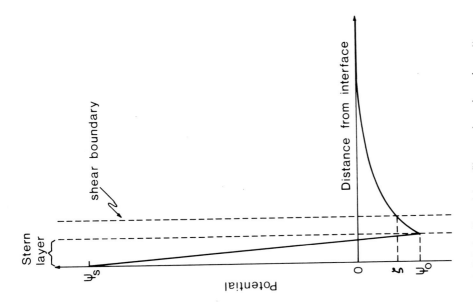

FIGURE 6. Potential profile at a charged membrane "over-loaded" with ions of the opposite sign.

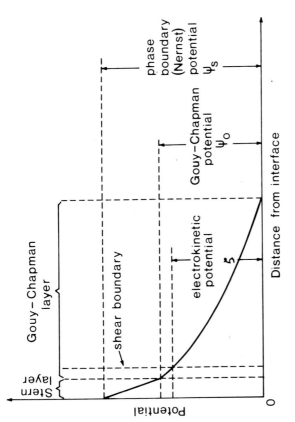

FIGURE 5. Potential profile at a charged surface. Although the total surface potential (Nernst potential, ψ_s) and the Gouy-Chapman potential (potential of the diffuse layer, ψ_0) are clearly differentiated in this scheme, for reason of simplicity and because little information is available on the structure of the Stern layer of biological membranes, no such differentiation will be made in the text and the notation ψ_s will merely apply to the diffuse layer potential. (From Jain, M. K., *The Bimolecular Lipid Membrane: A System*, Van Nostrand Reinhold, New York, 1972, chap. 2. With permission.)

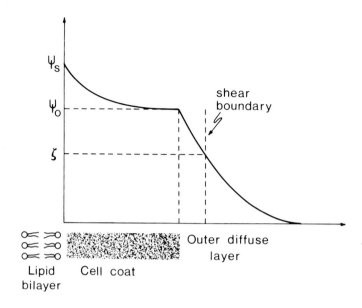

FIGURE 7.   Potential distribution at the cell membrane. (Modified from Dołowy, K., *Cell Adhesion and Motility*, Curtis A. S. G. and Pitts, J. D., Eds., Cambridge University Press, Cambridge, 1980, 39. With permission.)

complex structure[11-13] the charge distribution around a biological membrane is more complicated than in case of a simple solid-liquid interface.[14] Peripheral proteins and divalent cations complexed by polar head groups of membrane phospholipids may be regarded as corresponding to the Stern layer which, in this case, may extend over a few nanometers from the phospholipid bilayer. (Figure 7).

The surface charge of biological membranes is due to ionizable polar groups of phospholipids and integral and peripheral proteins (Table 1).

In most natural membranes the resulting net charge is negative at neutral pH.[8,9] Since the most abundant membrane phospholipids are neutral, it is generally assumed[8] that the negative charge of biological membranes is mainly due to negatively charged amino acid residues of membrane proteins, notably to their ionized free carboxylic groups. This assumption is compatible with the isoelectric point of biological membranes which amounts to 4 to 6 (cf. References 4, 8, 9) and is, therefore, one to three pH units higher than the isoelectric point of negatively charged phospholipids. However, negatively charged phospholipids may contribute significantly to the surface charge if they are present in larger quantities. This has been, for example, observed in erythrocyte membranes, whose internal side is more negative than the external one[15] in parallel with a higher content of acidic phospholipids.

The net negative charge density of biological membranes, calculated from their electrophoretic mobility (see Section III) usually amounts to 0.008 to 0.035 C/m^2 at neutral pH[9,16] which corresponds to one negative charge per 4.5 to 20 nm^2. This does not reflect, of course, the real density of charged groups at the membrane surface, since some of the negative groups are compensated by the positive ones. In artificial phospholipid membranes, e.g., those of liposomes, one polar head group occupies about 0.7 nm^2.[17-19] Based on this, the charge density of such membranes can be calculated from the composition of the phospholipid mixture, i.e., the proportion between negative and neutral phospholipids (see, for example, Fernández[19]). For high concentration of electrolytes (e.g., NaCl) in the surrounding medium, such calculation should be, however, corrected for a weak binding of monovalent cations[18] and anions.[20]

Table 1
MAIN CHARGED GROUPS OF BIOLOGICAL MEMBRANES

Anionic			Cationic		
Group		**Compounds**	**Group**		**Compounds**
$\begin{array}{c} O^- \\ \mid \\ -P-O^- \\ \parallel \\ O \end{array}$	Phosphate monoester	Phosphorylated proteins	$-NH_3^+$	Protonated amino group	Proteins, phospholipids (phosphatidyl-etanolamine)
$\begin{array}{c} \mid \\ -P-O^- \\ \parallel \\ O \end{array}$	Phosphate diester	Phospholipids	$-N^+(CH_3)_3$	Quaternary ammonium cation	Phospholipids (phos-phatidylcholine, sphingomyelin)
$\begin{array}{c} -C-O^- \\ \parallel \\ O \end{array}$	Carboxylic	Proteins, glycopro-teins, phospho-lipids (phosphati-dylserine)	$-NH-C\begin{array}{c} \nearrow NH_2^+ \\ \searrow NH_2 \end{array}$	Guanidine	Proteins

Surface charge density of biological and artificial membranes can be altered under experimental conditions by several factors. An obvious way of altering the surface charge is by changing pH of the surrounding medium.[16] Decreasing pH produces an increased protonation of ionizable polar groups of membrane proteins and phospholipids and, thus, reduces the negative surface charge, whereas making pH more alkaline decreases the protonation and makes the surface charge more negative.

The surface charge density can also be changed by divalent and polyvalent metal cations which can complex phospholipid head groups,[20] thus, neutralizing the negative charge of their phosphate moieties.[21-25] Under favorable conditions they can even make the net surface charge positive (see Section I).

Furthermore, the surface charge can be changed by ionic amphiphilic compounds. Many of them exhibit a high affinity towards the lipid core of the membrane where they become "anchored" by their hydrophobic moiety, whereas the charged hydrophilic groups protrude to the surface and, depending on their charge, either increase or decrease the net negative charge of the membrane. For example, we[25] have modified the surface charge of isolated mitochondria and microsomes by long-chain fatty acids and their CoA thioesters as well as by commercial detergents: dodecylsulfate, cetyltrimethylammonium bromide, and cetylpyridinium chloride. Schäfer et al.[26-29] and Byczkowski et al.[30] observed a decrease of the negative surface charge produced in mitochondrial and artificial membranes by aliphatic and aromatic guanidine derivatives. Other authors[31-35] modified the membrane surface potential by aliphatic long-chain amines, polyamines, local anesthetics, and some other related compounds of cationic character. A similar effect was also obtained with basic proteins.[22,36,37]

Chloride ion[20] and some other inorganic anions can be adsorbed to membranes. This explains the observation[38] that the negative surface charge density of biological membranes is often increased by increasing salt concentration (although the surface potential is decreased in accordance with Equation 6). Strongly lipophilic ions, like tetraphenylphosphonium, tetraphenylboron, or picrate, which become absorbed by the lipid core of the membrane,[39] will also shift their surface potential to more negative or more positive values.

There are indications that some compounds may change the surface potential of cells and organelles, not only when added to their suspension, but also if administered to the animal.

For example, Nałęcz et al.[40] shifted the surface charge of mouse liver mitochondria towards a less negative value by treating the animals with cuprizone (bis-cyclohexanone oxalyldihydrazone).

Digestion of the glycoprotein layer at the surface of various membranes by neuraminidase is another way of changing the surface potential[41-44] (for other references see Sherbet[16]). Other enzymic or chemical modifications of membrane lipids and proteins may be expected to alter the charge density and, hence, the surface potential as well. A well-documented effect is that of the phosphorylation of membrane proteins by endogenous or external protein kinases.[45,46]

Although all these effects have been obtained under experimental conditions and mainly by in vitro treatment of cells and cellular organelles, there is a good reason to expect that they also occur in vivo. In fact, divalent cations Mg^{2+} and Ca^{2+}, fatty acids and their CoA esters, as well as certain other amphiphiles of biological origin may change their concentration in cellular and extracellular media, thus affecting the surface charge of the membranes. Phosphorylation and dephosphorylation of proteins also occur in vivo and are subject to hormonal control. Moreover, phospholipid and protein composition of cell membranes can change within certain limits. Some of those changes may occur under specific physiological or pathological conditions.[16,47-50] On the other hand, the mechanism of action of certain drugs, e.g., the antidiabetic effect of biguanides and other guanidine derivatives, has been ascribed to their effect on the surface charge of biological membranes.[29,51]

III. METHODS OF MEASURING THE SURFACE POTENTIAL OF BIOLOGICAL MEMBRANES

The surface charge density of simple artificial membranes of known phospholipid composition, e.g., liposomes and black films, can be calculated from the area occupied by a single polar head group and its dissociation constant, and, hence, the surface potential can be obtained using Equation (6). This is, however, impossible for biological membranes and more complex model membranes where special measuring methods must be applied. They can be classified into the three following groups: (1) electrophoresis, (2) methods based on the affinity of specific membrane probes, and (3) methods based on changes of the probes bound to the membrane.

A. Electrophoresis

If the particle is large compared to the thickness of the diffuse layer, its electrophoretic mobility (U) is proportional to the surface potential and inversely proportional to the dynamic viscosity of the medium (η). As already mentioned in Section I, membraneous particles move in water solution together with an adhering film of the medium. Therefore, the electrophoretic mobility reflects the potential at the shear layer (ζ potential) and not the true surface potential (ψ_s). This dependence is described by the Smoluchowski equation[52]

$$U = \frac{\epsilon_0 \epsilon_r}{\eta} \zeta \qquad (12)$$

Its derivation assumes that the permittivity, conductivity, and viscosity are the same throughout the diffuse layer and the bulk solution and that the particle is nonconducting.

Since the ζ potential is not much different than the surface potential at the Stern layer (see Section I), electrophoresis is often used to estimate the surface potential of biological particles.[16,47-50,53,54]

The requirement of the Smoluchowski equation that the particle be much larger than the thickness of the diffuse layer holds for all cells and some cellular organelles suspended in a monovalent electrolyte of 0.1 *M* concentration or higher. However, for smaller particles a more general Henry's equation[16] has to be used. If the particle size is very small so that the diffuse layer becomes much larger that the particle diameter, then this equation becomes reduced to the following form:

$$U = \frac{2 \, \epsilon_0 \, \epsilon_r}{3 \, \eta} \, \zeta \qquad (13)$$

The same equation is also valid for larger particles whose electrical conductivity is high, i.e., close to that of the medium.

B. Methods Based on the Affinity of Probes to the Membrane

Many amphiphilic compounds partition between the membrane and the aqueous phase. If the compound is charged, its partitioning is affected by the surface potential, because the local concentration of the compound in the immediate vicinity of the membrane is different than that in the bulk solution. Based on the Boltzmann distribution (Equation 2), the following dependence can be formulated

$$K_d' = K_d^0 \exp(zF\psi_s/RT) \qquad (14)$$

where K_d^0 is the dissociation constant of the compound to the uncharged membrane (the so-called "intrinsic dissociation constant"), K_d' is the apparent dissociation constant to the membrane bearing a surface charge ψ_s, and z is the number of charges on the molecule of the compound. Equation (14) can be used to calculate the absolute surface potential if both K_d^0 and K_d' can be measured. In practice, the dissociation constant in the presence of high salt concentration can be assumed to approach the K_d^0 value[55] because of screening the surface potential by electrolytes (see Equation 6). However, even at 1 *M* concentration of a monovalent electrolyte the remaining surface potential may have a significant magnitude, especially at high surface charge density (Figure 2). On the other hand, the probe itself may also contribute to the surface charge of the membrane to which it becomes bound.[56] Therefore, Gibrat et al.[57] proposed two methods of analysis of experimental data which allow to overcome these difficulties and to calculate the intrinsic dissociation constant and, hence, the absolute surface potential. In a much simpler way charged affinity probes can be used to estimate relative values of the surface potential and, in particular, of changes of this potential according to the following formula:

$$\Delta\psi_s = \frac{RT}{zF} \ln \frac{K_d''}{K_d'} \qquad (15)$$

where $\Delta\psi_s$ is a difference, or a change, of the surface potential between two states of the membrane ($\Delta\psi_s = \psi_s'' - \psi_s'$), and K_d' and K_d'' are corresponding apparent dissociation constants.

Applicability of amphiphiles as probes of the surface potential requires that their partitioning between the membrane and the aqueous medium be easily and accurately determined. This condition is fulfilled by certain fluorescent compounds, e.g., 8-anilino-1-naphthalene sulfonate (ANS), 6-toluidyl-1-naphthalene sulfonate (TNS), and ethidium bromide whose fluorescence spectrum and/or quantum yield depend on the hydrophilicity or hydrophobicity of the environment.[58,59] In case of anionic probes, ANS and TNS, a blue shift of the emission maximum and a considerable increase of the quantum yield upon the transfer of the probe from aqueous into a hydrophobic medium allow to neglect their fluorescence in water and to assume all measured fluorescence practically to originate from the probe bound to the membrane.[58-60] Thus, apparent K_d values can easily be calculated from plots of the reciprocal of fluorescence against the reciprocal of probe concentration or fluorescence against fluorescence/probe concentration. Erroneous results can, however, be obtained if the probe is associated with the membrane in such a way that only a part of the bound probe is located in the hydrophobic region of the membrane, the other part being associated with hydrophilic regions and, therefore, exhibiting no increase of the fluorescence.[61] Application of fluorescent probes implies that they are associated with one side of the membrane only and do not penetrate across the membrane.[18] However, if the membrane is permeable for the probe so that it can be accumulated inside membraneous vesicles due to a transmembrane diffusion[62,63] or electrochemical[64] potentials, interpretation of the results in terms of the surface potential should be made with caution.[61]

Recently, methylene blue has been proposed as a probe for the surface potential.[65] This cationic dye partitions between the water phase and the membrane according to the surface potential. In the membrane, it forms dimers which can be distinguished from monomers by a reduced absorbance of visible light.

9-Aminoacridine is widely used as an indicator of pH gradient across membranes.[58] However, under specific conditions, e.g., in low-cation media, it can also monitor the surface potential of thylakoid[66,67] and mitochondrial[68] membranes, being accumulated at their surface according to the surface potential. Such accumulation causes a quenching of the fluorescence which can be measured.

Another group of surface potential probes based on affinity changes includes charged amphiphilic compounds containing a paramagnetic moiety in their hydrophilic part. Since the EPR signal of such compounds is altered when they are bound to membranes,[69-71] their partitioning between the aqueous phase and the membrane can easily be determined. Change of the surface potential is then calculated from the equation[70,71]

$$\Delta\psi_s = \frac{RT}{zF} \ln \frac{P'}{P''} \tag{16}$$

where P' and P'' are partition coefficients (water vs. membrane) for two different membrane states or conditions.

Since charged amphiphilic probes, when bound to the membrane, alter its surface charge, they must be used at very low concentrations[56] to obtain reliable figures for calculating their apparent K_d values. It should also be mentioned that some authors[27] proposed to use a change of the number of binding sites for ANS, and not a change of apparent K_d, as a measure of the change of surface charge density.

C. Methods Based on Changing Properties of Membrane-Bound Probes

To this group of probes belong compounds which change their optical properties (absorption spectrum or fluorescence) upon protonation/deprotonation. They can be regarded as membrane-bound pH indicators.[72-74] Since the local pH at the membrane surface is a function of the surface potential (Equation 11), the pH difference between the bulk solution (pH_∞) and the membrane surface (pH_0) enables us to calculate the actual absolute surface potential

$$\psi_s = \frac{2.3\ RT}{F} (pH_0 - pH_\infty) \tag{17}$$

A good candidate for such a probe appeared to be 4-heptadecylumbelliferone (4-heptadecyl-7-hydroxycoumarin).[19,73,74] Its insolubility in water and solubility in nonpolar media enable its complete incorporation into membranes. Then, its polar moiety protrudes to the surface and exhibits a fluorescence in the anionic (deprotonated) form only. By acid/base titration of membranous particles containing the probe, its apparent pK at the charged membrane (pK_{ch}) can be determined. This allows us to calculate the particle surface charge from the formula[19]

$$\psi_s = \frac{2.3\ RT}{F}\ (pK_n - pK_{ch}) \qquad (18)$$

where pK_n is the pK value of the probe in a neutral membrane.

Similar measurements have also been made using dansylated derivatives of phospholipids incorporated into lipid membranes.[75]

The advantage of probes of this kind is their ability to monitor absolute values of the surface potential and not only its difference or change.

A disadvantage of 4-heptadecylumbelliferone is, however, a virtual difficulty of incorporating this compound into natural membranes. Therefore, it has been applied so far to measure the surface potential of artificial phospholipid membranes only.[19]

IV. EFFECT OF THE SURFACE POTENTIAL ON ENZYME KINETICS

Since the concentration of anions and cations in the vicinity of a charged membrane differs from their concentration in the bulk solution (Section I), kinetic characteristics of membrane-associated enzymes should also be expected to depend on the surface potential in case the substrate is either cationic or anionic. Katchalski and co-workers[76-79] were the first to study a model system in which soluble proteolytic enzymes were immobilized on artificial poly-anionic or polycationic matrices, and differently charged synthetic oligopeptides were used as substrates. They observed that the affinity of the enzyme to its substrate in this system depended on the charge of the solid matrix as well as on the charge of the substrate molecule. It was found that, for cationic substrate analogs, apparent K_m was decreased and/or k_{cat} was increased when the enzyme was immobilized on a polyanionic support, whereas opposite changes occurred upon using a polycationic matrix. For anionic substrates, all these dependences were reversed.

If K_m^0 designates the Michaelis constant of an enzyme attached to an uncharged solid matrix ("true" K_m), then the apparent K_m (K_m') of this enzyme immobilized on a charged matrix could be deduced from the Boltzmann equation (Equation 2) as follows:[76]

$$K_m' = K_m^0\ \exp(zF\psi_s/RT) \qquad (19)$$

where ψ_s is the surface potential of the solid matrix and z is the charge of the substrate.

The same authors observed that the surface charge of the solid support not only altered the apparent K_m of immobilized enzymes, but it also shifted their pH optima.[76-79] This can also be explained on the basis of the Gouy-Chapman theory, since the surface potential changes the local concentration not only of charged substrate molecules, but also of hydrogen ions at the surface (Equation 11). In agreement with this, pH optima were displaced towards more alkaline values by negatively charged solid supports and towards more acidic pH by positively charged supports. These shifts of pH profiles were much smaller or disappeared when the ionic strength of the medium was increased. Similar changes of both the apparent K_m values and pH profiles were observed[80] with soluble polyanionic and polycationic derivatives of chymotrypsin, thus, fully confirming the assumption that these effects were due to intermolecular electrostatic interactions.

These studies do not imply that changes of enzyme kinetics on immobilization are solely due to electrical effects. In fact, several other factors resulting from immobilization may affect enzyme properties.[79,81] The importance of the pioneering work of Katchalski's group[79] is, however, that these authors drew, for the first time, the attention towards electric interactions at interfaces in regulating enzyme activities.

Studies in this line were subsequently made by other authors. For example, Remy et al.[82] found that immobilization of hexokinase on an artificial negatively charged membrane increased the apparent K_m for ATP (anionic substrate) and decreased the apparent K_a for the cationic activator Mg^{2+}. On the other hand, the apparent K_m of acetylcholinesterase for its cationic substrate was decreased. Beitz et al.[83] attached pyruvate decarboxylase to a positively charged support and found a decrease of apparent K_m for pyruvate. Moreover, there was an excellent agreement between the surface potential calculated from K_m values (Equation 19) and that computed from the positive charge density of the solid support.

Soluble enzymes immobilized on solid supports can thus be regarded as models for enzymes bound to biological membranes. A more systematic study on the kinetics of membrane-associated enzymes in relation to the surface potential was performed by the authors of the present chapter. In these investigations the surface potential was modified by addition of small amounts of surface-active agents or salts of divalent metal cations,[25] by changing pH of the medium,[25] by altering phospholipid composition of the membranes,[84] or by phosphorylating membrane proteins with the use of externally added or endogenous protein kinases.[45,46] It was observed[25,45,84] that enzymes reacting with negatively charged substrates, like arylsulfatase, glycerol-3-phosphate dehydrogenase, and glucose-6-phosphatase, were competitively inhibited by factors making the membrane more negative and activated by changing the membrane potential towards less negative values (Figure 8). In case of membrane-bound enzymes reacting with positively charged substrates, e.g., acetylcholinesterase, monoamine oxidase, and dimethylaniline oxidase, the opposite relationship was found, namely, the enzymes were inhibited competitively under conditions changing the membrane surface potential to less negative values and stimulated when the membrane became more negative[25,45,46,84] (Figure 9).

In these studies apparent K_m values were altered in agreement with predictions based on the work of Goldstein et al.,[76] whereas, in general, no significant change of the maximum reaction rate at saturating substrate concentration (V_{max}) was observed. Moreover, the effects on enzyme kinetics disappeared when the membranes became solubilized by detergents and reappeared after reconstitution of the membranes with externally added lipids.[25,84] It was also found[10] that the apparent K_m of microsomal pyrophosphate-glucose phosphotransferase for glucose (uncharged substrate) was not altered by changing the surface potential, whereas apparent K_m for pyrophosphate (anionic substrate) was increased or decreased when the microsomal membrane was made more or less negative, respectively.

Reconstitution experiments provide a further support to the view that the kinetics of membrane-bound enzymes are controlled by the surface potential. It has been observed[84] that apparent K_m depends on the phospholipid composition of the reconstituted membranes. For enzymes reacting with anionic substrates it is higher in negatively charged than in neutral phospholipids. The opposite is true for enzymes reacting with cationic substrates.

The surface potential can affect enzyme kinetics not only by altering the local concentration of substrates, but also that of charged activators. For example, the activity of enzymes which require Ca^{2+} or Mg^{2+} may be affected by the membrane surface potential since the local concentration of these cations largely depends on this potential (Figure 4). The inhibition of $(Na^+\text{-}K^+)$-ATPase by inorganic and organic cations[85,86] as well as positively charged amphiphiles[87] could also be explained by a diminution of the local concentration of Na^+, K^+, and Mg^{2+} due to a decrease of the negative surface potential. The activatory effect of the acidic phospholipid, phosphatidylserine,[88] is also compatible with this explanation.

On the other hand, some effects of divalent cations, which have been ascribed to their specific activatory action on enzymes, may be simply explained by diminution of the surface

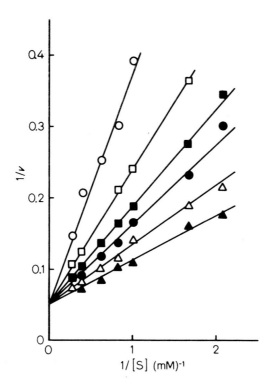

FIGURE 8. Effect of surface-active agents on arylsulfatase activity in microsomal membranes. Double reciprocal plots. ●, Control; ○, 100 μ*M* oleate; □, 100μ*M* dodecylsulfate; ■, 40 μ*M* palmitoyl-CoA; △, 150 μ*M* cetyltrimethylammonium bromide; ▲, 150 μ*M* cetylpyridinium chloride. Enzyme activity (V) is expressed in nmol *p*-nitrophenyl sulfate hydrolyzed per minute per milligram protein. (From Wojtczak, L. and Nałęcz, M. J., *Eur. J. Biochem.*, 94, 99, 1979. With permission.)

potential. This is likely to be the case with mitochondrial glycerol-3-phosphate dehydrogenase, the enzyme known to be activated by Ca^{2+} and Mg^{2+} and competitively inhibited by long-chain acyl esters of coenzyme A and long-chain free fatty acids.[89-92] All these substances were at first thought to be specific effectors of the enzyme. It was found, however, that glycerol-3-phosphate dehydrogenase was also activated by small amounts of cationic detergents and competitively inhibited by low concentrations of dodecylsulfate and other negatively charged amphiphiles.[25] Moreover, the cationic detergent cetyltrimethylammonium bromide, when added in stoichiometric amounts, was able to abolish the inhibition produced by palmitoyl-CoA. It seems, therefore, that these regulatory effects concerning glycerol-3-phosphate dehydrogenase may be sufficiently explained by electrostatic interactions at the membrane surface, without the need of postulating some more specific effects on the enzyme itself.

The fact that changes of the surface potential influence the activity of membrane-bound enzymes provides explanation of a number of earlier observed effects and may cast an additional light on regulatory processes occurring in biological membranes. For example, the activatory effect of high salt concentration on membrane enzymes reacting with anionic substrates, e.g., mitochondrial ATPase,[93] might be explained by decreasing the negative surface potential ("screening effect", Equation 6) and thus increasing the local concentration of the substrate at the membrane surface. More recently, Palmer and co-workers[94-96] have

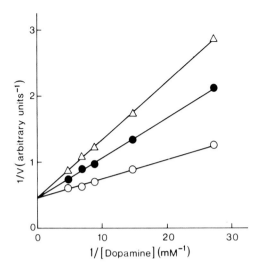

FIGURE 9.   Effect of surface-active agents on mitochondrial monoamine oxidase. Double reciprocal plots. ●, Control; ○, 69 μ*M* oleate; △, 200 μ*M* cetyltrimethylammonium bromide. (From Nałęcz, M. J. and Wojtczak, L., *Postepy Biochem. (Warsaw)*, 28, 191, 1982. With permission.)

clearly shown that the stimulatory effect of high ionic strength on the oxidation of external NADH by mitochondria from higher plants and microorganisms is, in fact, due to screening of the negative surface charge.

Some effects of pH on membrane enzymes can also be explained in terms of the role played by the surface potential. For example, Benga and Borza[97] and Douzou and Maurel[98] observed a shift of pH optimum of membrane-bound cytochrome oxidase by increasing strong electrolyte concentration (Figure 10). For better understanding this result, the following dependence has to be recalled. At high ionic strength, the surface potential of the inner mitochondrial membrane is largely screened and, therefore, the local pH at the membrane is not much different than that of the bulk solution. The observed pH optimum under these conditions corresponds roughly to the "intrinsic" pH optimum of the enzyme. However, at lower KCl concentrations, the surface potential resumes a substantial negative value, which results in lowering of the local pH at the membrane. This explains why the bulk pH must then be higher in order to keep the local pH at its optimum value for the enzyme (see Equation 11).

Another observation in this line is that made by ourselves[25] that the pH optimum of mitochondrial glycerol-3-phosphate dehydrogenase is shifted to more alkaline values when the surface charge of mitochondrial membranes is partly neutralized by an amphiphilic cation (Figure 11A). The mechanism of this effect is, however, somewhat different than that described before. Acidification of the medium decreases the surface charge density by decreasing the dissociation of anionic groups and increasing the protonation of cationic groups and thus diminishes the repulsion of the negatively charged substrate from the membrane. When the surface charge is neutralized by a cationic detergent, pH optimum is shifted to a more alkaline value. A support for such a mechanism is given by the observation that cetyltrimethylammonium and changes of the pH of the medium altered the apparent K_m of the enzyme, whereas V_{max} remained unchanged (Figure 11B). Since the latter value is the extrapolated enzyme activity at infintely high substrate concentration, it should not be affected by local repulsion or attraction.

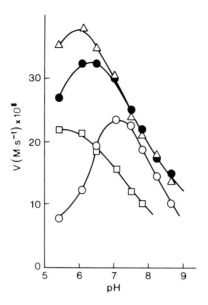

FIGURE 10. Effect of pH on cytochrome oxidase in the inner mitochondrial membrane. The enzyme activity was measured at the following KCl concentrations: ○, 10 mM; ●, 60 mM; △, 110 mM; □, 210 mM (From Douzou, P. and Maurel, P., *C. R. Acad. Sci. Paris*, 282 D, 2107, 1976. With permission.)

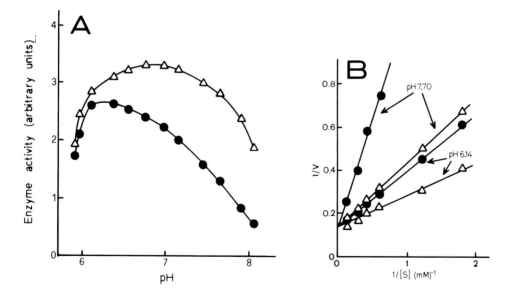

FIGURE 11. Effect of pH and cetyltrimethylammonium on the activity of glycerol-3-phosphate dehydrogenase in insect muscle mitochondria. (A) pH range from 5.95 to 8.05; (B) double reciprocal plots for two pH values. ●, No additions (control); △, with 34 μM cetyltrimethylammonium bromide. (From Wojtczak, L. and Nałęcz, M. J., *Eur. J. Biochem.*, 94, 99, 1979. With permission.)

The effects of detergents on membrane enzymes have been the subject of numerous studies.[99-102] Some of these effects may be mediated by changes of the surface potential evoked by sublytic concentrations of ionic detergents. The same can be postulated for other charged amphiphiles and polycations and polyanions. As mentioned before, our observations[25] on the action of charged surfactants on mitochondrial and microsomal enzymes are fully compatible with such interpretation (Figures 8 and 9). Inhibition of cytochrome oxidase by polycations[103] and activation by polyanions[104] concomitant with a shift of pH optimum, and inhibition by polyanions of monoamine oxidase[34] are other examples of this mechanism, since the substrates of both enzymes have a cationic character.

Inhibition by guanidine derivatives of mitochondrial respiration and its reversal by fatty acids[105] may be due to a similar mechanism as well. Finally, some effects of local anesthetics of cationic nature on liver adenylate cyclase can also be interpreted in terms of the surface charge.[106]

It has to be stressed, however, that many of the effects of detergents and other charged amphiphiles on enzymes cannot be attributed to the surface potential changes. This applies, in particular, to multienzyme complexes such as the respiratory chain,[34,107,108] photosynthetic systems,[9,109] and mitochondrial ATPase[110] where more complex relationships have been observed.

It is important to mention here a possible involvement of membrane lipids in electrostatic interactions influencing the activity of membrane-bound enzymes. The effect of lipids on enzymes located in biological membranes is usually discussed as specific molecular interactions between lipids and proteins, involving formation of functional complexes and maintaining an active conformation of enzymes, or in terms of lipid-dependent fluidity of the membrane, affecting rotational motion and lateral diffusion of membrane components.[111,112] However, besides these important aspects of lipid influence on membrane structure and functions, we also postulate that lipids may play a major role by regulating electrostatic interactions at the membrane surface. As already mentioned, phospholipid composition of reconstituted membranes determines kinetic properties of incorporated enzymes.[25,84] This suggests that lipid composition of biological membranes and, in particular, the content of acidic phospholipids may play an important role in regulating membrane enzymes.

It is interesting to note that factors such as pH, ionic strength, charged amphiphiles, and phospholipid composition may affect enzymes kinetics not only in systems of membrane-associated enzymes and their soluble substrates, but also in case of soluble enzymes acting on insoluble, e.g., membrane-bound, substrates. This has been, for example, observed with sialidase,[113] ribonuclease,[114] lysozyme,[115] and phospholipase.[116]

Another interesting variety of the regulatory function of the surface potential is its effect on the local concentration at the membrane of enzyme inhibitors. This is exemplified by mitochondrial NADH dehydrogenase which is inhibited by its product NAD^+.[117] At acidic pH which exists close to a negatively charged membrane (Equation 11), the latter compound forms a monovalent cation and, therefore, its concentration at the membrane vicinity depends on the surface potential. In agreement with this, we[25,118] observed that the inhibition by NAD^+ of NADH oxidation in submitochondrial particles was dependent on the surface potential.

Equation (19) could be used to determine the absolute value of the surface potential from enzyme kinetics if K_m^o, i.e., the "true" Michaelis constant, were known. Wojtczak et al.[119] have assumed that this is the case for enzymes which are set free from the membrane by solubilization. Although such assumption is not free from objections, the values which they obtained agree within reasonable limits with those calculated from electrophoretic mobility.

From Equation (19) one can derive a formula which enables to calculate a difference of the surface potentials ($\Delta\psi_s = \psi_s^{''} - \psi_s^{'}$) between two different states of a membrane or different conditions, if corresponding apparent Michaelis constants, $K_m^{'}$ and $K_m^{''}$, of a membrane-bound enzyme are known:[25]

$$\Delta\psi_s = \frac{RT}{zF} \ln \frac{K_m^{''}}{K_m^{'}} \tag{20}$$

Table 2
**EFFECT OF SURFACE-ACTIVE AGENTS ON THE SURFACE POTENTIAL,
AS CALCULATED FROM DISSOCIATION CONSTANT FOR 8-ANILINO-1-
NAPHTHALENE SULFONATE (EQUATION 15), PARTITIONING OF SPIN-
LABELED PROBE CAT$_{12}$ (EQUATION 16), AND CHANGE OF APPARENT K$_m$
OF TWO ENZYMES (EQUATION 20)**

Biological material and enzyme	Surfactant added	ψ_s (mV) calculated from		
		K$_d$ for ANS	Partition of CAT$_{12}$	Apparent K$_m$
Rat liver microsomes, arylsulfatase C	Cetyltrimethylammonium bromide, 150 μM	+ 10	+ 10	+ 8
	Cetylpyridinium chloride, 150 μM	+ 10		+ 14
	Oleate, 100 μM	− 12		− 18
	Na-dodecylsulfate, 100 μM		− 13	− 11
Rat liver mitochondria, monoamine oxidase	Cetyltrimethylammonium bromide, 71 μM	+ 9	+ 9	+ 8
	Cetylpyridinium chloride, 39 μM	+ 17		+ 22
	Oleate, 69 μM	− 16		− 19

From Nałęcz, M. J. and Wojtczak, L., *Postepy Biochem. (Warsaw)*, 28, 191, 1982. With permission.

A fairly good agreement of $\Delta\psi_s$ calculated in this way for a number of natural and reconstituted membranes with values for $\Delta\psi_s$ obtained from ANS binding and partitioning of the spin-labeled probe (Table 2) is the main argument for the applicability of this formula.

On the basis of what has been said, it is clear that the membrane surface potential is involved in modulation of activities of membrane-bound enzymes. It has to be stressed, however, that the model proposed by Katchalski et al.[76-80] and ourselves[10,25,84] is not always suitable to describe changes in enzyme kinetics supposed to be induced by changes in electrostatic interactions at the membrane surface. The main restriction of this model is the use of the Michaelis-Menten kinetics to describe the relationship between K$_m$ and ψ_s. Also, some oversimplifications, like the assumption that the surface charge is uniformly distributed over the membrane or that the charged substances involved (substrates, products, activators, and inhibitors) do not interact specifically with membrane components, are not always acceptable. On the other hand, the concept that electrostatic interactions are important factors in controlling enzyme activity can be extended over soluble systems as well.[80,104,120,121]

V. EFFECT OF THE SURFACE POTENTIAL ON MEMBRANE TRANSPORT

The effect of the surface charge on membrane permeability and transport has long attracted attention. Fundamental research in this line has been made by McLaughlin et al.[122] They observed that the permeability of bimolecular phospholipid membranes for the cationic nonactin-K$^+$ complex was higher with negatively charged phospholipids, phosphatidylserine and phosphatidylglycerol, than with neutral phosphatidylethanolamine. The reverse relationship was observed with respect to the permeability for the negatively charged I$_5^-$ complex. Similar observations were made subsequently by other authors[26,123,124] for liposomal membranes and black lipid films. An interesting model is represented by asymmetric membranes, i.e., membranes whose phospholipid composition on either side is different. It was found[125] that the complex formation between valinomycin present in such membrane and K^+ from the aqueous phase, and, hence, the conductance of the membrane, was higher when K^+ was added at the negatively charged side of the membrane than at its neutral side.

Ionophore-mediated permeability is also affected by factors changing the surface potential in natural membranes. For example, valinomycin-mediated transport of K^+ through mitochondrial membranes is competitively inhibited by magnesium ions[126] and cationic local anesthetics.[127]

All these effects are obvious consequences of the fact that the local concentration of transported ion in the immediate vicinity of the membrane depends on the surface potential (Section I). A survey of studies on the transport phenomena in a variety of biological membranes allows to conclude that many of the observed effects can be explained in these terms. The passive permeability to, and the transport of, inorganic ions are, perhaps, best examples. As already described (Section IV), the effect of salts, charged amphiphiles, and phospholipids on (Na^+-K^+)-ATPase is, at least partly, due to changing the surface potential.[85-88] It should be then expected that the same factors may respectively increase or decrease the transport of Na^+ and K^+ across the membrane, mediated by this enzyme. The effect of pH on Na^+ and K^+ conductance of frog nerves has also been attributed[128] to changes of the surface potential. Since local anesthetics are mostly positively charged amphiphiles (however, some of them may be negatively charged), their effect on nerve excitability[129] could also be explained by blocking Na^+ and K^+ transport.

Extensive studies have been carried out on inorganic ion transport through the yeast plasma membrane by Borst-Pauwels and co-workers.[21,130-135] They found that factors changing the surface potential in either direction either increased or decreased the uptake by yeast cells of Rb^+, Ca^{2+}, Sr^{2+}, phosphate, and sulfate. Inhibition of the inorganic cation uptake in yeast by positively charged compounds exhibiting a high affinity to membranes, guanidine derivatives and ethidium bromide, was also observed by Peña.[136,137]

Because many carrier-mediated transport processes can be described by the Michaelis-Menten kinetics, relationship presented in Section IV between a change of the surface potential and that of enzyme kinetics (Equations 19 and 20) should also hold for membrane transport. Theoretical considerations of this problem have been presented by Theuvenet and Borst-Pauwels[138] and a quantitative experimental approach for Rb^+ uptake by yeast has recently been made by the same authors.[134]

An increased permeability of the inner mitochondrial membrane to K^+ and other alkali metal cations is produced by long-chain fatty acids and their CoA esters[139] (for review see also Wojtczak[140]). On the other hand, Mg^{2+} associated with the membrane maintains its permeability to these monovalent cations at a low level[141,142] (review by Wojtczak[143]) and counteracts the effect of fatty acids.[139] These effects are best understood in terms of the surface potential changes[143] which either increase or decrease the local concentration of monovalent cations at the membrane surface, thus increasing or decreasing the rate of their transport.

Inhibitory effect of Mg^{2+} and spermine on Ca^{2+} uptake by mitochondria[32] can also be discussed in terms of the surface potential changes.

The inner membrane of mitochondria contains transport systems for important metabolites, like inorganic phosphate,[144] di- and tricarboxylic respiratory substrates,[145] and adenine nucleotides.[146,147] Numerous studies have shown that all these carrier systems are strongly affected by factors expected to alter the membrane surface potential. A strong inhibition of the adenine nucleotide carrier by free fatty acids[148,149] and acyl-CoA esters[150-153] is well recognized. Although these compounds may have a specific effect on the adenine nucleotide translocator,[140,153,154] there are indications that the inhibition is also due to the increased surface potential. This is indicated by the fact that the inhibition can be partly overcome by factors decreasing the negative surface potential, as cationic detergents,[154] Mg^{2+}, and high

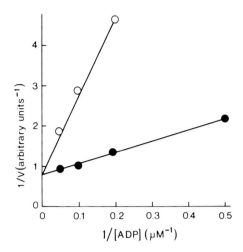

FIGURE 12.   Effect of dodecylsulfate on the transport of ADP through the inner mitochondrial membrane. Double reciprocal plot. ●, Control; ○, Na-dodecylsulfate 60 nmol/mg mitochondrial protein. (From Duszyński, J. and Wojtczak, L., *FEBS Lett.*, 40, 72, 1974. With permission.)

ionic strength.[155] On the other hand, anionic detergents[154] (Figure 12), lipophilic anions,[156] and polyanions[157,158] also inhibit this transport in a competitive way. The dependence of the adenine nucleotide translocator on the surface potential has recently been directly documented by Krämer.[159] He showed that the effect of surface charges was weak on the cytosolic side of the inner mitochondrial membrane, whereas it was rather strong on the matrix side.

Also, the carrier systems for di- and tricarboxylate anions in the inner mitochondrial membrane are inhibited by long chain acyl-CoA.[160] Similarly, as in case of the adenine nucleotide translocase, this inhibition is competitive and strongly depends on the chain length of the fatty acid moiety.[153] The latter dependence probably reflects variations in the affinity of various acyl-CoA esters towards mitochondrial membranes. The lipophilic anion tetraphenylboron[156] and a synthetic polyanion[158] also inhibit the carrier-mediated di- and tricarboxylic anion transport in mitochondria. On the other hand, the cationic detergent cetylpyridinium chloride in sublytic concentrations potentiates mitochondrial transport of pyruvate by decreasing its apparent K_m value[10] (Figure 13). In addition, it has been shown[161,162] that mitochondrial anion translocators are activated (lowering of apparent K_m without a change of V_{max}) by increasing bulk electrolyte concentration and decreasing pH. Moreover, mitochondrial phosphate carrier is also inhibited by acyl-CoA esters,[153] free fatty acids,[163] polyanions,[158] and tetraphenylboron.[156] All these observations strongly suggest that all mitochondrial carrier systems for anionic metabolites are highly susceptible to variations of the surface potential. It is also interesting to note that K_i values of acyl-CoA esters[153] and of the polyanion of König et al.[157,158] are the lowest for the adenine nucleotide translocator and the tricarboxylate carrier, higher for dicarboxylates, and the highest for phosphate. This corresponds to the decreasing valency of corresponding anions of these metabolites at neutral pH. As explained in Section I (Equations 1 and 2), the effect of the surface potential on the local concentration of an ion at the interface depends on its valency (see also Figure 4).

It has, however, to be stressed that factors changing the surface potential of membranes may also affect their molecular structure, e.g., fluidity of the phospholipid core. Therefore, their effect on membrane permeability is not always explainable solely in terms of the surface

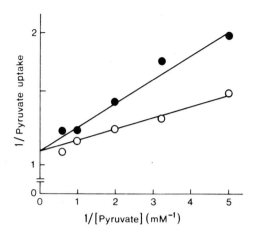

FIGURE 13. Effect of cetylpyridinium on pyruvate transport in mitochondria. Double reciprocal plot. The uptake of pyruvate by rat liver
mitochondria was measured by the rapid centrifugation technique and expressed in nmol/mg mitochondrial protein during the time of passage of
mitochondria through a layer containing [^{14}C]pyruvate (approximately 20
sec). ●, Control; ○, cetylpyridinium chloride 12 nmol/mg mitochondrial
protein. (From Nałęcz, M. J. and Wojtczak, L., *Postepy Biochem. (Warsaw)*, 28, 191, 1982. With permission.)

potential. This is especially true for complex systems such as, for example, proton pump
of the inner mitochondrial membrane. While Papa et al.[127] observed an increase of the
respiration-driven proton transport by the cationic local anesthetic dibucaine, Schäfer
et al.[26-28] found an inhibition of this transport by guanidine derivatives, also bearing a
positive charge. In both cases, however, oxidative phosphorylation was impaired. Apparently, the mitochondrial apparatus for energy conservation depends in a more complex way
on the surface potential than the carrier systems for cations and metabolites.

VI. CONCLUDING REMARKS

The surface charge and surface potential of biological membranes can influence their
permeability, transport processes, and enzymic function. Since the surface potential can be
changed by electrolyte composition of the surrounding medium, its pH, the presence of
polyvalent ions, and charged amphiphilic compounds, both of endogenous origin and introduced into the organism, e.g., as drugs, it may be an important controlling factor in cellular
metabolism. It has to be remembered, however, that the surface potential, as measured by
membrane probes or electrophoretic mobility, reflects an average distribution of charges on
the membrane surface. If charges are not uniformly spread over the membrane, the potential
sensed by membrane enzymes and carriers may differ considerably from its average value.
This fact implies a considerable limitation in experimental and theoretical approach to the
problem. On the other hand, it should also be stressed that electrostatic interactions are of
an unspecific character and thus may often form only a background for other, more specific,
regulatory mechanisms. Changes of membrane fluidity and hydratation, the mobility of
membrane components, and sensitivity to specific modulators are some of the phenomena
which, together with electrostatic interactions, may eventually give a complete picture of
the extremely complicated system referred to as "the biological membrane".

REFERENCES

1. **Gouy, M.,** Sur la constitution de la charge électrique a la surface d'un électrolyte, *J. Phys. (Paris),* 9, 457, 1910.
2. **Chapman, D. L.,** A contribution to the theory of electrocapillarity, *Philos. Mag.,* 25, 475, 1913.
3. **Overbeek, J. Th. G.,** Electrochemistry of the double layer, in *Colloid Science,* Vol. 1, Kruyt, H. R., Ed., Elsevier, Amsterdam, 1952, chap. 4.
4. **Delahay, P.,** *Double Layer and Electrode Kinetics,* John Wiley & Sons, New York, 1965.
5. **Shaw, D. J.,** *Introduction to Colloid and Surface Chemistry,* Butterworths, London, 1966.
6. **Jain, M. K.,** *The Bimolecular Lipid Membrane: A System,* Van Nostrand Reinhold, New York, 1972, chap. 2.
7. **Träuble, H.,** Membrane electrostatics, in *Structure of Biological Memebranes,* Abrahamson, S. and Pascher, I., Eds., Plenum Press, New York, 1977, 509.
8. **McLaughlin, S.,** Electrostatic potentials at membrane-solution interfaces, in *Current Topics in Membrane Transport,* Vol. 9, Bonner, F. and Kleinzeller, A., Eds., Academic Press, New York, 1977, 71.
9. **Barber, J.,** Membrane surface charges and potentials in relation to photosynthesis, *Biochim. Biophys. Acta,* 594, 253, 1980.
10. **Nałęcz, M. J. and Wojtczak, L.,** Ladunek powierzechniowy blon biologicznych i jego funkce regulacyjne, *Postepy Biochem. (Warsaw),* 28, 191, 1982.
11. **Malhotra, S. K.,** Organization, composition, and biogenesis of animal cell membranes, in *Membrane Structure and Function,* Vol. 1, Bittar, E. E., Ed., John Wiley & Sons, New York, 1980, chap. 1.
12. **Danielli, J. F.,** Experiment, hypothesis, and theory in development of concepts of cell membrane structure 1930—1970, in *Membrane and Transport,* Vol. 1, Martonosi, A. N., Ed., Plenum Press, New York, 1982, 3.
13. **Sjöstrand, F. S.,** Models for molecular architecture of cell membranes, in *Structure and Properties of Cell Membranes,* Vol. 1, Benga, Gh., Ed., CRC Press, Boca Raton, Fla., 1985, chap. 1.
14. **Dołowy, K.,** A physical theory of cell-cell and cell-substratum interactions, in *Cell Adhesion and Motility,* Curtis, A. S. G. and Pitts, J. D., Eds., Cambridge University Press, Cambridge, 1980, 39.
15. **Lin, G. S. B., Macey, R. I., and Mehlhorn, R. J.,** Determination of the electric potential at the external and internal bilayer-aqueous interfaces of the human erythrocyte membrane using spin probes, *Biochim. Biophys. Acta,* 732, 683, 1983.
16. **Sherbet, G. V.,** *The Biophysical Characterisation of the Cell Surface,* Academic Press, London, 1978.
17. **Small, D. M.,** Phase equilibria and structure of dry and hydrated egg lecithin, *J. Lipid Res.,* 8, 551, 1967.
18. **Eisenberg, M., Gresalfi, T., Riccio, T., and McLaughlin, S.,** Adsorption of monovalent cations to bilayer membranes containing negative phospholipids, *Biochemistry,* 18, 5213, 1979.
19. **Fernández, M. S.,** Determination of surface potential in liposomes, *Biochim. Biophys. Acta,* 646, 23, 1981.
20. **Grasdalen, H., Eriksson, L. E. G., Westman, J., and Ehrenberg, A.,** Surface potential effects on metal ion binding to phosphatidylcholine membranes. ^{31}P NMR study of lanthanide and calcium ion binding to egg-yolk lecithin vesicles, *Biochim. Biophys. Acta,* 469, 151, 1977.
21. **Theuvenet, A. P. R. and Borst-Pauwels, G. W. F. H.,** Surface charge and the kinetics of two-site mediated ion-translocation. The effects of UO_2^{2+} and La^{3+} upon Rb^+-uptake into yeast cells, *Bioelectrochem. Bioenerg.,* 3, 230, 1976.
22. **Stollery, J. G. and Vail, W. J.,** Interactions of divalent cations or basic proteins with phosphatidylethanolamine vesicles, *Biochim. Biophys. Acta,* 471, 372, 1977.
23. **Puskin, J. S. and Coene, M. T.,** Na^+ and H^+-dependent Mn^{2+} binding to phosphatidylserine vesicles as a test of the Gouy-Chapman-Stern theory, *J. Membr. Biol.,* 52, 69, 1980.
24. **Ohki, A. and Kurland, R.,** Surface potential of phosphatidylserine monolayers. II. Divalent and monovalent ion binding, *Biochim. Biophys. Acta,* 645, 170, 1981.
25. **Wojtczak, L. and Nałęcz, M. J.,** Surface charge of biological membranes as a possible regulator of membrane-bound enzymes, *Eur. J. Biochem.,* 94, 99, 1979.
26. **Schäfer, G. and Rieger, E.,** Interaction of biguanides with mitochondrial and synthetic membranes. Effect on ion conductance of mitochondrial membranes and electric properties of phospholipid bilayers, *Eur. J. Biochem.,* 46, 613, 1974.
27. **Schäfer, G. and Rowohl-Quisthoudt, G.,** Influence of electric potential on mitochondrial ADP-phosphorylation, *FEBS Lett.,* 59, 48, 1975.
28. **Schäfer, G. and Rowohl-Quisthoudt, G.,** Influence of surface potentials on the mitochondrial H^+ pump and on lipid-phase transitions, *J. Bioenerg.,* 8, 73, 1976.
29. **Schäfer, G.,** Some new aspects on the interaction of hypoglycemia-producing biguanides with biological membranes, *Biochem. Pharmacol.,* 25, 2015, 1976.

30. **Byczkowski, J. Z., Salamon, W., Harlos, J. P., and Porter, C. W.,** Actions of bis(guanylhydrazones) on isolated rat liver mitochondria, *Biochem. Pharmacol.,* 30, 2851, 1981.

31. **Huunan-Seppälä, A. J. and Nordling, S.,** Effect of local anaesthetics and related agents on mitochondrial volume oscillations and surface charge density, in *Energy Transduction in Respiration and Photosynthesis,* Quagliariello, E., Papa, S., and Rossi, C. S., Eds., Adriatica Editrice, Bari, 1971, 317.

32. **Åkerman, K. E. O.,** Effect of Mg^{2+} and spermine on the kinetics of Ca^{2+} transport in rat-liver mitochondria, *J. Bioenerg. Biomembr.,* 9, 65, 1977.

33. **Eriksson, L. E. G. and Westman, J.,** Interaction of some charged amphiphilic drugs with phosphatidylcholine vesicles. A spin label study of surface potential effects, *Biophys. Chem.,* 13, 253, 1981.

34. **Byczkowski, J. Z., Zychlinski, L., and Porter, C. W.,** Inhibition of the bioenergetic functions of isolated rat liver mitochondria by polyamines, *Biochem. Pharmacol.,* 31, 4045, 1982.

35. **Schlieper, P. and Steiner, R.,** Drug-induced surface potential changes of lipid vesicles and the role of calcium, *Biochem. Pharmacol.,* 32, 799, 1983.

36. **Tomasiak, M., Tomasiak, A., and Rzeczycki, W.,** The interaction of polycations with mitochondria, *Bull. Acad. Polon. Sci. Sér. Sci. Biol.,* 28, 1, 1980.

37. **Kantcheva, M. R., Popdimitrova, N. G., and Stoylov, S. P.,** Microelectrophoretic properties of purple membrane particles, *C. R. Acad. Bulg. Sci.,* 35, 633, 1982.

38. **Ohsawa, K., Ohshima, H., and Ohki, S.,** Surface potential and surface charge density of the cerebral-cortex synaptic vesicle and stability of vesicle suspension, *Biochim. Biophys. Acta,* 648, 206, 1981.

39. **Skulachev, V. P., Jasaitis, A. A., Navickaite, V. V., Yaguzhinsky, L. S., Liberman, E. A., Topali, V. P., and Zofina, L. M.,** Five types of uncouplers for oxidative phosphorylation, in *Mitochondria — Structure and Function,* Ernster, L. and Drahota, Z., Eds., Academic Press, London, 1969, 275.

40. **Nałęcz, M. J., Wroniszewska, A., Famulski, K. S., and Wojtczak, L.,** Changes of the mitochondrial surface potential during cuprizone-induced formation of megamitochondria, *Eur. J. Cell Biol.,* 27, 289, 1982.

41. **Eylar, E. H., Madoff, M. A., Brody, O. V., and Oncley, J. L.,** The contribution of sialic acid to the surface charge of the erythrocyte, *J. Biol. Chem.,* 237, 1992, 1962.

42. **Cook, G. M. W., Heard, D. H., and Seaman, G. V. F.,** The electrokinetic characterization of the Ehrlich ascites carcinoma cell, *Exp. Cell Res.,* 28, 27, 1962.

43. **Seaman, G. V. F. and Vassar, P. S.,** Changes in the electrokinetic properties of platlets during their aggregation, *Arch. Biochem. Biophys.,* 117, 10, 1966.

44. **Bosmann, H. B., Myers, M. W., Dehond, D., Ball, R., and Case, K. R.,** Mitochondrial autonomy. Sialic acid residues on the surface of isolated rat cerebral cortex and liver mitochondria, *J. Cell Biol.,* 55, 147, 1972.

45. **Famulski, K. S., Nałęcz, M. J., and Wojtczak, L.,** Effect of the phosphorylation of microsomal proteins on the surface potential and enzyme activities, *FEBS Lett.,* 103, 260, 1979.

46. **Famulski, K. S., Nałęcz, M. J., and Wojtczak, L.,** Phosphorylation of mitochondrial membrane proteins: effect of the surface potential on monoamine oxidase, *FEBS Lett.,* 157, 124, 1983.

47. **Ambrose, E. J., James, A. M., and Lowick, J. H. B.,** Differences between the electrical charge carried by normal and homologous tumor cells, *Nature (London),* 177, 576, 1956.

48. **Ruhenstroth-Bauer, G., Fuhrmann, G. F., Granzer, G., Kübler, W., and Rueff, F.,** Elektrophoretische Untersuchungen an normalen und malignen Zellen, *Naturwissenschaften,* 49, 363, 1962.

49. **Boltz, R. C., Jr., Todd, P., Gaines, R. A., Milito, R. P., Docherty, J. J., Thompson, C. J., Notter, M. F. D., Richardson, L. S., and Mortel, R.,** Cell electrophoresis research directed toward clinical cytodiagnosis, *J. Histochem. Cytochem.,* 24, 16, 1976.

50. **Bohn, B., Thies, C., and Brossmer, R.,** Cell surface charge, sialic acid content and metabolic behaviour of two tumor sublines. A comparative study, *Eur. J. Cancer,* 13, 1145, 1977.

51. **Schäfer, G.,** On the mechanism of action of hypoglycemia-producing biguanides. A reevaluation and a molecular theory, *Biochem. Pharmacol.,* 25, 2005, 1976.

52. **Smoluchowski, M.,** Contribution à la théorie de l'endosmose électrique et de quelques phénomènes corrélatifs, *Bull. Intern. Acad. Sci. Cracovie Classe Sci. Math. Nat.,* No. 3, 182, 1903.

53. **Haydon, D. A.,** The surface charge of cells and some other particles as indicated by electrophoresis. I. The zeta potential surface charge relationships, *Biochim. Biophys. Acta,* 50, 450, 1961.

54. **Haydon, D. A.,** The surface charge of cells and some other small particles as studied by electrophoresis. II. The interpretation of the electrophoretic charge, *Biochim. Biophys. Acta,* 50, 457, 1961.

55. **Haynes, D. M.,** A fluorescent indicator of ion binding and electrostatic potential on the membrane surface, *J. Membr. Biol.,* 17, 341, 1974.

56. **Yguerabide, J. and Foster, M. C.,** Fluorescence spectroscopy of biological membranes, in *Membrane Spectroscopy,* Grell, E., Ed., Springer-Verlag, Berlin, 1981, 199.

57. **Gibrat, R., Romien, C., and Grignon, C.,** A procedure for estimating the surface potential of charged or neutral membranes with 8-anilino-1-naphthalenesulphonate probe. Adequacy of the Gouy-Chapman model, *Biochim. Biophys. Acta,* 736, 196, 1983.

58. **Azzi, A.,** The application of fluorescent probes in membrane studies, *Q. Rev. Biophys.,* 8, 237, 1975.
59. **Slavík, J.,** Anilinonaphthalene sulfonate as a probe of membrane composition and function, *Biochim. Biophys. Acta,* 694, 1, 1982.
60. **Haynes, D. H. and Staerk, H.,** A fluorescent probe of membrane surface structure, composition and mobility, *J. Membr. Biol.,* 17, 313, 1974.
61. **Robertson, D. E. and Rottenberg, H.,** Membrane potential and surface potential in mitochondria: fluorescence and binding of 1-anilinonaphthalene-8-sulfonate, *J. Biol. Chem.,* 258, 11039, 1983.
62. **Gaines, N. and Dawson, A. P.,** Transmembrane electrophoresis of 8-anilino-1-naphthalene sulfonate through egg lecithin liposome membranes, *J. Membr. Biol.,* 24, 237, 1975.
63. **Haynes, D. H. and Simkowitz, P.,** 1-Anilino-8-naphthalene sulfonate: a fluorescent probe of ion and ionophore transport kinetics and trans-membrane asymmetry, *J. Membr. Biol.,* 33, 63, 1977.
64. **Williams, W. P., Layton, D. G., and Johnson, C.,** An analysis of the binding of fluorescence probes in mitochondrial systems, *J. Membr. Biol.,* 33, 21, 1977.
65. **Nakagaki, M., Katoh, I., and Handa, T.,** Surface potential of lipid membrane estimated from the partitioning of methylene blue into liposomes, *Biochemistry,* 20, 2208, 1981.
66. **Searle, G. F. W., Barber, J., and Mills, J. D.,** 9-Aminoacridine as a probe of the electrical double layer associated with chloroplast thylakoid membranes, *Biochim. Biophys. Acta,* 461, 413, 1977.
67. **Chow, W. S. and Barber, J.,** 9-Aminoacridine fluorescence changes as a measure of surface charge density of the thylakoid membrane, *Biochim. Biophys. Acta,* 589, 346, 1980.
68. **Møller, I. M., Chow, W.-S., Palmer, J. M., and Barber, J.,** 9-Aminoacridine as a fluorescent probe of the electrical diffuse layer associated with the membrane of plant mitochondria, *Biochem. J.,* 193, 37, 1981.
69. **Castle, J. D. and Hubbell, W. L.,** Estimation of membrane potential and charge density from the phase equilibrium of a paramagnetic amphiphile, *Biochemistry,* 15, 4818, 1976.
70. **Quintanilha, A. T. and Packer, L.,** Surface potential changes on energization of the mitochondrial inner membrane, *FEBS Lett.,* 78, 161, 1977.
71. **Mehlhorn, R. J. and Packer, L.,** Membrane surface potential measurements with amphiphilic spin labels, in *Methods in Enzymology,* Vol. 56, Fleischer, S. and Packer, L., Eds., Academic Press, New York, 1979, 515.
72. **Montal, M. and Gitler, C.,** Surface potential and energy-coupling in bioenergy-conserving membrane systems, *J. Bioenerg.,* 4, 363, 1973.
73. **Fromherz, P.,** A new method for investigation of lipid assemblies with a lipoid pH indicator in monomolecular films, *Biochim. Biophys. Acta,* 323, 326, 1973.
74. **Fromherz, P. and Masters, B.,** Interfacial pH at electrically charged lipid monolayers investigated by the lipoid pH-indicator method, *Biochim. Biophys. Acta,* 356, 270, 1974.
75. **Vaz, W. L. C., Nicksch, A., and Jähnig, F.,** Electrostatic interactions at charged lipid membranes. Measurement of surface pH with fluorescent lipoid pH indicators, *Eur. J. Biochem.,* 83, 299, 1978.
76. **Goldstein, L., Levin, Y., and Katchalski, E.,** A water-insoluble polyanionic dervative of trypsin. II. Effect of the polyelectrolyte carrier on the kinetic behavior of the bound trypsin, *Biochemistry,* 3, 1913, 1964.
77. **Goldstein, L.,** Water-insoluble derivatives of proteolytic enzymes, in *Methods in Enzymology,* Vol. 19, Perlmann, G. E. and Lorand, L., Eds., Academic Press, New York, 1970, 935.
78. **Goldstein, L. and Katchalski, E.,** Use of water-insoluble enzyme derivatives in biochemical analysis and separation, *Fresenius' Z. Anal. Chem.,* 243, 375, 1968.
79. **Katchalski, E., Silman, I., and Goldman, R.,** Effect of the microenvironment on the mode of action of immobilized enzymes, *Adv. Enzymol.,* 34, 445, 1971.
80. **Goldstein, L.,** Microenvironmental effects on enzyme catalysis. A kinetic study of polyanionic and polycationic derivatives of chymotripsin, *Biochemistry,* 11, 4072, 1972.
81. **Laidler, K. J. and Bunting, P. S.,** The kinetics of immobilized enzyme systems, in *Methods in Enzymology,* Vol. 64, Purich, D. L., Ed., Academic Press, New York, 1980, 227.
82. **Remy, M. H., David, A., and Thomas, D.,** Insolubilization and charge effects on crosslinked enzyme polymers. Kinetic studies in solution and in gelified membranes, *FEBS Lett.,* 88, 332, 1978.
83. **Beitz, J., Schellenberger, A., Lasch, J., and Fischer, J.,** Catalytic properties and electrostatic potential of charged immobilized enzyme derivatives. Pyruvate decarboxylase attached to cationic polystyrene beads of different charge densities, *Biochim. Biophys. Acta,* 612, 451, 1980.
84. **Nałęcz, M. J., Zborowski, J., Famulski, K. S., and Wojtczak, L.,** Effect of phospholipid composition on the surface potential of liposomes and the activity of enzymes incorporated into liposomes, *Eur. J. Biochem.,* 112, 75, 1980.
85. **Ahrens, M.-L.,** Electrostatic control by lipids upon the membrane-bound $(Na^+ + K^+)$-ATPase, *Biochim. Biophys. Acta,* 642, 252, 1981.
86. **Ahrens, M.-L.,** Electrostatic control by lipids upon the membrane-bound $(Na^+ + K^+)$-ATPase. II. The influence of surface potential upon the activating ion equilibria, *Biochim. Biophys. Acta,* 732, 1, 1983.

87. **Abdelfattah, A.-S. A. and Koch, R. B.,** Inhibition of dog brain synaptosomal Na$^+$ − K$^+$ ATPase and K$^+$-stimulated phosphatase activities by long chain *n*-alkyl-amine and -piperidine, and N″-alkyl-nicotinamide derivatives, *Biochem. Pharmacol.,* 30, 3195, 1981.

88. **Kimelberg, H. K.,** Alterations in phospholipid-dependent (Na$^+$ + K$^+$)-ATPase activity due to lipid fluidity. Effects of cholesterol and Mg^{2+}, *Biochim. Biophys. Acta,* 413, 143, 1975.

89. **Hansford, R. G. and Chappell, J. B.,** The effect of Ca^{2+} on the oxidation of glycerol phosphate by blowfly flight-muscle mitochondria, *Biochem. Biophys. Res. Commun.,* 27, 686, 1967.

90. **Bukowiecki, L. J. and Lindberg, O.,** Control of *sn*-glycerol 3-phosphate oxidation in brown adipose tissue mitochondria by calcium and acetyl-CoA, *Biochim. Biophys. Acta,* 348, 115, 1974.

91. **Houštěk, J. and Drahota, Z.,** The regulation of glycerol-3-phosphate oxidase of rat brown-adipose tissue mitochondria by long-chain free fatty acids, *Mol. Cell. Biochem.,* 7, 45, 1975.

92. **Ścisłowski, P. W. D.,** The effect of some glycolytic intermediates and long-chain acyl-CoA esters on rat skeletal muscle mitochondrial α-glycerophosphate dehydrogenase, *Mol. Cell. Biochem.,* 18, 93, 1977.

93. **Verdouw, H. and Bertina, R. M.,** Affinities of ATP for the dinitrophenol-induced ATPase, *Biochim. Biophys. Acta,* 325, 385, 1973.

94. **Johnston, S. P., Møller, I. M., and Palmer, J. M.,** The stimulation of exogenous NADH oxidation in Jerusalem artichoke mitochondria by screening of charges on the membranes, *FEBS Lett.,* 108, 28, 1979.

95. **Møller, I. M. and Palmer, J. M.,** Charge screening by cations affects the conformation of the mitochondrial inner membrane. A study of exogenous NAD(P)H oxidation in plant mitochondria, *Biochem. J.,* 195, 583, 1981.

96. **Møller, I. M., Schwitzguébel, J.-P., and Palmer, J. M.,** Binding and screening by cations and the effect on endogenous NAD(P)H oxidation in *Neurospora crassa* mitochondria, *Eur. J. Biochem.,* 123, 81, 1982.

97. **Benga, Gh. and Borza, V.,** Differences in reactivity of cytochrome oxidase from human liver mitochondria with horse and human cytochrome c, *Arch. Biochem. Biophys.,* 169, 354, 1975.

98. **Douzou, P. and Maurel, P.,** Le contrôle ionique des réactions biochimiques, *C. R. Acad. Sci. (Paris),* 282 D, 2107, 1976.

99. **Helenius, A. and Simons, K.,** Solubilization of membranes by detergents, *Biochim. Biophys. Acta,* 415, 29, 1975.

100. **Tzagoloff, A. and Penefsky, H.,** Extraction and purification of lipoprotein complexes from membranes, in *Methods in Enzymology,* Vol. 22, Jakoby, W. B., Ed., Academic Press, New York, 1971, 219.

101. **Razin, S.,** Reconstitution of biological membranes, *Biochim. Biophys. Acta,* 265, 241, 1972.

102. **Kagawa, Y.,** Reconstitution of oxidative phopshorylation, *Biochim. Biophys. Acta,* 265, 297, 1972.

103. **Person, P. and Fine, A. S.,** Reversible inhibition of cytochrome system components by macromolecular polyions, *Arch. Biochem. Biophys.,* 94, 392, 1961.

104. **Maurel, P., Douzou, P., Waldmann, J., and Yonetani, T.,** Enzyme behaviour and molecular environment. The effects of ionic strength, detergents, linear polyions and phospholipids on the pH profile of soluble cytochrome oxidase, *Biochim. Biophys. Acta,* 525, 314, 1978.

105. **Davidoff, F.,** Effect of guanidine derivatives on mitochondrial function. II. Reversal of guanidine-derivative inhibition by free fatty acids, *J. Clin. Invest.,* 47, 2344, 1968.

106. **Rubalcava, B., Grajales, M. O., Cerbon, J., and Pliego, J. A.,** The role of surface charge and hydrophobic interaction in the activation of rat liver adenylate cyclase, *Biochim. Biophys. Acta,* 759, 243, 1983.

107. **Mehlhorn, R. J. and Packer, L.,** Inactivation and reactivation of mitochondrial respiration by charged detergents, *Biochim. Biophys. Acta,* 423, 382, 1976.

108. **Itoh, S.,** Membrane surface potential and the reactivity of the system II primary electron acceptor to charged electron carriers in the medium, *Biochim. Biophys. Acta,* 504, 324, 1978.

109. **Gómez-Puyou, A. and Tuena de Gómez-Puyou, M.,** Site and mechanism of action of cations in energy conservation, in *Perspectives in Membrane Biology,* Estrada-O., S. and Gitler, C., Eds., Academic Press, New York, 1974, 303.

110. **Tuena de Gómez-Puyou, M., Gómez-Puyou, A., and Beigel, M.,** On the mechanism of action of alkylguanidines in oxidative phosphorylation: their action on soluble F$_1$, *Arch. Biochem. Biophys.,* 173, 326, 1976.

111. **Farías, R. N., Bloj, B., Morero, R. D., Siñeriz, F., and Trucco, R. E.,** Regulation of allosteric membrane-bound enzymes through changes in membrane lipid composition, *Biochim. Biophys. Acta,* 415, 231, 1975.

112. **Sandermann, H., Jr.,** Regulation of membrane enzymes by lipids, *Biochim. Biophys. Acta,* 515, 209, 1978.

113. **Barton, N. W., Lipovac, V., and Rosenberg, A.,** Effects of strong electolytes upon the activity of *Clostridium perfringens* sialidase toward sialyllactose and sialoglycolipids, *J. Biol. Chem.,* 250, 8462, 1975.

114. **Douzou, P. and Maurel, P.,** Ionic regulation in genetic translation systems, *Proc. Natl. Acad. Sci. U.S.A.,* 74, 1013, 1977.

115. **Maurel, P. and Douzou, P.,** Catalytic implications of electrostatic potentials: the lytic activity of lysozyme as a model, *J. Mol. Biol.,* 102, 253, 1976.

116. **Willman, C. and Hendrickson, H. S.,** Positive surface charge inhibition of phospholipase A_2 in mixed monolayer systems, *Arch. Biochem. Biophys.,* 191, 298, 1978.

117. **Hatefi, Y. and Stempel, K. E.,** Isolation and enzymatic properties of the mitochondrial reduced diphosphopyridine nucleotide dehydrogenase, *J. Biol. Chem.,* 244, 2350, 1969.

118. **Nałecz, M. J. and Wojtczak, L.,** Effect of monovalent cations on the inhibition by NAD^+ of NADH oxidation in submitochondrial particles, *Biochem. Biophys. Res. Commun.,* 80, 681, 1978.

119. **Wojtczak, L., Famulski, K. S., Nałęcz, M. J., and Zborowski, J.,** Influence of the surface potential on the Michaelis constant of membrane-bound enzymes. Effect of membrane solubilization, *FEBS Lett.,* 139, 221, 1982.

120. **Fukuyama, H. and Yamashita, S.,** Activation of rat liver choline kinase by polyamines, *FEBS Lett.,* 71, 33, 1976.

121. **Douzou, P. and Maurel, P.,** Ionic control of biochemical reactions, *Trends Biochem. Sci.,* 2, 14, 1977.

122. **McLaughlin, S. G. A., Szabo, G., and Eisenman, G.,** Divalent ions and the surface potential of charged phospholipid membranes, *J. Gen. Physiol.,* 58, 667, 1971.

123. **de Gier, J., Haest, C. W. M., van der Neut-Kok, E. C. M., Mandersloot, J. G., and van Deenen, L. L. M.,** Correlations between liposomes and biological membranes, in *Mitochondria, Biomembranes,* van den Bergh, S. G., Borst, P., van Deenen, L. L. M., Riemersma, J. C., Slater, E. C., and Tager, J. M., Eds., North-Holland, Amsterdam, 1972, 263.

124. **Apell, H.-J., Bamberg, E., and Läuger, P.,** Effects of surface charge on the conductance of the gramicidin channel, *Biochim. Biophys. Acta,* 552, 369, 1979.

125. **Caspers, J., Landuyt-Caufriez, M., Deleers, M., and Ruysschaert, J. M.,** The effect of surface charge density on valinomycin-K^+ complex formation in model membranes, *Biochim. Biophys. Acta,* 554, 23, 1979.

126. **Ligeti, E. and Fonyó, A.,** Competitive inhibition of valinomycin-induced K^+-transport by Mg^{2+}-ions in liver mitochondria, *FEBS Lett.,* 79, 33, 1977.

127. **Papa, S., Guerrieri, F., Simone, S., and Lorusso, M.,** Action of local anaesthetics on passive and energy-linked ion translocation in the inner mitochondrial membrane, *J. Bioenerg.,* 3, 553, 1972.

128. **Drouin, H.,** Surface charges at nerve membranes, *Bioelectrochem. Bioenerg.,* 3, 222, 1976.

129. **Seeman, P.,** The membrane actions of anaesthetics and tranquilizers, *Pharmacol. Rev.,* 24, 583, 1972.

130. **Theuvenet, A. P. R. and Borst-Pauwels, G. W. F. H.,** Kinetics of ion translocation across charged membranes mediated by two-site transport mechanism. Effect of polyvalent cations upon rubidium uptake into yeast cells, *Biochim. Biophys. Acta,* 426, 745, 1976.

131. **Roomans, G. M. and Borst-Pauwels, G. W. F. H.,** Interaction of cations with phosphate uptake by *Saccharomyces cerevisiae.* Effects of surface potential, *Biochem. J.,* 178, 521, 1979.

132. **Roomans, G. M., Theuvenet, A. P. R., van den Berg, Th. P. R., and Borst-Pauwels, G. W. F. H.,** Kinetics of Ca^{2+} and Sr^{2+} uptake by yeast. Effect of pH, cations and phosphate, *Biochim. Biophys. Acta,* 551, 187, 1979.

133. **Roomans, G. M., Kuypers, G. A. J., Theuvenet, A. P. R., and Borst-Pauwels, G. W. F. H.,** Kinetics of sulfate uptake by yeast, *Biochim. Biophys. Acta,* 551, 197, 1979.

134. **Theuvenet, A. P. R. and Borst-Pauwels, G. W. F. H.,** Effect of surface potential on Rb^+ uptake in yeast. The effect of pH, *Biochim. Biophys. Acta,* 734, 62, 1983.

135. **Borst-Pauwels, G. W. F. H.,** Ion transport in yeast, *Biochim. Biophys. Acta,* 650, 88, 1981.

136. **Peña, A.,** Studies with guandidines on the mechanism of K^+ transport in yeast, *FEBS Lett.,* 34, 117, 1973.

137. **Peña, A.,** Effect of ethidium bromide on Ca^{2+} uptake by yeast, *J. Membr. Biol.,* 42, 199, 1978.

138. **Theuvenet, A. P. R. and Borst-Pauwels, G. W. F. H.,** The influence of surface charge on the kinetics of ion-translocation across biological membranes, *J. Theor. Biol.,* 57, 313, 1976.

139. **Wojtczak, L.,** Effect of fatty acids and acyl-CoA on the permeability of mitochondrial membranes to monovalent cations, *FEBS Lett.,* 44, 25, 1974.

140. **Wojtczak, L.,** Effect of long-chain fatty acids and acyl-CoA on mitochondrial permeability, transport and energy-coupling processes, *J. Bioenerg. Biomembr.,* 8, 293, 1976.

141. **Wehrle, J. P., Jurkowitz, M., Scott, K. M., and Brierley, G. P.,** Mg^{2+} and the permeability of heart mitochondria to monovalent cations, *Arch. Biochem. Biophys.,* 174, 312, 1976.

142. **Duszyński, J. and Wojtczak, L.,** Effect of Mg^{2+} depletion of mitochondria on their permeability to K^+: the mechanism by which ionophore A23187 increases K^+ permeability, *Biochem. Biophys. Res. Commun.,* 74, 417, 1977.

143. **Wojtczak, L.,** Control by magnesium ions of the monovalent cation/proton exchange in mitochondria, in *Membrane Dynamics and Transport of Normal and Tumor Cells,* Trón, L., Damjanovich, S., Fonyó, A., and Somogyi, J., Eds., Akadémiai Kiadó, Budapest, 1984, 239.

144. **Fonyó, A., Ligeti, E., Palmieri, F., and Quagliariello, E.,** Carrier mediated transport of phosphate in mitochondria, in *Biomembranes: Structure and Function,* Gárdos, G. and Szász, I., Eds., Akadémiai Kiadó, Budapest, 1975, 287.

145. **Meijer, A. J. and van Dam, K.,** The metabolic significance of anion transport in mitochondria, *Biochim. Biophys. Acta,* 346, 213, 1974.

146. **Vignais, P. V.,** Molecular and physiological aspects of adenine nucleotide transport in mitochondria, *Biochim. Biophys. Acta,* 456, 1, 1976.

147. **Vignais, P. V.,** Molecular aspects of structure-function relationships in mitochondrial adenine nucleotide translocator, in *Structure and Properties of Cell Membranes,* Vol. 2, Benga, Gh., Ed., CRC Press, Boca Raton, Fla., 1985, chap. 7.

148. **Wojtczak, L. and Załuska, H.,** The inhibition of translocation of adenine nucleotides through mitochondrial membranes by oleate, *Biochem. Biophys. Res. Commun.,* 28, 76, 1967.

149. **Wojtczak, L., Bogucka, K., Sarzała, M. G., and Załuska, H.,** Effect of fatty acids on energy metabolism and the transport of adenine nucleotides in mitochondria and other cellular structures, in *Mitochondria — Structure and Function,* Ernster, L. and Drahota, Z., Eds., Academic Press, London, 1969, 79.

150. **Pande, S. V. and Blanchaer, M. C.,** Reversible inhibition of mitochondrial adenosine diphosphate phosphorylation by long chain acyl coenzyme A esters, *J. Biol. Chem.,* 246, 402, 1971.

151. **Lerner, E., Shug, A. I., Elson, C., and Shrago, E.,** Reversible inhibition of adenine nucleotide translocation by long chain fatty acyl coenzyme A esters in liver mitochondria of diabetic and hibernating animals, *J. Biol. Chem.,* 247, 1513, 1972.

152. **Harris, R. A., Farmer, B., and Ozawa, T.,** Inhibition of the mitochondrial adenine nucleotide transport system by oleyl CoA, *Arch. Biochem. Biophys.,* 150, 199, 1972.

153. **Morel, F., Lauquin, G., Lunardi, J., Duszyński, J., and Vignais, P. V.,** An appraisal of the functional significance of the inhibitory effect of long chain acyl-CoAs on mitochondrial transports, *FEBS Lett.,* 39, 133, 1974.

154. **Duszyński, J. and Wojtczak, L.,** Effect of detergents on ADP translocation in mitochondria, *FEBS Lett.,* 40, 72, 1974.

155. **Duszyński, J. and Wojtczak, L.,** Effect of metal cations on the inhibition of adenine nucleotide translocation by acyl-CoA, *FEBS Lett.,* 50, 74, 1975.

156. **Meisner, H.,** Inhibition of metabolite anion uptake in mitochondria by tetraphenylboron, *Biochim. Biophys. Acta,* 318, 383, 1973.

157. **König, T., Kocsis, B., Mészáros, L., Nahm, K., Zoltán, S., and Horváth, I.,** Interaction of a synthetic polyanion with rat liver mitochondria, *Biochim. Biophys. Acta,* 462, 380, 1977.

158. **König, T., Stipani, I., Horváth, I., and Palmieri, F.,** Inhibition of mitochondrial substrate anion translocators by a synthetic amphipathic polyanion, *J. Bioenerg. Biomembr.,* 14, 297, 1982.

159. **Krämer, R.,** Interaction of membrane surface charges with the reconstituted ADP/ATP-carrier from mitochondria, *Biochim. Biophys. Acta,* 735, 145, 1983.

160. **Halperin, M. L., Robinson, B. H., and Fritz, I. B.,** Effect of palmitoyl CoA on citrate and malate transport by rat liver mitochondria, *Proc. Natl. Acad. Sci. U.S.A.,* 69, 1003, 1972.

161. **Meisner, H., Palmieri, F., and Quagliariello, E.,** Effect of cations and protons on the kinetics of substrate uptake in rat liver mitochondria, *Biochemistry,* 11, 949, 1972.

162. **Palmieri, F., Quagliariello, E., and Klingenberg, M.,** Kinetics and specificity of the oxoglutarate carrier in rat-liver mitochondria, *Eur. J. Biochem,* 29, 408, 1972.

163. **Wojtczak, L. and Załuska, H.,** Effect of fatty acids on the transmembrane potential and phosphate transport in liver mitochondria, in *Function and Molecular Aspects of Biomembrane Transport,* Quagliariello, E., Palmieri, F., Papa, S., and Klingenberg, M., Eds., Elsevier/North-Holland, Amsterdam, 1979, 309.

INDEX

A

F

G

H

U

V

X

Y

Z